SOLAR CENSUS

the directory for the 80s

aatec publications   p.o. box 7119   ann arbor, michigan   48107

Copyright © 1980 by **aatec publications**
p.o. box 7119, ann arbor, michigan 48107

Library of Congress Catalog Card No. 80-68910
ISBN 0-937948-00-4

All Rights Reserved
Manufactured in the United States of America

Cover design by Carl Benkert

CONTENTS

MANUFACTURERS 1
DESIGN—ARCHITECTURE AND ENGINEERING . 151
RESEARCH AND DEVELOPMENT 299
EDUCATION AND INFORMATION 363
Alphabetical Index 391
Tradename Index 411
Subject Index 415
Geographical Index 431

SOLAR CENSUS
Organization

SOLAR CENSUS was compiled from questionnaires submitted by solar professionals throughout the United States. The entries were divided into four major sections based on the primary activities designated by the respondents: Manufacturers, Design— Architecture and Engineering, Research and Development, and Education and Information.

The directory is thoroughly cross-referenced to reflect multiple interest areas. Cross-reference listings appear at the end of each section.

The entries were alphabetized within each major section (using the letter-by-letter method), and then assigned a numerical entry code to facilitate index use.

For the user's convenience, **SOLAR CENSUS** contains four indexes: Alphabetical, Tradename, Subject and Geographical.

- The Alphabetical Index includes all entries cited within the directory, by name and numerical code.
- The Tradename Index cross-references product tradenames to their corresponding OEM listings.
- The Subject Index contains over 200 solar specialty areas, keyed by numerical code to their respective entries within **SOLAR CENSUS**.
- The Geographical Index repeats the name, address and entry number of all entries, and organizes them alphabetically by state.

SOLAR CENSUS is a biannual publication of **aatec publications**. **aatec** specializes in technical publications in the energy field.

MANUFACTURERS

1. **A.A.I. CORPORATION**
   P.O. Box 6767
   Baltimore, Maryland 21204

   Mr. H. A. Wilkenington, Manager/Energy Systems
   (301) 628-3486

   A.A.I. manufactures a single-axis tracking collector, with a 24:1 concentration ratio, a fixed receiver and line focusing for industrial process heat, hot water, and space heating and cooling applications.

   A.A.I.'s Energy Program Department has been involved in numerous major solar projects. All systems for the seven projects listed below were conceived, designed and developed by A.A.I. Preliminary fabrication and tests of prototype models were performed on in-house funds. The actual field installations were part of the government's solar demonstration projects and applications experiments and as such were funded primarily by the Department of Energy.

   Projects include: Timonium Elementary School Heating and Cooling Project; A.A.I. Solar Test Facility; Padonia Elementary School Heating and Cooling Project; Industrial Solar Hot Water as Applied to the Curing of Concrete Blocks; Modular Solar Roof for the Disney World Central Energy Plant Office

Building; Photovoltaic Concentrator for Puerto Rico; Design of a 200/1 Photovoltaic Concentrator for Sandia Labs.

2. **ABACUS CONTROLS INC.**
   80 Readington Road
   Somerville, New Jersey 08876

   F. Curtis Lambert
   (201) 526-6010

   Abacus Controls Inc. is a manufacturer of inverters under the trade name Sunverter. These devices convert direct current electricity to alternating current. The Sunverter models can return solar energy power to the utility lines.

3. **ACROSUN INDUSTRIES, INC.**
   1024 W. Maude Avenue
   Sunnyvale, California 94086

   Jim D. Smith
   (408) 738-2442

   AcroSun manufactures a liquid flat plate solar collector that consists of an all-copper absorber plate, glass cover plate, anodized bronze aluminum frame with aluminum back plate, and is fully insulated. It is the highest performing nonselective collector certified by California.

4. **ACUREX SOLAR CORPORATION**
   485 Clyde Avenue
   Mountain View, California 94042

   Jorgen Vindum
   (415) 964-3200 Ext. 3342

   The Acurex Model 3001 Concentrating Collector is a reflecting parabolic trough solar collector designed to heat fluids to temperatures between 140F and 600F. In this temperature range, it is highly cost-effective compared to flat plate and other concentrating collectors. Typical applications include industrial process hot water and steam, space cooling, organic Rankine cycle systems, and photovoltaic power generation.

The Acurex Solar Tracking System increases the efficiency and reliability of concentrating single-axis tracking collector fields. The system combines three control elements: Direct Insolation Monitor, Shadow-Band Sensor, and Tracker Motor Control.

Acurex has developed a new low-cost, high-efficiency photovoltaic concentrator that uses the technology of sealed-beam automobile headlights. This new concentrator is capable of focusing the sunlight falling on each photovoltaic cell at very high concentrator ratios, requiring fewer photovoltaic cells and resulting in higher cell efficiency.

Recent projects and applications include: a solar industrial process steam system at the Johnson and Johnson plant in Sherman, TX; a deep-well solar irrigation system near Coolidge, AZ; a shallow-well irrigation system near Albuquerque, NM; a solar industrial process hot water system at the Campbell Soup plant, Sacramento, CA; photovoltaic concentrator application experiment at Wilcox Memorial Hospital, Hawaii; solar photovoltaic power system with the Sacramento Municipal Utility District; and a 500-kWe solar thermal-electric power system for the province of Almeria, Spain.

5. **ADVANCED ENERGY SYSTEMS INC. (SUN STONE)**
   P.O. Box 194, 126 Water Street
   Baraboo, Wisconsin 53913

   Bob Arnold, President
   (608) 356-6844

Advanced Energy Systems Inc. has manufactured, distributed and installed the Sun Stone solar energy equipment system for four years; also installation of heat pumps and energy recovery equipment such as Z-ducts. Our plant builds most of the components from raw stock, but we do buy other components factory-direct to complete our systems. Each one is custom-fitted to the home, building or factory for its own special needs, but the system is simple, easily installed and durable.

Our panels are government-tested and approved for state and federal rebates. A digital controller runs each system. Our systems are manufactured with various forms of storage: concrete walls or floors, storage bins with rock, salts or our main storage form—water in plastic packs. All components are of

24-gauge steel, tempered glass, with industrial insulation. Ducts are built in the panels and require runs only from collector to storage. The University of Wisconsin provides weather data, degree days and computer readouts for all systems.

6. **ADVANCED ENERGY TECHNOLOGY, INC.**
   121 C Albright Way
   Los Gatos, California 95030

   Bob Altieri
   (408) 866-7686

   AET manufactures flat plate collectors available single- or double-glazed, with all-copper waterways, unitized aluminum case. Complete design for water, space, swimming pool/spa heating.

7. **ADVANCE ENERGY TECHNOLOGIES INC.**
   P.O. Box 387
   Solartown
   Clifton Park, New York 12065

   Ed O'Hanlon
   (518) 371-2140

   Advance Energy Technologies Inc. is basic in food preservation, energy conservation, and solar systems. We manufacture total system buildings of the "Zeroenergy" type—manufactured the nation's first solar, low-energy-use, superinsulated building in 1973. Have systems and components for HVAC and building construction market—residential, commerical and industrial. Also package a kit for easy-to-construct Zeroenergy solar buildings.

   Advance products include: solar hot water systems; solar Homeside™ liquid collectors; solar Homeside™ air collectors; Sol·R·Wal™ heating systems; Solar Bank™ thermal storage systems; monolithic insulation systems; refrigerated warehouses and environmentally controlled boxes; commercial buildings; Zeroenergy® Shelters; solar greenhouses; roof insulation systems; insulated pipe; heating and ventilating equipment; walk-in coolers and freezers; Zerotherm™ Insulation; modular insulated panels; and Homeside collector mounts.

8. **ADVANCE TECHNOLOGY ENGINEERING**
   P.O. Box 176
   Canoga Park, California 91305

   L. Hoffman/D. Warne
   (213) 703-0239

   Manufacture do-it-yourself kits for swimming pool heating, low-temperature industrial and DHW (when used with heat exchanger) applications, for distribution in Southern California. We provide all technical assistance and materials required to solarize swimming pools.

9. **AEOLIAN KINETICS**
   P.O. Box 100
   Providence, Rhode Island 02901

   Brad Krevor
   (401) 421-5033

   Aeolian Kinetics is an interdisciplinary group of professionals with expertise in electronics, computer science, architecture and the physical sciences. The synthesis of Aeolian Kinetics' hardware and software systems is a result of the group's common goal to develop and manufacture research tools which can help us to understand our complex environment.

   Aeolian Kinetics manufactures data acquisition systems for the wind and solar energy community, including:

   —The MS 778 Digital Data Recorder System: a low-cost, remote site battery-powered data acquisition system for wind and solar research.

   —The PDL-24 Programmable Data Logger: a portable, 24-channel intelligent data acquisition system designed for the Solar Energy Research Institute's Passive Solar Monitoring Program: a fully integrated system of hardware and software.

   —The Wind Prospector 4000: a unique wind survey instrument: combination wind odometer/speedometer.

10. **AFG INDUSTRIES INC.**
P.O. Box 929
Kingsport, Tennessee 37662

Ralph Carter/Brenda Doane
(615) 245-0211

Manufacturer of solar glass—glazing for active and passive applications.
- Sunadex: colorless, iron-free, transmits more energy than other flat glass currently manufactured for commercial glazing.
- Solatex: low iron oxide content, designed for flat plate collectors and to non-vision areas of passive systems.
- Clearlite: higher iron oxide content than other AFG solar glasses, utilized when lower "first cost" takes precedence over "life cycle" costing.

11. **ALDERMASTON INC.**
P.O. Box 34
Locust Valley, New York 11560

Mr. Bru
(516) 676-6198

Aldermaston develops and assembles various solar-operated units that are suitable for gifts. This includes music boxes, kinetics, and the first solar-operated AM/FM pocket radio ever offered to the public.

12. **ALLIED INDUSTRIES INTERNATIONAL**
East Highway 275 Industrial Area
Fremont, Nebraska 68025

Dean J. Morrison
(402) 721-9812

Manufacturer of solar collectors and storage units (air and water) complete or in kit form; fiberglass holding tanks, all sizes; hot tubs (solar heated). In business since 1928. Mail order and direct sales.

MANUFACTURERS 7

13. **ALPHA SOLARCO INC.**
    Suite 2230, 1014 Vine Street
    Cincinnati, Ohio 45202

    M. Uroshevich
    (513) 621-1243

    Manufactures collectors, concentrators and complete systems.

    Econosol flat plate collectors are used for industrial, commercial and agricultural space and water heating, and as preheaters for process heat application. Integral multi-fluid passageways (33) of the absorber panels allow the collected heat to be quickly transformed from the panel into the system. The all-copper Econosol panel has a black chrome selective finish, tempered low-iron solar glass cover, frames of extruded aluminum (custom-finished when required), and high-temperature, no-binder insulation.

    The Suntrek Series of concentrators are used when temperatures above 160F are required. Tracking, acquisition and automatic return-to-start are controlled by a solid-state electronic unit. A single array is made of six parabolic troughs with a total of 104 square feet of reflective surface (aluminum or glass mirrors). Black chrome selective finish receiver tubes are polished prior to plating for reduced heat loss protection.

    Solaqua-58 is a totally prepackaged system of redwood construction, utilizing Alpha Solarco Econosol flat plate collectors, premium nonfreezing transfer liquid, all-copper absorber panel, black chrome selective finish, tempered solar glass and high-temperature insulation.

    Solaqua-95 is a low-cost combination space and water solar heating system.

14. **ALTERNATIVE ENERGY RESOURCES, INC.**
    1155K Larry Mahan Drive
    El Paso, Texas 79925

    Robert V. Butterfield
    (915) 593-1927

    Manufactures air and liquid collectors and DHW systems.

15. **ALUFOIL PACKAGING CO., INC.**
    Route 3, Box 15, 1800 33rd Street
    Fort Madison, Iowa 52627

    Glenn C. Wooldridge
    (319) 372-5512

    Manufacturers of foil containers and other products made of foil. The extent of our involvement in solar products is manufacturing component parts for other manufacturers, inventors and experimenters working from their prints or drawings. However, we do have stock containers that various manufacturers have incorporated in their plans for solar equipment. We can also supply aluminum foil in various sizes and coatings.

16. **THE AMCOR GROUP LTD.**
    350 Fifth Avenue, Suite 1907
    New York City, New York 10001

    N. Zohar/D. Gazith
    (212) 736-7711

    Manufacturer of flat plate solar collector, complete thermosyphonic solar water heating systems, including do-it-yourself systems. Also, driver modules and antifreeze devices for thermosyphonic solar systems.

17. **AMERICAN ENERGY SAVERS, INC.**
    912 St. Paul Road, Box 1421
    Grand Island, Nebraska 68801

    David E. Lentz, President
    (308) 382-1831

    Manufacturer of Reinke wind turbine systems. Also, distributor of solar systems and wood burning systems.

18. **AMERICAN HELIOTHERMAL CORPORATION**
    720 South Colorado Boulevard, Suite 450
    Denver, Colorado 80222

    C. A. Van Nortwick
    (303) 753-0921

MANUFACTURERS 9

American Heliothermal Corporation acquired Miromit, Ltd. of Israel—a 20-year-old manufacturer of the Tabor selective-surface solar collector—in 1978. Since 1975 American Heliothermal has supplied Miromit solar equipment and engineering technology for a number of residential, commercial and industrial projects.

American Heliothermal offers a complete line of solar energy products, from systems and subsystems to various components. Systems utilize a specially designed flat plate liquid collector to absorb solar radiation. Energy is transferred either directly or through a heat exchanger into a water storage tank.

The AHC Solar Appliance is a compact, integrated package that has been simplified and streamlined for DHW application. The easily installed system is engineered and tested to meet government requirements. It consists of standard components to minimize cost and maintenance while assuring reliability. The Solar Appliance is a closed-loop system to prevent corrosion and scaling, and to permit use of an antifreeze and distilled water mixture in freezing climates.

AHC also manufactures Miroblack™, a coating which is less costly to apply to absorber plates than competitive electroplated selective surfaces currently available in the U.S.

Among AHC's current projects is the research and testing of the "solar roof." In the solar roof concept, the solar collectors themselves form the roof, thus substantially cutting costs.

19. **AMERICAN HOME SOLAR ENERGY SYSTEMS, INC.**
    23142 Alcalde Drive
    Laguna Hills, California 92653

    S. W. Crosland
    (714) 951-8507

    Manufacturer of NOVA™ all-copper flat plate solar collectors featuring American Home's patented all-weather EPDM frame which has an anticipated life-span of 50+ years.

20. **AMERICAN SOLAR HEAT CORPORATION**
    7 National Place
    Danbury, Connecticut 06810

    Joseph Heyman
    (203) 748-5554

    American Solar Heat Corporation manufactures solar collectors and differential thermostats.

    Amsolheat's collector cover is made of specially formulated fiberglass which has all the advantages of glass without the disadvantages of weight and fragility. The absorber is copper foil, coated with a special black surface. The grid is made of copper to insure durability and superior heat transferences. To insulate the panel and to retain heat, the collector uses polyurethane foam.

    Amsolheat's differential thermostat, Sunmeter, gives a constant readout of the collector temperature and gives the temperature of the water in the storage tank. It can be used with any solar heating system.

21. **AMERICAN SOLAR KING CORPORATION**
    P.O. Box 7399
    Waco, Texas 76710

    Richard Lloyd, Vice-President
    Donna Graham
    (817) 776-3860

    American Solar King Corporation is a manufacturer of solar collectors and marketer of solar hot water, space heating, and air conditioning systems. The company manufactures two flat plate models and an evacuated tube model. The company presently has over 300 distributors and dealers in almost every state in the United States and international distributors in Korea, Saudi Arabia and Italy.

    Components can be ordered for domestic hot water, heat pump, absorption cooling, space heating, agricultural and industrial systems. These components, together with American Solar King collectors, make it possible for customers to install complete solar heating/cooling systems. While the company does not directly install the systems it markets, it provides a training school for employees of its distributors.

22. **AMERICAN SUN CORPORATION**
 2913 Ponce de Leon Boulevard
 Coral Gables, Florida 33143

 James Dudek
 (305) 661-2501

 Design, engineer, manufacture, sell and install solar energy systems for hot water, space heating and cooling, and direct electricity.

23. **AMERICAN TIMBER HOMES/SOLARTRAN DIVISION**
 Escanaba, Michigan 49829

 John Walbridge
 (906) 786-4550

 American Timber Homes/Solartran Division designs and manufactures panelized solar homes and commercial buildings. We supply both passive and active systems. For active systems, we do all of the necessary engineering and supply the solar hardware with the buildings.

24. **AMETEK, POWER SYSTEMS GROUP**
 1025 Polinski Road
 Ivyland, Pennsylvania 18974

 Jack Nelson
 (215) 441-8770

 Manufacturer of solar collectors and systems, including the SunJammer high-thermal-efficiency flat plate collector designed for commercial, industrial and residential solar heating and cooling applications.

 At the present time AMETEK collectors power the world's largest solar swimming pool heating system at the International Swim Center at Santa Clara, CA, and provide cooling, heating and hot water at the adjacent 27,000 square foot recreation center.

12 SOLAR CENSUS

25. **ANDERSEN CORPORATION**
Bayport, Minnesota 55003

Ernie Betker
(612) 770-7293

Andersen Corporation manufactures windows and gliding doors for passive solar applications.

26. **APPLIED SOLAR ENERGY CORPORATION**
P.O. Box 1212
Industry, California 91746

George Home
(213) 968-6581

Manufacturer of silicon solar cells for space and terrestrial applications: terrestrial one-sun cells and concentrator cells. Also manufacture solar cell modules and stand-alone remote power systems.

27. **APPROTECH SOLAR PRODUCTS**
770 Chestnut Street
San Jose, California 95110

Kent Dogey
(408) 297-6527

Manufacturer of flat plate liquid and air collectors, solar ovens and distillation units. Solaristocrat—high-thermal performance, liquid flat plate panel used for hot water and space heating/cooling. Solo—an all metal, low-temperature flat plate collector used for swimming pool applications. Solaroaster—concentrating air collector for cooking. Heliostill—for solar-powered distillation of liquids with vaporization temperatures below 250F.

Areas of R&D include biomass conversion, eutectic thermal storage, hydrogen fuel generation, and designs for wind and wave capture.

MANUFACTURERS 13

28. **APTEC CORPORATION**
    1637 Pontius Avenue
    Los Angeles, California 90025

    Howard Helfman
    (213) 478-4031

    Manufacturer of Heliophase™ water heating systems utilizing phase-change fluid (freon) with automatic temperature control using saturation pressure.

29. **AQUASOLAR, INC.**
    1232 Zacchini Avenue
    Sarasota, Florida 33578

    (813) 366-7080

    Manufacturer of the Aquasolar Pool Heater, a low-temperature high-volume heater, with an open-air tube collector system using 1.5-inch internal-finned black tubes placed in a serpentine pattern. Easily installed, new or retrofit applications.

30. **ARCO SOLAR INC.**
    20554 Plummer Street
    Chatsworth, California 91311

    Cheryl Burton
    (213) 998-2482

    Manufacturing encompasses all stages of solar cell manufacture beginning with purified polysilicon raw material to finished solar cells, and finishing with laminated weather-resistant modules.

    Extensive laboratory capacity for basic research concerning materials and processes.

    A full line of accessory products also supplied, including charge controllers, regulators, alarms, power panels, control boxes, array support structures, specially designed refrigerators, fluorescent lights, and water pumps.

31. **ARKLA INDUSTRIES INC.**
    P.O. Box 534
    Evansville, Indiana 47704

    John F. Burr
    (812) 424-3331

    Development and manufacture of hot water activated chillers for solar air conditioning. Also manufacture a complete system for solar space heating, space cooling, and domestic hot water.

    Arkla's Solaire 36 water chiller is designed primarily for solar-operated air conditioning applications, but can also be used in small industrial process cooling applications. The unit is rated at 3 tons, but design flexibility allows for operation over a wide range of cooling capacities. The Solaire 300 water chiller can produce from 7.5 tons to 26.5 tons of cooling capacity.

    The Arkla Solaire 36P engineered package is designed to provide total year-round comfort air conditioning, heating and domestic hot water. The primary feature of the system is the air conditioner which derives its operating energy from solar-heated hot water.

32. **ASTRO RESEARCH CORPORATION**
    6390 Cindy Lane
    Carpinteria, California 93013

    James R. Wilson
    (805) 684-6641

    Manufacturer of portable towers which may be used for wind measuring equipment. Larger sizes support wind-driven generators.

33. **ATR ELECTRONICS, INC.**
    300 E. 4th Street
    St. Paul, Minnesota 55101

    Albert Goffstein
    (612) 222-3791

    Manufacture solid-state inverters.

MANUFACTURERS

34. **AUTOMATIC SOLAR COVERS, INC.**
    1970 Gladwick
    Compton, California 90220

    Marc Smith
    (213) 639-7800

    Manufacture Poolsaver® automatic solar cover that acts as a solar collector, thermal insulator and safety device.

35. **AZTECH INTERNATIONAL, LTD.**
    2417 Aztec Road, NE
    Albuquerque, New Mexico 87107

    Manufacture low-temperature electric radiant ceiling heaters as backup for active and passive solar systems.

    Also manufacture air collectors for localized distribution in immediate market area.

36. **BASIC ENVIRONMENTAL ENGINEERING INC.**
    21 W 161 Hill Avenue
    Glen Ellyn, Illinois 60137

    Rey Familar
    (312) 469-5340

    Manufacture solid waste boilers which burn pollution-free without collectors. Have radiant water tubes in fire box and bare tubes in convection section. Waste products do not have to be shredded to burn in our units.

37. **BELL & GOSSETT**
    8234 North Austin Avenue
    Morton Grove, Illinois 60053

    (312) 677-4030

    Manufacture Solar Heat Transfer Modules, pre-engineered packages of selected pumps, heat exchangers and differential control centers that automatically control space and DHW systems.

16 SOLAR CENSUS

38. **BERRY SOLAR PRODUCTS**
    P.O. Box 327, Woodbridge at Main
    Edison, New Jersey 08817

    Calvin C. Beatty
    (201) 548-3800

    Electroplating selective black chrome on copper with nickel underlay called SolarStrip® for absorber plate materials. Through an affiliated company, manufacture laminated insulation back plates for collectors.

    West Coast Representative: Engelhardt Associates, 2449 Via Anacapa, Palos Verdes Estates, California 92074 (213) 377-1619

39. **BIO-ENERGY SYSTEMS, INC.**
    221 Canal Street
    Ellenville, New York 12446

    Dennis Moore
    (914) 647-6700

    Bio-Energy Systems, Inc. is the manufacturer of SolaRoll® solar and radiant heating systems. SolaRoll components are flexible, durable extrusions of EPDM. The primary components of a SolaRoll system consist of a uniquely designed heat exchanger/absorber mat and a series of specially engineered framing strips which are used for glazing applications. It is the first solar energy system to be recommended to the U.S. Department of Energy by the National Bureau of Standards. SolaRoll systems are low in cost and easy to install.

40. **BLUE WHITE INDUSTRIES**
    14931 Chestnut Street
    Westminster, California 92683

    Clyde King (Technical)/Robert Gledhill (Sales)
    (714) 893-8529

    We manufacture a line of variable area and pitot tube flowmeters for the solar industry. These products are marketed under the name "CalQFlo."

MANUFACTURERS 17

41. D. W. BROWNING CONTRACTING CO.
475 Carswell Avenue
Holly Hill, Florida 32017

Robert S. Reed, President
(904) 252-1528

D. W. Browning manufactures liquid collectors for DHW systems. The outstanding features of our collector are Inherent Freeze Protection (IFP), salt resistance, and low price.

42. BUDCO
6 Cadwell Road
Bloomfield, Connecticut 06002

C. Pirrello
(203) 242-6180

Design and manufacture of SOLARGY™ modules for use with low-temperature collectors and sources for space heating/cooling and hot/chilled water. Also available with automatic logic for operation with blower only with input temperatures of 90F (18C)+. Useful for all solar collector/storage applications and for energy conservation and recovery.

43. CALMAC MANUFACTURING CORPORATION
Box 710
Englewood, New Jersey 07631

John Armstrong
(201) 569-0420

Manufacture the SUNMAT-I liquid collector (unglazed) for outdoor pool heating, the SUNMAT-II single-glazed flat plate liquid collector for structure use, and the SUNMAT-70 for indoor pool heating. (Collector is glazed for winter operation.)

18 SOLAR CENSUS

44. CANFIELD GROUP
1000 Brighton Street
Union, New Jersey 07083

Karl Lazan, Engineer
(201) 588-5050

Canfield manufactures both hard and soft solders needed for the manufacture of solar units. Special high-temperature solders for heat transfer from plate to tube. Safe fluxes for proper metal joining.

45. CARDOR COMPANY
743 Bay East Drive
Traverse City, Michigan 49684

Carlus Kotila
(616) 941-7426

Cardor Company manufactures the C.O.E. (Conservation of Energy) water heating system—a collector system plus thermal conduction flue element—for utilizing solar energy in the summer and waste heat from the stove or chimney flue in the winter.

46. CAROLINA SOLAR SYSTEMS
527 Hillsborough Street
Raleigh, North Carolina 27602

Rick Dixon
(919) 828-4328

Primary emphasis to date has been design, manufacture and installation of solar water heaters.

47. CELLULAR PRODUCT SERVICES, INC. (CPS, INC.)
3125A North El Paso
Colorado Springs, Colorado 80907

Phil McClain
(303) 475-9443

CPS, Inc. manufactures insulation material for collectors and storage tanks.

MANUFACTURERS

48. **CENTURY FIBERGLASS PRODUCTS**
    XERXES FIBERGLASS, INC.
    1210 North Tustin Avenue
    Anaheim, California 92807

    Roland Fribourghouse
    (714) 630-0012

    Manufacturer of fiberglass-reinforced plastic tanks for particular solar system needs, e.g., shopping centers and industrial solar systems.

49. **CHAMPION HOME BUILDERS CO.**
    Route 2, Box 321
    Plainfield, Indiana 46168

    (317) 272-2996

    Manufacture Vertafin collectors—flat plate, single or double glazing, aluminum fin absorber plate. Manufacture and package complete DHW systems; solar furnace systems.

50. **CHEMAX MANUFACTURING CORPORATION**
    211 River Road
    New Castle, Delaware 19720

    (302) 328-2440

    Manufacturer of heat transfer cement for solar heating and cooling applications.

51. **CHEMPLAST INC.**
    150 Dey Road
    Wayne, New Jersey 07470

    C. Starke
    (201) 696-4700

    Chemplast is a manufacturer and distributor of fluorocarbon products. We supply the solar industry with films used in inner and outer glazing of solar collectors.

52. CHICAGO SOLAR CORPORATION
2001 Highway 3 North, P.O. Box 839
Faribault, Minnesota 55021

T. Crombie
(507) 332-2251

Chicago Solar Corporation has been in business since 1977 manufacturing the patented inflatable solar collector. Our system is marketed under the trade name SOLAR PAK II.

The complete system consists of the 96 square foot inflatable collector, the window-mounted air handling unit containing a 350-cfm blower, flexible connecting ducts, and all installation instructions and mounting hardware. It is designed to be installed in one afternoon by the average do-it-your-selfer. We have also developed a grain drying system using heavy-duty air handlers, special frames, and two collectors. The grain drying system has been adapted to use on small commercial buildings and warehouses.

We are also in the process of developing a unique water-air storage system. Just as our collector is extremely lightweight and easy to ship, the storage system will be designed for easy shipment unassembled. Future projects will include a line of evaporative coolers to be used in conjunction with our air handlers.

53. CLAIREX ELECTRONICS
560 South Third Avenue
Mt. Vernon, New York 10550

(914) 664-6602

Clairex Electronics is a manufacturer of photovoltaic cells.

54. CMI SOLARGLAS™
11015 Cumpston Street
North Hollywood, California 91601

Cory Isaacson
(213) 764-7880

Manufacturer of a 100% glass collector made from borosilicate and tempered glass. The collector comes in one-square-foot interlocking modules that yield a very attractive and durable installation. The collector can also be used in place of conventional roofing materials.

55. **COATING LABORATORIES INCORPORATED**
    3133 East Admiral Place
    Tulsa, Oklahoma 74110

    Edward B. Kaplan
    (918) 932-5991

    Coating Laboratories, Inc. is the manufacturer of Plasticool® solar heat reflective coating, a passive air conditioning that cools without the use of energy.

56. **COLE SOLAR SYSTEMS, INC.**
    440A East St. Elmo Road
    Austin, Texas 78745

    Warren Cole
    (512) 444-2565

    Manufacturer of solar water heating and swimming pool heating systems, incorporating the following Cole collector models:

    Model 18 features a high-transparency tempered glass cover and an all-copper absorber with a plated black chrome selective surface. The enclosure is made of galvanized steel with bronze enamel finish. The collectors weigh about 85 pounds each.

    Model 40 features copper tubing throughout and aluminum fins spot-welded onto the copper tubes. The top cover is a high-transparency acrylic, and the second cover is FEP Teflon®, a material with excellent sunlight stability and the most transparent of all solar glazing materials. Both cover materials have a 20+-year life expectancy. A self-generating dessicant system eliminates moisture condensation on the covers. The enclosure is made of galvanized steel with bronze enamel finish. Each collector weighs about 150 pounds.

    Model 31 is similar to Model 40 except that the collectors are single-glazed with a tempered glass cover.

22 SOLAR CENSUS

57. **COLUMBIA CHASE CORPORATION**
    Solar Energy Division
    55 High Street
    Holbrook, Massachusetts 02343

    Don Ranney
    (617) 767-0513

    Manufacturer of Redi-Mount liquid collectors for domestic hot water, space and pool heating applications.

58. **COMPOOL CORPORATION**
    333 Fairchild Drive
    Mountain View, California 94043

    (415) 964-2201

    Manufacturer of the SC-22 controller for solar heating of swimming pools and spas.

59. **CON-EGY**
    P.O. Box 740, 2nd & Main
    Maple Hill, Kansas 66507

    Sid Hoobler
    (913) 663-2264

    Con-Egy provides a sales and service facility for solar, wood and conventional hydronic heating systems.

60. **CONSERDYNE CORPORATION**
    4437 San Fernando Road
    Glendale, California 91204

    Howard Kraye
    (213) 246-8408

    Manufacture and install monitoring equipment. Custom design and install solar thermal systems—hot water, space heating and swimming pools.

61. CONTEMPORARY SYSTEMS, INC.
The CSI Solar Center, Route 12
Walpole, New Hampshire 03608

Harry Wolhandler
(603) 756-4796

CSI provides technical solar engineering and design support for passive, hybrid and active solar space heating applications. In addition, modular premanufactured components provide complete systems kits, which are easily installed by contractors onsite using typical building skills (a manual is included). In the Northeast, onsite technical assistance is provided to the contractor. A complete packaged system delivery program, plus modular system components, allow quality solar applications to be provided by conventional building architects, engineers, designers and contractors.

CSI provides a modular integrated system which can be custom-tailored through a choice of many component options. Featuring the Series V Warm Air Collectors and the CSI/SOLAR® and No-Frills® lines of air handling and control systems, systems can range from 200 to 1000 square feet. The complete warm air space heating system includes all the components necessary for complete onsite installation, including: all collector mounting and ducting accessories, insulated flexible solar ducting, and miscellaneous sheet metal connectors and adaptors.

Features: Integrated mounting collectors which replace roof or wall sections and provide a weatherproof exterior surface with no roof penetrations. Clients may choose custom-length collectors between 10 and 16 feet and have a choice of air handling and control systems, from 400 to 2000 cfm at 1.5 in sp.

62. CONTROLEX ENGINEERING
P.O. Box 473
Birmingham, Michigan 48012

Thomas Hamilton

Controlex Engineering is a young company whose goal is to apply electronic technology to energy management. Our

24 SOLAR CENSUS

first product line is a complete line of standard and custom-made temperature sensors. Future products, currently under development, are: temperature controllers, digital and analog temperature readout instruments, power factor controller and a humidity controller.

We specialize in low-volume products with an eye toward unusual applications for both the OEM and the end user/experimenter.

63. **CONTROLS/INC.**
 1509 Woodlawn Avenue
 Logansport, Indiana 46947

 Mr. Max A. Nelson
 (219) 722-1167

 Manufacturer of solid-state electronic controls for original equipment manufacturers.

64. **COPPERSMITH'S**
 P.O. Box 907
 Cypress, California 90630

 Bernard J. Mottershead
 (714) 761-2758

 Manufacturer of solar panels, custom panels, specialty tubing and heat exchangers. All materials used are approved. Solar panels are designed with nonrestrictive flow pattern through the waterways because of Coppersmith's own header manufacture; of multiple tube design for Thermosiphon solar hot water systems.

65. **COVER POOLS, INC.**
 117 West Fireclay Avenue
 Salt Lake City, Utah 84107

 (801) 262-2724

 Manufacturer of automatic solar safety covers for swimming pools.

MANUFACTURERS 25

66. **CREATIVE ENERGY PRODUCTS**
    1053 Williamson Street
    Madison, Wisconsin 53703

    Randy Korda
    (608) 256-7696

    We manufacture and sell a sewing manual and do-it-yourself kit for Window Warmers®. The sewing instructions are sold separately from the kit. The kit includes all components for the insulated shades except for the outer fabric and wood for the side clamps.

67. **CREIGHTON SOLAR CONCEPTS, INCORPORATED**
    Kennedy Boulevard
    Woodbine Airport, Building No. 4
    Woodbine, New Jersey 08270

    Vance R. Creighton, Chairman of the Board
    (609) 861-2442

    Creighton Solar Concepts, Inc. is an energy-oriented company which markets products and services that are designed to save money through energy conservation. It is our practice to sell only to the geographic markets near one of our outlets. We are forging ahead in the research and development area to develop a strong background in the energy production market.

    The highly efficient Creighton solar collector panel is the main product of manufacture in our operations. It is tested and approved by Lockheed of Palo Alto, CA, Polytechnic Institute of New York, and HUD. We also manufacture kits for constructing attached solar greenhouses of single-glazed Lexan®. Redwood, cedar and aluminum frames are available.

    Some of our other products and services include: wind generators, hydroplaces, passive engineering, geothermal stratifiers, photovoltaic cells, heliostats, diaelectric inversion, ocean thermal energy conversion, pyrolytic incinerators.

26 SOLAR CENSUS

68. CRESCENT ENGINEERING CO., INC.
12118 S. Western Avenue
Gardena, California 90249

Jesse O. House
(213) 323-9060

Manufacturer of solar water heating systems for both potable water and swimming pools.

69. CSI SOLAR SYSTEMS DIVISION
12400 49th Street
Clearwater, Florida 33520

L. H. Sallen
(813) 577-4228

Manufacturer of liquid collectors, photovoltaic cells and panels; electric power generation systems; wind generation systems.

70. DALE AND ASSOCIATES INC.
Distributor Division
1401 Cranston Road
Beloit, Wisconsin 53511

Dale Swanson
(608) 362-1495

Manufacturer of air moving fans for solar collectors; continuous air circulation for hard to heat areas.

71. L. M. DEARING ASSOCIATES, INC.
12324 Ventura Boulevard, P.O. Box 1744
Studio City, California 91604

(213) 769-2521

Manufacturer of the Dearing SOLARCAP™ modular pool blanket system for public pools. Floating, insulating cover, reel and portable benches provide fast reel-off for swimming, quick flotation for solar energy collection by day, and heat conservation by night.

72. DECO PRODUCTS
506 Sanford Street
Decorah, Iowa 52101

(319) 382-4264

Manufacturer of hardware for solar panels.

73. del SOL CONTROL CORPORATION
11914 U.S. 1
Juno, Florida 33408

Rodney E. Boyd
(305) 626-6116

Manufacturer of controls and instruments. The Model 02B solar combination was designed specifically for use with the March Pump and Motor Model 809-115VAC. It may be used, however, for on/off control of any motor using 115 volt AC, up to 2 amperes. Higher volt/amp ratings are available on special order. The multiple sensor system ensures against heat loss. Accessories available through del Sol.

74. DELTA H SYSTEMS
Route 3
Sterling, Colorado 80751

Larry L. Jackson
(303) 522-4300

Manufacturer of the Rota Flow II solar air controller. It comes prewired for 115 VAC operation. An air-liquid heat exchanger and 30-ampere power relay for pump control are standard. Four thermistors for temperature sensing of collectors, storage and DHW are included with the logic circuit. Any use point temperature may be selected; however, 80 or 90F is standard. A single thermostat with simple contact closure controls air distribution; however, additional logic is available for two-level or two thermostat operation for backup control.

Delta H also does custom design and manufacture of solar air collectors, greenhouses, passive and active design, grain dryers, etc., locally.

75. **DELTA SOLAR SYSTEM COMPANY**
    ENERGY SAVING SHOP
    2930 S. Creyts Road
    Lansing, Michigan 48917

    Al Ashari
    (517) 626-6902
    (517) 322-0313

    Consult, design and construct collector systems. Also, retail sales of energy saving materials (solar collectors, systems, plumbing materials, wood stoves, etc.).

76. **DENCOR ENERGY COST CONTROLS, INC.**
    2750 South Shoshone
    Englewood, Colorado 80110

    Maynard L. Moe, President
    (303) 761-2553

    Manufacturer of electronic control systems primarily for OEM sales. Dencor solar control systems will perform the decision-making functions required to operate a solar heating system at optimum efficiency. These controllers are compatible with standard thermostats and are designed to permit easy installation.

77. **DEPOSITION TECHNOLOGY, INC.**
    7670 Trade Street
    San Diego, California 92121

    Nathan Meckel
    (714) 578-4711

    DTI presently contributes to the solar industries through special coatings and laminations. For example, we manufacture our own line of solar window films. In addition to our coating and laminating operations, we also metallize reflective mirror surfaces.

78. DE SOTO INC.
1700 South Mount Pleasant Road
Des Plaines, Illinois 60018

Kenneth Lawson
(312) 391-9434

DeSoto Inc. is one of the largest manufacturers of sophisticated chemical coatings in the U.S.

ENERSORB®A is a durable flat-black nonselective solar absorbing coating. Its specially chosen pigmentation provides a highly efficient energy-absorbing surface for a variety of flat plate solar collectors. The two-component urethane resin system provides durability and adhesion when applied to properly prepared surfaces as aluminum, copper and ferrous alloys. Used in conjunction with Super Koropon® epoxy primer, it has exceptional corrosion resistance and withstands the effects of weathering, rapid temperature variations, water immersion, salt fog, humidity, and chemicals such as ethylene glycol, sealants and caulks. Its very low gloss surface is particularly effective in capturing sunlight at low incident angles. Whereas many collector surfaces lose efficiency upon aging, ENERSORB®A's solar absorptance remains high over a long period.

ENERSORB©B is a spray-applied black fluorocarbon-based coating which is intended primarily for use on aluminum solar collector panels. It is an extremely durable nonselective black solar absorbing coating for chemically cleaned and pretreated aluminum extrusions, sheet and fabricated collector plates, resistant to long-term degradation and long periods at high stagnation temperatures.

79. DEVICES & SERVICES CO.
3501-A Milton Avenue
Dallas, Texas 75205

Tom Ashley/Charles Moore
(214) 368-5749

Manufacture solar instruments to measure solar energy, absorptivity, reflectivity, transmissivity and emissivity: Alphatometer, Emissometer, Pyranometer, Scaling Digit Voltmeter, and Solar

Spectrum Reflector, plus two Solar Energy Educational Packages. Develop custom instrumentation for solar and other systems. R&D on selective absorber coatings.

80. **DIXIE ROYAL HOMES, INC.**
    460 East 15th Street
    P.O. Box 805
    Cookeville, Tennessee 38501

    Kris Ballal
    (615) 528-1589

    Design, manufacture and market four models of passive and passive hybrid factory-manufactured housing. These homes are from approximately 1000 to 1500 square feet. Marketing area is Tennessee and surrounding states.

81. **DIY-SOL, INC.**
    29 Highgate Road
    Marlboro, Massachusetts 01752

    Felix Rapp
    (617) 481-0359

    DIY-Sol is a leading supplier to the rapidly growing site-built solar heating industry and to do-it-yourself homeowners. The DIY-Sol heating system uses a two-pass air heating collector, thermosiphon DHW preheater and rock storage bin. These are available in pre-engineered kit form from DIY-Sol.

    Our XSOLTHERM solar energy absorbers and THERMALATOR control systems are used in more than 50 solar heating systems totaling over 15,000 square feet of collector area.

82. **DODGE PRODUCTS, INC.**
    P.O. Box 19781
    Houston, Texas 77024

    Claire Dodge
    (713) 467-6262

    Dodge Products manufactures solar measuring devices.

Portable Solar Meter Model 776 is hand-held, used to measure direct plus diffuse solar radiation. The meter is a two bearing micro-ammeter. The sensor is a space-grade silicon cell carefully loaded to insure good linearity of response versus intensity.

Solar Sensor SS-100 is a silicon cell powered transducer designed to provide an electrical signal proportional to solar insolation. It is used to provide an input signal for meters, recorders and integrators.

The DM-25 Desk Meter and the PM-45 Panel Meter are both powered directly from the Solar Sensor SS-100. No batteries are used. The meter scale is calibrated in heat, electrical and scientific units; it may be placed up to several hundred feet from the sensor.

The SR-77 Solar Recorder provides a continuous strip chart recording of solar intensity variation over the day. The instrument prints 21,600 data points in a 12-hr solar day. The recording speed is one inch per hour and provides 756 hours of continuous recording. Calibration is available in $Btuh/ft^2$, mW/cm^2 or Langleys/hr, and is referenced to a laboratory pyranometer.

The Solar Integrator Model SI-377 measures the amount of solar energy that has fallen on a surface over a period of time, from a few minutes to nine months. Solid-state with CMOS integrated circuits and a 4-digit LED display.

Signal Processor SP-6 converts a varying input signal into a constant averaged value easy to record and interpret. Can obtain averaged values of: solar radiation, wind velocity, liquid or gas flow, temperature, electrical current, voltage or power.

83. **DOUCETTE INDUSTRIES**
P.O. Box 1641
York, Pennsylvania 17405

(717) 845-8746

Design and manufacture Vented Double Wall Solar Heat Exchangers.

84. **DOW CORNING CORPORATION**
 2200 Salzburg Road
 P.O. Box 1767
 Midland, Michigan 48640

 (517) 496-4000

 Manufacture Silicone/Urethane Roof System, a spray-in polyurethane foam insulation coated with Dow Corning Silicone Elastomeric Membrane. Also manufacture Silicone Rubber Sealants.

85. **E. I. du PONT de NEMOURS AND COMPANY, INC.**
 10th and Market Streets
 Wilmington, Delaware 19898

 (302) 774-7637

 Manufacture Teflon®solar film and Tedlar®PVF, low-cost, lightweight inner glazings for flat plate collectors. Pyranometers have measured transmissivities of 89-90% total incident solar energy for Tedlar®PVF. It may be shrink-wrapped or bonded by adhesives.

86. **DWYER INSTRUMENTS INC.**
 P.O. Box 373
 Michigan City, Indiana 46360

 Dave Decker
 (219) 872-9141

 Manufacture a broad range of measuring and control instruments for low-pressure air and gases, including low-pressure gauges and pressure switches, low-flow flowmeters for air or water.

87. **DYNA TECHNOLOGY, INC.**
 WINCO DIVISION
 7850 Metro Parkway
 Minneapolis, Minnesota 55420

 Len Attema, Vice-President Marketing
 (612) 853-8400

Manufacture the WINCHARGER® wind electric 12-volt battery charger that starts charging in a 7-mph breeze and reaches its maximum charge of 14 amps in 23-mph winds. Standard equipment includes: insulated instrument panel, 6-ft propeller, 200-W generator, enclosed collector ring, air brake governor, and 10-ft WINCHARGER® tower.

88. **EARTH SERVICES INC.**
    Route 30, Box 99
    Pawley, Vermont 05761

    (802) 325-3093

    Manufacture the Solar Module, a single unit, preassembled, to handle control, pumping, air elimination, fill/drain, and monitoring in closed-loop DHW systems. Includes solid-state electronic differential thermostat, temperature sensors, expansion tank, air purger, vent, pressure relief valve, fill/drain check valve, pump isolation valves, thermometer and pressure gauge.

89. **EASCO ALUMINUM**
    P.O. Box 73
    North Brunswick, New Jersey 08902

    Michael Slom
    (201) 249-6867

    EASCO Aluminum extrudes, fabricates and finishes aluminum extrusions used in the solar industry as component parts, i.e., collector frames, absorbers, parabolic reflectors and support racks. Having been involved directly in the industry for a number of years, EASCO offers assistance with product design. Both standard and custom extrusions are available through our plants in New Jersey, North Carolina and Arizona.

90. **EDMUND SCIENTIFIC CO.**
    7897 Edscorp Building
    Barrington, New Jersey 08007

    (609) 547-3488

Research, develop and manufacture solar cells and panels. Photovoltaic units, including the Solarvent,℠ photovoltaic-powered fan.

91. **edSON SOLAR SYSTEMS**
    7600 Capricorn Drive
    Citrus Heights, California 95610

    G. V. Edson
    (916) 967-2600

    Manufacture multi-use open or closed heating systems (pool, space and water). edSON solar panels are of metal construction: all piping is copper, connections are high-temperature Silfos welds.

92. **EDWARDS ENGINEERING CORP.**
    101 Alexander Avenue
    Pompton Plains, New Jersey 07444

    Robert Pennabere
    (201) 835-2808

    Edwards Engineering Corp. manufactures heat exchange equipment, complete packages or components: boilers, chillers, finned tubing, radiation terminal units, circulators, controls and heat exchangers.

93. **EISLER ENGINEERING COMPANY**
    750 South 13th Street
    Newark, New Jersey 07103

    Charles Eisler, Jr.
    (201) 243-5310

    Manufacturer of automatic assembly machines for the glass working trade since 1920. In recent years the design has been expanded to make the equipment responsive to the manufacturing needs of vacuum-type solar heat collectors.

MANUFACTURERS 35

94. **E & K SERVICE CO.**
    16824 - 74th NE
    Bothell, Washington 98011

    James Ewbank
    (206) 488-2863

    E & K Sol-R panels are suited for a variety of applications in space heating and domestic hot water systems. Can be used with heat pump air conditioners and for swimming pool heating.

95. **ELECTRA SOL LABS INC.**
    2326 Fieldingwood Road
    Maitland, Florida 32751

    Dr. Harold R. Dessan
    (305) 339-0511

    Manufacturer of liquid collectors. Active in solar research and development.

96. **ELECTROLAB INC.**
    2103 Mannix
    San Antonio, Texas 78217

    Karl A. Senghaas
    (512) 824-5364

    Custom design and fabrication for photovoltaic-powered systems for OEMs.

97. **ELKHART PRODUCTS CORPORATION**
    1255 Oak Street
    P.O. Box 1008
    Elkhart, Indiana 46515

    Charles B. Pearman
    (219) 264-3181

    Manufacture custom-fabricated nonferrous tubular components.

36 SOLAR CENSUS

98. ELMWOOD SENSORS INC.
1655 Elmwood
Cranston, Rhode Island 02907

(401) 781-6500

Manufacture Models 3100 and 3150 SPST Hermetic Thermostats, high limit or control devices to activate heaters that aid in eliminating freeze-up in solar panels. Both units open or close on a temperature rise, operate in an ambient temperature range of -30 to 350F, and offer maximum resistance to shock and vibration. A variety of standard mountings are offered, as well as special designs to customer specifications.

99. ENERGY ASSOCIATES
101 Townsend
San Francisco, California 94107

Terry Beynart
(415) 777-1811

Manufacturer/distributor of solar components and systems. Companion company of SOLAR DESIGNS.

100. ENERGY CONTROL SYSTEMS
3324 Octavia Street
Raleigh, North Carolina 27606

Dr. G. G. Reeves
(919) 851-2310

Energy Control Systems manufactures air collectors that are double-glazed and form the roof when mounted on rafters. We also manufacture an Energy Processor™ which is a computer-controlled air handler with air-water heat exchanger, pump and valves to give a complete heating, cooling and domestic hot water system.

101. ENERGY DESIGN CORPORATION
P.O. Box 34294
Memphis, Tennessee 38134

John Connelly
(901) 382-3000

Energy Design Corporation is engaged in the development, manufacturing and marketing of premium quality equipment and systems for efficient, economical conversion of solar radiation to thermal energy for domestic hot water, space heating and cooling of residential, commercial, industrial and institutional buildings; for agricultural and industrial processes requiring heat; for refrigeration; and for operation of heat engines. Energy Design Corporation is a wholly owned subsidiary of Steelcraft Corporation and was established in 1976 to independently continue Steelcraft's inhouse solar energy collector program.

Energy Design manufactures two high-efficiency solar fluid heaters, and related accessory components: the XE-300 vacuum tube solar energy collector; and the HP-150 flat plate solar energy collector. Both are factory assembled into ready-to-use modules.

102. ENERGY EQUIPMENT SALES
412 Longfellow Boulevard
Lakeland, Florida 33801

Ron Yachabach
(813) 688-8373

Represents 18 manufacturers of solar-related equipment. Florida coverage only.

103. THE ENERGY FACTORY
1550 N. Clark
Fresno, California 93703

Tom Kristy
(209) 441-1833

The Energy Factory was founded in 1974 in Fresno, CA, as a solar design and manufacturing firm and is committed to the constant search for a better environment through solar energy applications. In addition to manufacturing and research and development, the company provides a comprehensive support team with the expertise and flexibility to maximize the client's goal through the following services: feasibility studies, environmental impact studies, financial

38 SOLAR CENSUS

analysis and packaging, solar planned communities, real estate promotional planning, product implementation, and energy conservation design.

The TEF solar heating and cooling system is a totally integrated solar package that features the latest technological and design innovations for the ultimate in comfort and convenience. Developed through extensive research over the past seven years, the system combines passive solar principles with space age computer technology.

The TEF solar space heating and cooling system, a passive-hybrid type that operates very much like a conventional forced air system, is currently installed and operating in over 30 residential applications and one commercial office-warehouse complex.

104. ENERGY HARVESTER
11807 Bernardo Terrace
San Diego, California 92128

R. H. Zakhariya, General Manager
(714) 485-8454

Energy Harvester is a manufacturer of high-performance, high-efficiency solar energy systems. Modular solar energy systems that are natural, 100% dependent on the sun for their operational needs to serve the purpose of clean and quiet energy source. The SPP-4 is the smallest standard mini solar power plant, composed of four vacuum tube concentrating collectors, photovoltaic power module, circulating pump, and installation hardware. The collectors are light in weight, available in various colors to blend with the environment. No controls are required to operate this solar power plant. The SPP-8 has almost twice the capabilities of the SPP-4.

105. ENERGY MATERIALS INC.
2622 S. Zuni Street
Englewood, Colorado 80465

Dave Sibila
(303) 934-2444

Energy Materials Inc. manufactures THERMALROD 27 — an alternative to rock and hydronic thermal storage systems. THERMALROD is a new phase-change storage device. The phase-change salt is Dow's specially formulated calcium chloride hexahydrate. The compound is encased in a high-density polyethylene tube which is permanently sealed on the ends by a heat fusion process. Principal applications relate to direct heat storage such as passive solar heating and waste heat reclamation.

106. **ENERGY SYSTEMS, INC.**
4570 Alvarado Canyon Road
San Diego, California 92120

Craig Caster
(714) 280-6660

Energy Systems, Inc. (ESI) has over 7 years experience in the solar industry, holding two primary patents: aluminum extruded spring tempered fin absorber; and mechanical bonding equipment. ESI collectors (9 models) are specified and in use on federal, California state and city buildings and projects.

107. **ENERGY TRANSFER SYSTEMS INC.**
5001 W. Waters Avenue
Tampa, Florida 33614

Clyde Bouse
1-800-237-5057

Manufacturing and marketing of complete solar systems in the United States. Module construction of ETS solar collectors permits installation and testing prior to cover module installation; conversion from single to double glazing without disassembly; simple disassembly for alterations.

108. **ENERSPAN, INC.**
14168 Poway Road
Poway, California 92064

J. Coxsey
(714) 748-1810

40 SOLAR CENSUS

Enerspan, Inc. manufactures flat plate collectors and complete do-it-yourself systems for pools, spas and domestic hot water.

109. ENERTECH CORPORATION
P.O. Box 420
Norwich, Vermont 05055

David Foster
(802) 649-1145

Enertech is the largest manufacturer of wind generators in America. Enertech manufactures the Enertech 1500, which is a utility interface machine. The 1500 produces electricity for one quarter of the cost of a battery charging system. We are developing at 500-kilowatt wind generator for the Department of Energy. The 15-kilowatt machine is ten times the size of the 1500 and will be on the market in the near future.

Enertech is one of the oldest and largest solar companies in the Vermont and New Hampshire area. We have half a decade of experience with solar applications, including total solar heating design and engineering.

110. ENGEL INDUSTRIES, INC.
8122 Reilly Avenue
St. Louis, Missouri 63111

Arthur B. Heuer, Vice-President, Sales
(314) 638-0100

Manufacturer of roll forming units to fabricate absorber panels and collector enclosures. Do not sell component parts.

111. ENTROPY LIMITED
5735 Arapahoe
Boulder, Colorado 80303

Jerry Moon, Sales Manager
(303) 443-5103

Entropy Limited was organized in 1974 to further the development of a major advance in solar thermal technology, the application of heat pipe concepts (vapor energy or phase-change energy) to solar energy systems. The company's Sunpump solar energy systems feature a concentrating, non-tracking collector that delivers energy passively at a constant temperature—the boiling point of water. The patented Sunpump collector has been in production since 1977; the Sunpump/Suncycle domestic water package, using Sunpump collector modules and a Suncycle heat exchanger also developed by the company, has been available since 1979. All systems manufactured by Entropy Limited are sold nationally through a network of dealers and manufacturer's representatives.

112. ENVIRONMENTAL ENERGY MANAGEMENT & MANUFACTURING CORPORATION (E^2M^2)
2722 Temple Avenue
Signal Hill, California 90806

S. R. Fulton
(213) 427-0991

Dedicated to an energy conservation/preservation philosophy, E^2M^2 has manufactured over 50,000 square feet of collectors, with the unique capability of manufacturing nonstandard sizes as well as specific design. Factory 10-year guarantee on standard collectors, quality controlled manufacture has provided E^2M^2 with a no-mechanical defect record for the 2.5 years Sunsorbers have been in the field.

We are involved in all phases of flat plate design and application and are growing rapidly in conservation audits and Title 34 calculations. We give strong engineering support and provide solar seminars for aiding new-to-solar tradesmen.

113. EPPLEY LABORATORY, INC.
12 Sheffield Avenue
Newport, Rhode Island 02840

George L. Kirk
(401) 847-1020

Manufacturer of pyrheliometers and radiometers.

114. ERGENICS
Division of MPD Technology Corporation
A Wholly Owned Subsidiary of INCO LTD.
681 Lawlins Road
Wyckoff, New Jersey 07481

Gregory J. Egan
(201) 891-9103

Selective surface technology for passive and active solar components.

MAXORB Solar Foil is a thin nickel foil with a black selective surface designed for high-efficiency solar energy collection. The black surface is produced by a new process which gives the solar foil an outstanding combination of high absorptance, low emittance and high resistance to humidity and thermal degradation. It is supplied in continuous strips, uncoated or precoated with a high-temperature, pressure sensitive adhesive.

NOVAMET 150 is an inorganic coating in a water-base binder for high-temperature solar absorbers. It can be applied by spraying or brush painting and dries to a hard, durable coating which is resistant to high temperatures.

SKYSORB Stainless Steel Solar Collector is the product of years of research and testing at INCO. It is made of type 304 austenitic stainless steel, well known for its durability and resistance to corrosion. The stainless steel collector frame is made from type 301 annealed stainless steel and is treated by a proprietary coloring process to produce designer colors. The absorber plate is also colored to achieve maximum collection of solar energy while minimizing heat losses.

Every component in the collector can withstand temperatures up to 400F. Insulation used is Owens Corning fiberglass. Glazing is tempered glass with 91.3% transmissivity. Silicone gasketing is used as a weather seal. All internal plumbing connections are made of stainless steel.

SKYSORB solar collectors can be used for medium temperature hydronic heating applications including domestic hot water, space heat and process heat. A special swimming pool collector is also available.

115. ERIE MANUFACTURING CO.
4000 S. 13th Street
Milwaukee, Wisconsin 53221

Robert W. Couffer
(414) 483-0524

Erie Manufacturing produces control valves for the heating and air conditioning industry, as well as for the solar systems industry. Erie also manufactures water softener controls and timers for the residential market. It also produces a complete line of emergency oxygen controls.

116. ERIE SCIENTIFIC COMPANY
Division of Sybron Corporation
Portsmouth Industrial Park
Portsmouth, New Hampshire 03801

Albert H. Stahmer, Jr.
(603) 431-8410

Manufacture diversified flat glass parts, microscope slides and coverglasses, specializing in thin lightweight glass.

117. E-TECH, INC.
3570 American Drive
Atlanta, Georgia 30341

Susan Lewis
1-800-241-7755

Manufacturer of the Efficiency II™ water heating heat pump. The Efficiency II heat pump water heater heats water at the rate of 13,000 Btu per hour. But it consumes only 1230 watts of electricity—less than one-third as much as a resistance element would use. It connects to conventional electric water heaters, which serve as storage tanks only. The company plans to market a unit with its own storage tank.

118. FABRICO MANUFACTURING CORPORATION
4222 South Pulaski Road
Chicago, Illinois 60632

Charles Carboneau
(312) 890-5358

Manufacturer of SUN DOME pool enclosures, greenhouses, and pond liners.

119. FAFCO, INC.
235 Constitution Drive
Menlo Park, California 94025

Terri Walton
(415) 321-3650

Manufacturer of solar systems featuring FAFCO's unglazed polyolefin solar collector for pool heating and the FAFCO IV glazed plastic panels for space heating. The FAFCO IV solar collector panel is the world's first glazed plastic design manufactured and sold in high volume. Design of this product is based upon ten years of product research and technology refined through field experience of more than 30,000 FAFCO solar heating system installations worldwide.

120. FALBEL ENERGY SYSTEMS CORP.
114 Manhattan Street
Stamford, Connecticut 06902

Gerald Falbel
(203) 357-0626

Falbel manufactures Delta liquid collectors which concentrate both direct and diffuse radiation onto aluminum absorber plates, which cover only one-third of the collection area, eliminating the need for tracking devices. The smaller absorber plate of the Falbel Delta solar collectors results in a more economical, lighter, very efficient collector, particularly for medium temperature applications such as space heating and hot water.

The Falbel SolaRoof™ is an integrated, built-in-place solar collector system using air or water as the heat transfer medium. This collector is intended to cover the entire surface of a southerly facing roof and is mounted directly on the roof rafters. In the configuration, which uses air for heat transport, blackened galvanized steel air ducts are incorporated in the SolaRoof structure. If water is the heat transport medium, high temperature EPDM synthetic rubber tubes are bonded to the SolaRoof structure. In the water type construction no anti-freeze is required in the collector loop since the rubber tubes will not be damaged by repeated freezing. An air-type Falbel SolaRoof may be used in conjunction with water storage or the more usual rock storage.

121. **FILON DIVISION OF VISTRON CORP.**
12333 Van Ness Avenue
Hawthorne, California 90250

(213) 757-5141

Filon Division of Vistron Corporation manufactures fiberglass reinforced plastic panels used for glazing, flat plate solar collectors, hot air collectors and greenhouses. These panels may also be used in various types of passive solar applications.

122. **FIRST SOLAR INDUSTRIES, INC.**
P.O. Box 303
Plymouth, Connecticut 06782

Raymond Flanagan
(203) 283-0223

We are a manufacturer of air-type solar collectors. We also handle various other collectors with sales and installation of solar domestic hot water systems, and engineering of systems for residential and commercial application.

123. **FISCHER SUN COOKER**
302 Center Street
Redwood City, California 94061

Guy Fischer/Sandy Fischer
(415) 368-7930

46 SOLAR CENSUS

Manufacture and sales of the Fischer Sun Cooker, which incorporates both a parabolic reflector and an enclosed oven for easy adjustment and oven access, easier temperature control, comfortable cooking height unaffected by wind.
Also involved in R&D, solar concentrators.

124. **FLAGALA CORPORATION**
9700 W. Alt. 98
Panama City, Florida 32407

H. G. Swicord
(904) 234-6550

Manufacturer of components and complete systems; air- and liquid-type collectors, and liquid storage units.

125. **FORD PRODUCTS CORPORATION**
Ford Products Road
Valley Cottage, New York 10989

Ronald F. Cataldo
(914) 358-8282

Manufacturer of solar water heaters featuring stone-lined tanks. Also, oil, electric, wood/coal-fired DHW backup systems; and combined collection/storage tanks for passive/greenhouse DHW systems.

126. **FOUR SEASONS SOLAR PRODUCTS CORP.**
910 Route 110
Farmingdale, New York 11735

(516) 694-4400

Manufactures kits for the construction of attached greenhouses of single- or double-glazed glass, Lexan® or acrylic, with aluminum frames.

127. **FREE ENERGY SYSTEMS INC.**
Holmes Industrial Park
Holmes, Pennsylvania 19043

Mark Sanderson
(215) 583-4780

Free Energy Systems Inc. manufactures photovoltaic solar modules for boats, recreational vehicles and the home. Panels are designed to be mounted directly on the deck of the boat or the roof of the RV. The primary use of the panel is to keep the battery completely trickled charged at 100% charge at all times. The solar charges are hermetically sealed in surgical-grade silicon rubber to keep them salt-water and weather-proof.

Free Energy also can design complete alternative energy systems to reduce energy costs while protecting the home from the effects of power failure.

Free Energy also manufactures a photovoltaic pumping system for DHW, a solar-powered fluorescent light system, a line of toys and educational matter, and the Free Energy 17, a 17-foot sailboat which utilizes solar electric drive for auxilliary power. Free Energy is also involved with the remote power industry.

128. **FRIEDRICH & DIMMOCK, INC.**
P.O. Box 230
Millville, New Jersey 08332

(609) 825-0305

Friedrich & Dimmock is a glass capillary manufacturer. Also, distributor of glass tubing and rods.

129. **FUSION INCORPORATED**
4658 East 355th Street
Willoughby, Ohio 44094

Brad A. Hoffman
(216) 946-3300

Fusion manufactures brazing and soldering alloys in paste form. These consist of powdered alloy, appropriate flux, and special paste binders blended together. The mixture may be stored in a pressurized reservoir, for automatic application to parts via a simple dispenser gun. Upon heating, the liquid flux is first released, cleaning the joint area of oxidation. The powdered filler metal then liquefies and flows into the joint area, cooling to form a strong, void-free brazed or soldered joint. The material may be applied in small dots, or in continuous stripes of any length. The latter is especially suitable for applications such as soldering long fluid tubes to copper sheet such as found in flat plate collectors. To facilitate the use of these paste alloys, Fusion also designs and fabricates a wide variety of automated equipment to apply paste alloy, heat, cool, etc. for high-volume production.

130. GANTEC CORPORATION
P.O. Box 88447
Emeryville, California 94662

David Gantz/Bill Gantz
(415) 658-9824

Gantec Corporation manufactures high-quality semirigid plastic mirrors for use as solar concentrators. The advantages of this material are: (1) ease of handling, (2) recyleability, and (3) weight reduction. Gantec Corporation can assist in engineering problems and demonstrates a modular approach to solar energy harvest. The Reflec-tec™ panels are 51 x 21 x 0.030 in. and will take any curve that is required, i.e., parabolic or spherical.

131. GENERAL ENERGY DEVELOPMENT CORPORATION
661 Highland Avenue
Needham Heights, Massachusetts 02194

Robert A. Harrow
(617) 444-4770

The company is photovoltaic and thermal systems oriented, specializing in the applications area. The company has inhouse capabilities for manufacturing collector hardware and balance of system components. We manufacture a line of universal panel racks for mounting collectors.

132. GENERAL ENERGY DEVICES, INC.
P.O. Box 5679
Clearwater, Florida 33515

(813) 461-2557

Manufacturer of Solatron flat plate liquid collector DHW, space heating, and swimming pool heating systems.

133. **GENERAL INSTRUMENT CORPORATION**
 3811 University Boulevard, W26
 Jacksonville, Florida 32217

 Daniel Alvo
 (904) 731-0233

 Supply thermometers and thermo-hydrometers (combination temperature and pressure in one unit) for all types of solar systems.

134. **GENERAL SOLAR SYSTEMS DIVISION**
 GENERAL EXTRUSIONS INC.
 P.O. Box 2687
 Youngstown, Ohio 44507

 Larry Groves
 (216) 783-0270

 Developed limited tracking concentrator for IPH use, temperature range of 150—210 process requirements. Well-suited for absorption chillers.

 Now introducing an air flat plate collector for space heat and hot water applications. Flush-mount style, much design work devoted to improving weather seal and reducing installation time. All sales direct at this time.

135. **THE GLASS-LINED WATER HEATER COMPANY**
 13000 Athens Avenue
 Cleveland, Ohio 44107

 Ralph R. Mendelson, President
 (216) 521-1377

 Manufacturer of liquid storage units.

136. W. J. GRAHAM CO. INC.
1759 E. Borchard
Santa Ana, California 92705

Mal Tardiff
(714) 972-9200

W. H. Graham Co. Inc. manufactures a full line of polyethylene closed cell foam pipe insulation. The process is unique and provides a product which is fully competitive thermally with traditional products. We guarantee our product against attack from ultraviolet rays for 10 years with no painting or shielding necessary.

137. GREEN MOUNTAIN HOMES, INC.
Royalton, Vermont 05068

James Kachadorian
(802) 763-8384

Green Mountain Homes manufactures pre-fabricated solar homes, combining architectural appeal and passive solar engineering, and incorporating their patented thermal mass heat exchanger, SOLAR-SLAB(TM). All homes may be purchased in kit form and constructed by owner or contractor.

138. GRISWOLD CONTROLS
2803 Barranca
Irvine, California 92714

Ray Renaud
(714) 559-6000

Griswold manufactures constant automatic flow controllers that maintain a specified flow rate where there is need for accurate, dependable energy saving and management.

139. GRUMMAN ENERGY SYSTEMS, INC.
10 Orville Drive
Bohemia, New York 11716

(516) 244-2700

Manufacturers of the complete line of Sunstream collectors.

MANUFACTURERS 51

140. GRUNDFOS PUMPS CORPORATION
2555 Clovis Avenue
Clovis, California 93612

Ole Mathiasen
(209) 299-9741

Manufacturer of maintenance-free domestic circulators for closed system heating applications—hydronic heating, solar thermal heating, and heat recovery.

141. GULF THERMAL CORPORATION
645 Central Avenue, P.O. Box 1273
Sarasota, Florida 33578

Dudley Slocum/John O'Neil
(813) 957-0106

Gulf Thermal manufactures single-glazed, 2 and 4 outlet flat plate hydronic collectors. Our absorber plate is all-copper flow tube in aluminum fin. We will be offering an all-copper absorber with selective surface soon.

142. HALEAKALA SOLAR RESOURCES, INC.
865-A Ahua Street
Honolulu, Hawaii 96819

Marc Anthony/Robyn Newton
(808) 833-2561

Involved in residential hot water systems; also townhouse and commercial systems. Manufacturer of flat plate collector utilizing the Olin Brass absorber, fiberglass box, Owens Corning insulation, ASG glazing. Full service contracting company. Design and consult with architects and engineers for passive and active systems.

143. HALSTEAD & MITCHELL CO.
P.O. Box 1110
Scottsboro, Alabama 35768

Troy Barkley
(205) 259-1212

Halstead & Mitchell manufactures the SunCeiver, a liquid-type solar collector for DHW applications.

144. HANAT SALES COMPANY
2220 West Hassell Road, Suite 304
Hoffman Estates, Illinois 60195

Richard J. Hanat, President
(312) 882-7436

We are a five-man sales organization covering Illinois, Indiana, Wisconsin and the Upper Peninsula of Michigan. Our specialization is products involved in energy creation, energy conservation, energy reclamation and environmental ecology.

145. HAWTHORNE INDUSTRIES, INC.
Solar Energy Division
3114 Tuxedo Avenue
West Palm Beach, Florida 33405

J. B. Carr, Technical Supervisor
(305) 684-8400

Hawthorne Industries manufactures differential thermostat controls for the solar industry. A complete line of 100% solid-state controls includes both on/off and proportional models for direct, indirect, draindown and drainback systems. Several different configurations of these controls are available—everything from the control circuit board only (ideal for OEMs) to plug-in units (ideal for do-it-yourselfers) to hardwired conduit-mount units for industrial applications.

All Hawthorne controls are subjected to a seven-day "burn-in" in which the controls are placed in a heated environment and electrically stressed for seven days before shipping.

A new addition to the line is the H-1640 Digital Temperature Monitor. Capable of monitoring up to eight stations, this device has an LED readout display capable of displaying temperatures from -100 to +400F. with a degree of accuracy unrivalled by other low-cost temperature monitoring equipment. The newly introduced "1600" series of controls is scheduled to complete UL certification.

146. HEDLAND PRODUCTS
2200 South Street
Racine, Wisconsin 53404

David Wehling
(414) 639-6770

Manufacture in-line flow meters for oil, water and other liquids. Can be mounted horizontally or vertically. Monitor fluid flow rates including pump performance, flow regulator settings or hydraulic system performance. For use in mobile or industrial oil hydraulic circuits as well as water in lubrication, coolant or transfer lines. Either permanent or temporary installation. Can be applied to pressure lines, return or drain lines. A moving indicator on the meter provides direct reading and eliminates need for electrical connections or readout devices.

147. HELIODYNE, INC.
770 South 16th Street
Richmond, California 94804

Christel D. Bieri
(415) 237-9614

Manufacturer of Heliodyne flat plate collectors featuring bronze anodized aluminum extrusion frames and all-copper absorber plates. Also, Helio-Pak™ solar mechanical and control modules. Complete systems for space heating and cooling, domestic water heating, pool heating, process heating.

148. HELIODYNE, INC.
2629 Charles Street
Rockford, Illinois 61108

Milt Kling
(815) 397-4208

Manufacturer of air collector panels and controls.

149. HELIOS INTERNATIONAL CORP.
SUNSTAR GROUP
2120 Angus Road
Charlottesville, Virginia 22901

John B. Lange
(804) 977-3719

Helios International Corp. has operated outside the U.S. and Canada since 1976. It has licensees and distributors in Chile, Argentina, Uruguay, Brazil, Portugal, Spain, Saudi Arabia and Lebanon.

SunStar Group utilizes Helios International's technology and serves markets in the U.S. and Canada. It has two manufacturing licensees: Virginia and British Columbia.

Manufactures total DHW systems utilizing concentrating, liquid, vacuum tube and flat plate collectors, and liquid and phase-change storage.

150. HELIO THERMICS INC.
1070 Orion Street, Donaldson Center
Greenville, South Carolina 29605

Randy Granger
(803) 299-1300

The Helio Thermics solar heating system contains all the necessary components, including backup heat: the air handler/furnace/control system, the collector cover material, and the solar hot water package. An optional high efficiency wood stove/fireplace that allows ducting into your regular ducting system to provide backup heat is also available. The system will provide 70% or more of your space heating and at least 50% of your hot water requirements.

Helio Thermics will engineer the system to fit your exact requirements, whether residential, commercial, industrial— new or retrofit.

151. HELIOTROPE GENERAL
3733 Kenora Drive
Spring Valley, California 92077

A. K. Abernathy
(714) 460-3930

Manufacture the Delta-T®Differential Temperature Thermostat for solar space and water heating, featuring the Temp-Trak Controller with digital temperature readout of collector and storage. 40 UL-listed models available; each model series available in four ranges of differential. Also manufacture electronic thermometers and strip chart recorders, and the Helio-Matic I®, valve or pump controller of solar heat collection in pool or spa. Accessory items available also. At Heliotrope General, two of our principal goals are: 1. educating the public about solar use, and 2. making the best solar controls on the market.

152. HELIX SOLAR SYSTEMS
Division of American Creative Engineering
P.O. Box 2038
City of Industry, California 97146

Robert S. Giovannucci
(213) 961-0471

First fully integrated solar systems company in California. We engineer, design, manufacture, market, install and service only our own products and systems: KGL®collectors, Hi-Cal®heat exchangers and XC-200®solar fluid.

153. H&H PRECISION PRODUCTS
Division of Emerson Electric Co.
25 Canfield Road
Cedar Grove, New Jersey 07009

John Mangan
(201) 239-1331

Manufacture precision engineered flow control products: two- and three-way thermal modulation valves and mixing valves; water regulation valves.

154. H&H TUBE AND MANUFACTURING CO.
4000 Town Center
Suite 485
Southfield, Michigan 48075

(313) 355-2500

Manufacturer of brass and copper tubing.

155. HITACHI CHEMICAL CO. AMERICA, LTD.
437 Madison Avenue
New York, New York 10022

T. Hamajima
(212) 838-4804

Our solar system has been on the market for ten years with great success, but mainly in Japan where the products are manufactured.

156. HI-TECH, INC.
3600 16th Street
Zion, Illinois 60099

Tom Kinney
(312) 746-2447

Hi-Tech manufactures a complete line of magnetic drive circulators for the solar industry, consisting of models ranging from a 4-ft to a 35-ft lift.

157. HOLLIS OBSERVATORY
One Pine Street
Nashua, New Hampshire 03060

Mr. Joseph F. Litwin, Director
(603) 882-5017

Hollis Observatory is a manufacturer of instruments which measure solar radiation and of environmental monitoring systems.

158. HOT STUFF
   406 Walnut
   La Jara, Colorado 81140

   Andrew Zaugg
   (303) 274-4069

   Involved in the development of simple control systems and the manufacture of low-cost, durable, versatile and efficient air handling components. Manufacture very tightly sealing dampers and the electronic assemblies to control them; also manufacture round motorized dampers that consume power only when changing position.

159. HY-CAL ENGINEERING
   12105 Los Nietos Road
   Santa Fe Springs, California 90670

   A. E. Bowman, Marketing Manager
   (213) 698-7785

   Manufacture sensors and signal conditioning to measure temperature, humidity and heat. Sensors and systems include: MgO compacted thermocouples, platinum resistance temperature sensors, micro-interface systems, linearized RTD bridge amplifier, passive RTD bridge network, calorimeters and pyrheliometers.

160. HYDRO-FLEX CORPORATION
   2101 N.W. Brickyard Road
   Topeka, Kansas 66618

   Max Campbell
   (913) 233-7484

   Manufacture flexible fittings made from bronze, stainless or Teflon®—used for expansion and contraction and panel misalignment. Also manufacture heat exchanger coils (bronze).

58 SOLAR CENSUS

161. HYPERION, INC.
4860 Riverbend Road
Boulder, Colorado 80303

Leonard Rozek
(303) 449-9544

Manufacture flat plate solar collectors, featuring the Selective Vee corrugated absorber with black iron oxide surface, low iron glazing, all-copper passageways, and a durable silicone overcoat.

162. HYPERION INCORPORATED
1300 South Maybrook Drive
Maywood, Illinois 60153

Keith Zar
(312) 343-9514

Distributor for FAFCO, Inc. and Solar Thermal Systems. Sell and install complete systems for swimming pool and domestic water heating. Also, distributor for Solar International—sell solar cell panels for marine applications.

163. ILI, INC.
5965 Peachtree Corners East
Norcross, Georgia 30071

W. T. Hudson
(404) 449-5900

Design and manufacture total systems for residential, commercial and industrial applications. Projects include a number of HUD homes in Georgia, Brandon Life Clinic in Brandon, FL, and the Savannah Science Museum.

164. ILSE ENGINEERING, INC.
7177 Arrowhead Road
Duluth, Minnesota 55811

John F. Ilse
(218) 729-6858

MANUFACTURERS 59

Ilse is an original manufacturer of liquid collectors/storage systems.

165. **IMPAC CORP.**
Division of Decker Manufacturing Co.
P.O. Box 365
Keokuk, Iowa 52632

Marty Fox
(319) 524-3304

Manufacture the Isis solar panel air collector. Also design active applications for residential, commercial, industrial and retrofit uses.

166. **INDEPENDENT ENERGIES INC.**
Route 131
P.O. Box 398
Schoolcraft, Michigan 49087

Dennis Adams
(616) 679-5024

Manufacturer of Timberline Woodstoves and Heliopass passive collectors.

Distributor for Timberline Woodstoves, Valley Forge forced air and hydronic, Jensen forced air, Strato-Jet & Strato Therm destratification device, SolaRoll, "The Flue" & Bel Vent chimneys.

Dealer for Sunspot Solar DHW, Cawley Lemay Woodstoves, and other associated items.

167. **INDEPENDENT ENERGY, INC.**
42 Ladd Street
P.O. Box 860
East Greenwich, Rhode Island 02818

Leonard I. Tinkoff
(401) 884-6990

Manufacturer of basic and advanced solar controls for DHW, greenhouses, space heating, pool heating, hot tub heating, and heat pumps. Controls are standard on most solar systems produced by major U.S. manufacturers. Microcomputer-based, 100% tested. Available with LED digital readouts of system performance.

168. INFINITY ENERGY SYSTEMS
Division of The SIRI Corporation
7015 North Sheridan Road
Chicago, Illinois 60626

S. T. S. Khalsa
(312) 338-2227

IES designs and manufactures microcomputer process control systems for energy applications. These applications can range from extremely accurate tracking mechanisms to temperature or pH parameter control for alcohol production/distillation. Using state-of-the-art computer technology, IES has developed sophisticated cost- and energy-effective control systems. IES is also available for custom design and production for any small computer application.

169. INSOLARATOR
a product of SPECIALTY MANUFACTURING INC.
7926 Convoy Court
San Diego, California 92111

Frank B. Ames
(714) 292-1857

Established 1971, one of the oldest established manufacturers in U.S. today. Insolarator is a fully patented, exclusively designed solar collector for pools, spas, DHW and space heating systems.

INTERNATIONAL SOLAR TECHNOLOGIES, INC. See (430A)

170. INTERTECHNOLOGY/SOLAR CORPORATION
100 Main Street
Warrenton, Virginia 22186

William J. Graves, Marketing Manager Commercial Division
(703) 347-7900

InterTechnology/Solar Corporation has been actively involved in all aspects of solar technology: low-temperature thermal, industrial process heat, biomass, ocean thermal energy conversion, photovoltaic power systems, solar thermal energy systems, and geothermal energy systems. ITC/Solar has grown to an organization consisting of some 40 persons who are specialists in the fields of systems engineering and marketing. Most of ITC/Solar's work has been on research and development contracts in the area of alternative energy strategies.

ITC/Solar has generated numerous pieces of software in the solar area. The company maintains a software library of programs regularly required for a variety of data manipulation tasks. Also, ITC/Solar manufactures a flat plate collector which is distributed nationwide.

171. **ISTA ENERGY SYSTEMS CORPORATION**
29 South Union Avenue
Cranford, New Jersey 07016

Bernhard Amstutz
(201) 272-2822

Ista Energy Systems is the U.S. subsidiary of Ista International which is headquartered in Mannheim, West Germany. We manufacture a complete line of energy conservation products such as: Btu meters and flowmeters, thermostatic valves, outdoor reset controls, water refining products (backwash filters and chemical metering pumps), and flowmeters for oil and water.

172. **ITT FLUID HANDLING DIVISION**
4711 Golf Road
Skokie, Illinois 60076

Robert L. Archer
(312) 677-4030

Manufacturer of pumps, heat exchangers, valves and other components including completely assembled module to provide all the mechanical components for use with liquid solar collectors for domestic water heating, heating and heating/cooling of buildings and process heating.

62 SOLAR CENSUS

173. **W. L. JACKSON MANUFACTURING CO.**
1205 East 40th Street
P.O. Box 11168
Chattanooga, Tennessee 37401

Jerry Balch
(615) 867-4700

Manufacturer of a series of domestic solar water heating systems and storage vessels for such systems.

174. **JACK JANOFSKY & ASSOCIATES**
dba Energy Conservaton Rep.
148 South Crescent Heights Blvd.
P.O. Box 48405
Los Angeles, California 90048

Jack Janofsky
(213) 931-5562

We sell energy conservation products such as the heat reflective coating Plasticool (note: Plasticool sales to industry are backed with a free computerized evaluation report), manufactured by Coating Laboratories, Tulsa, OK; infra-red gas heaters, space heating for industry by Enerco, Cleveland, OH (can be used as stand-by heaters for rock storage bins; thermal barrier air curtains for doors and pass-through windows by Mars Air Doors, El Segundo, CA.

175. **JETEL COMPANY**
2811 North 24th Street
Phoenix, Arizona 85008

Jay
(602) 956-6190

Build, modify or repair controls (electronic), charging systems and inverters, and quality control instrumentation. Also, electronics consulting.

176. JOHNSON CONTROLS
Control Products Division
2221 Camden Court
Oakbrook, Illinois 60521

Wendell Wolka
(312) 654-4900

Johnson Controls provides differential temperature controllers and sensors.

177. JORDAN'S SOLAR EQUIPMENT SALES
671 Newcastle Road
P.O. Box 758
Newcastle, California 95658

Martin Donahue
(916) 663-2045

Manufactures/designs systems incorporating the American Solar King "Decade 80" glazed copper collector into DHW systems, as well as into hydronic space heating/DHW preheat/pool and spa heating systems, hot tub/DHW preheat systems, and solar source heat pump systems. Jordan's Solar Equipment is central distributor for American Solar King in northern California.

178. KAELIN INDUSTRIES, INC.
2210 North Grand Avenue
Evansville, Indiana 47711

Harold Bender
(812) 423-7200

Manufacturer of air-type solar collectors.

179. KALWALL CORPORATION
Solar Components Division
P.O. Box 237
Manchester, New Hampshire 03105

(603) 668-8186

64 SOLAR CENSUS

Manufacture passive solar heating components: Sun-Lite® glass fiber reinforced polymer for collector covers; insulated Sun-Lite® Glazing Panels for solar greenhouses or windows; Sunwall® wall and roof systems; solar storage tubes; Solar-Kal air heaters.

180. KATHABAR SYSTEMS
Ross Air Systems Division
Midland-Ross Corporation
Box 791
New Brunswick, New Jersey 08903

Jos. Pash
(201) 356-6000

Kathabar equipment integrated with alternate energy sources and terminal units provide a better environment with energy, space and cost savings.

A Kathabar System consists of a conditioner for modifying the air temperature and humidity; a regenerator for transferring moisture to an exhaust air stream; and Kathene® solution circulating system for interconnection. Air is dehumidified as it passes through a cool Kathene spray in the conditioner. This spray is either over a cooling coil, Spray Cel® Series; or through a heat exchanger and over a pack surface, Kathapac Series. The concentration of the Kathene is controlled by the regenerator. An exhaust air stream removes the water from a warm Kathene spray in the Cn Series by a heat coil or in the Kathapac Series with a heat exchanger and packing. Simple control of Kathene temperatures and concentration determines the extent of dehumidification.

181. KEM ASSOCIATES, INC.
153 East Street
New Haven, Connecticut 06507

Colman C. Kramer, President
(203) 865-0584

We are the largest manufacturers' rep in New England, with a complete line of backup systems: Peerless Heating, Myson, Wayne Home Equipment and New Yorker Steel Boiler.

182. KINGSTON INDUSTRIES CORPORATION
205 Lexington Avenue
New York, New York 10016

Mark Sherwood
(212) 889-0190

Kingston Industries manufactures King-Lux aluminum reflector sheet for solar application.

183. KIRKHILL RUBBER COMPANY
300 East Cypress
Brea, California 92621

Don L. Reid, Vice-President, Sales
(714) 529-4901

Kirkhill Rubber can provide the collector manufacturer with molded or extruded seals for all types of design requirements. Seals can be produced from the full range of elastomers including silicones and fluoroelastomers.

184. KOMARA COMPANY
205 South M Street
Dinuba, California 93618

Paul Komara
(209) 591-1703

Manufacturer of polyethylene and polybutylene pipe and tubing.

185. LAFONT CORPORATION
1319 Town Street
Prentice, Wisconsin 54556

Howard C. Heikkinen
(715) 428-2881

LaFont manufactures a line of Hydronic Fireplace Inserts which are used in series with hydronic and solar hydronics systems. The Hydronic Inserts maximize the heating capacity of a fireplace, transferring heat generated by burning firewood to circulating water which is fed into an existing boiler, or installed in series with solar-heated water, to help maintain temperatures overnight or on cloudy days. The LaFont Insert is intended for use in a constant-flow system.

186. **WM. LAMB & CO.**
10615 Chandler Boulevard
North Hollywood, California 91605

Wm. Lamb
(213) 980-6248

Manufacture solar-powered ceiling and attic fans, photovoltaic panels and storage battery systems, and swimming pool heaters and pumps. Researching other solar applications, solar-powered television sets.

187. **LAMCO, INC.**
5923 North Nevada
Colorado Springs, Colorado 80907

Dan L. McPherson
(303) 599-8744

LamCo, Inc. is a manufacturer of solar heating systems. Our system is an active one, utilizing aluminum collector plates, and freon-12 with vegetable oil, which is pumped through a patented design collector plate. The LamCo Solar System is able to heat residential and commercial facilities along with greenhouses, pools, spas, domestic hot water and space.

188. **LASCO INDUSTRIES**
Division of Philips Industries Inc.
3255 East Miraloma Avenue
Anaheim, California 92806

Harold Hartman
(714) 993-1220

Manufacturer of Lascolite Crystalite-T, fiberglass-reinforced plastic panels, corrugated and flat, with weather-protective surfaces of DuPont's Tedlar® polyvinylfluoride film; highly translucent (95% visible light transmission); used in glazing active and passive collectors, solar greenhouses and providing solar illumination in metal-clad buildings, commercial and agricultural.

189. **THE L.A. SOLAR MART**
3145 Glendale Boulevard
Los Angeles, California 90039

Suzan Hawkins
(213) 665-4159

A total energy conserving service—do-it-yourself assistance (films, library). Design, component manufacture, installation, computer calculations.

190. **LENNOX INDUSTRIES INC.**
P.O. Box 400450
Dallas, Texas 75240

Dave Peters
(214) 783-5427

Lennox Industries Inc. is a leading manufacturer of residential and commercial heating, air conditioning and solar equipment.

191. **LETRO THERMOMETER, INC.**
8311 Airport Road
Redding, California 96002

Jeri L. Campbell
(916) 365-5424

Manufacturer of thermometers for solar applications. Small diameter stem and powerful low mass temperature element provides fast response to temperature change. Hermetically sealed, climate-proof instruments. SL-1D recommended for use on swimming pool solar installations. SL-2D recommended for use on solar hot water system installations.

192. LEWIS & ASSOCIATES
Solar Site Selector
105 Rockwood Drive
Grass Valley, California 95945

Sheri Lewis
(916) 272-2077

Lewis & Associates manufactures the Solar Site Selector which calculates solar day/hours, percent of occlusion or shading patterns, and a prime solar percentage or percent of direct solar radiation available for any given site from a silk-screened grid (solar window) that approximates the relative insolation values along its sunpaths. The instrument has recently been re-engineered for additional energy auditor use in the national RCS program. The instrument is used by most professionals in or relating to the solar field, and has been recommended by SERI, Pacific Gas and Electric, as well as many solar professionals.

193. LIBBEY-OWENS-FORD COMPANY
Solar Energy Systems
1701 East Broadway Street
Toledo, Ohio 43605

Lloyd E. Bastian, Manager
(419) 247-4355

Manufacture LOF SunPanel™ Solar Collectors (all-copper, flat plate). Offers computer simulation and analysis services to architects and designers.

194. LI-COR
4421 Superior Street
Box 4425
Lincoln, Nebraska 68504

Mike O'Hara, Sales Engineer
(402) 467-3576

LI-COR, involved in light measuring applications, applies this same experience and background to provide accurate and dependable solar radiation measurements for site evaluation and system monitoring. Particular instruments are the: LI-1776 Solar Monitor, LI-175 Solar Meter/Integrator, LI-200SB Pyranometer and 2010S Miniature Shadow Band.

195. **LODI SOLAR CO. INC.**
16 North Cherokee Lane
Lodi, California 95240

Wally Emery
(209) 334-2132

Lodi Solar Collectors are made of the finest materials available on the market today. They reach temperatures as high as 325F and produce the highest percentage of energy, even on a cloudy day. Lodi Solar Company guarantees their solar system will perform as represented or Lodi Solar Company will modify it at no charge to the customer. Lodi Solar Company offers a 10-year warranty on all collectors and free 3-year maintenance by a factory-trained technician.

196. **H. D. LUTHER MANUFACTURING CO.**
5297 Southway Avenue S.W.
Canton, Ohio 44706

H. D. Luther, President
(216) 477-8634

Manufacturer of material handling equipment. Design and build to customers' requirements. Machine job shop.

197. **MacBALL INDUSTRIES, INC.**
1820 Embarcadero
Oakland, California 94606

W. McKiroy
(415) 534-0274

Manufacture the heat saving insulated swimming pool blanket.

198. MARCH MANUFACTURING INC.
1819 Pickwick Avenue
P.O. Box 87
Glenview, Illinois 60025

Dean Werner
(312) 729-5300

Manufacture a complete line of pumps for the solar industry.

199. MARK ENTERPRISES INC.
30 Hazel Terrace
P.O. Box 3659
Woodbridge, Connecticut 06525

H. B. Finer
(203) 389-5598

Manufacture and market heat transfer modules and synthetic fluids.

200. MARTIN PROCESSING, INC.
P.O. Box 5068
Martinsville, Virginia 24112

Mr. Jim Aker
(703) 629-1711

Martin Processing, Inc. manufactures solar control window film and dyed weatherable polyester (P.E.T.). Also have capabilities to vacuum metalize almost any substrate to insulation, i.e., polyethylene.

MASS ENERGY SYSTEMS, INC. *See (430C)*

201. MATRIX, INC.
537 South 31st Street
Mesa, Arizona 85204

Don Pershing
(602) 832-1380

Matrix manufactures Sol-A-Meters, low-cost, rugged, weatherproof solar radiometers intended for use under the most severe field conditions. They are completely self-contained, and do not require any external power supply. All models incorporate the silicon photovoltaic cell as a sensor. Spectral response is from 0.35 to 1.15 microns. 100% response to a change in signal takes less than one millisecond. All Sol-A-Meters are calibrated for the entire solar spectrum by comparison with thermopile-type radiometers in bright sunshine on clear days. Cosine effect and airmass effect corrections are made during calibration. Standard temperature compensation is from 40 to 140F. Each instrument is furnished with a certificate of calibration showing radiation in both English and metric units.

202. MECHANICAL SEALS CORPORATION
2540 West 237th Street
Torrance, California 90505

Murray/Orlet
(213) 539-5511

MCS manufactures LOK™RING tube joints for hermetic sealing.

203. MEGATHERM
803 Taunton Avenue
East Providence, Rhode Island 02914

Charles E. Mellor
(401) 438-3800

Manufacturer of Megatherm Thermal Storage Systems for space heating and domestic hot water applications. For new construction, energy retrofit, replacement programs, supplemental heating for solar installations—residential and commercial.

204. MEHRKAM ENERGY DEVELOPMENT CO.
R.D. 2
Hamburg, Pennsylvania 19526

(215) 562-8856.

Manufacturer of complete wind energy systems.

205. MICROCONTROL SYSTEMS, INC.
6579 North Sidney Place
Milwaukee, Wisconsin 53209

Harshad Shah
(414) 351-0281

Manufacturer of the MICROL™ Energy Management System, an advanced microprocessor-based controller which has been configured for performing the task of monitoring and controlling the energy consumption of an industrial or commercial environment.

206. MID PACIFIC SOLAR CORPORATION/ SOLAR DISC CORPORATION
540 Lagoon
Honolulu, Hawaii 96819

14202 Ventura Boulevard
Sherman Oaks, California 91403

Genus Colvin
(808) 833-6075/833-0001 (Hawaii)
(213) 501-2988/501-4377 (California)

Complete active solar heating systems for space, water, pool, spa and special design installations. All systems using Solar Disc Corporation photovoltaic modules for 100% total solar utilization. Have installed over 1500 systems in Hawaii and California.

207. MIDWEST COMPONENTS INC.
Port City Industrial Park
Box 787
Muskegon, Michigan 49443

(616) 777-2602

Manufacturer of temperature sensors.

208. **MILLVILLE WINDMILLS, INC.**
(TA-SOLAR, a division of Millville Windmills, Inc.)
10335 Old 44 Drive
Millville, California 96062

Devon Tassen, President
(916) 547-4302

Manufacture and distribute wind generation systems for supplemental energy for residential, agricultural and industrial needs. Also manufacture TA-SOLAR®tracking flat plate collectors.

209. **MINNESOTA, MINING AND MANUFACTURING (3M)**
Decorative Products Division
223-1 South 3M Center
St. Paul, Minnesota 55144

David L. Hyman
(612) 733-0127

Manufacture 0.004 aluminized acrylic sheeting with pressure-sensitive adhesive backing. Available 24" wide x 50 yards minimum. Also manufacture Scotchcal Brand 5400 and/or FEK 244 (where critical focus is a requirement). 86% spectral reflectivity.

210. **MINNESOTA TRITEC INC.**
Route No. 3
Box 524
Delano, Minnesota 55328

John E. Bolkcom
(612) 955-1591

Manufacture air collectors, controls and heat exchangers, and total DHW systems.

211. MONOSOLAR, INC.
a subsidiary of Monogram Industries, Inc.
100 Wilshire Boulevard, Suite 600
Santa Monica, California 90401

Dr. Robert L. Rod
(213) 451-8151

Monogram/Monosolar is a conglomerate with interests in materials used in solar systems, photovoltaic and thermoelectric R & D, and the building of homes and shopping centers.

212. MONROE MODULAR HOMES, INC.
Route 5, Industrial Park
Madisonville, Tennessee 37354

James E. Wallace, Jr., Vice-President
(615) 442-3905

Manufacturer of modular single-family solar homes.

213. MONSANTO COMPANY
800 North Lindbergh Boulevard
St. Louis, Missouri 63166

E. K. Brakebill
(314) 694-2153

Monsanto manufactures a line of Therminol® heat transfer fluids for use in concentrating collector systems and in thermal storage applications. These fluids have operating capabilities over a temperature range of -60F to 750F.

214. MOR-FLO INDUSTRIES, INC.
18450 South Miles Road
Cleveland, Ohio 44128

John Fabrizio
(216) 663-7300

Manufacture Solarstream™ systems for DHW applications. The original Solarstream™ System is a closed system featuring a patented heat exchanger, Reynolds Aluminum collectors and

and top-mounted acessories for ease of installation. Combination electric/solar packages or water heaters alone are available in 66-, 82- and 120-gallon sizes, with or without controls.

Also manufacture the Solarstream Direct-Flo™ open system which features the SSTA Solar Storage Tank, Copper Tubeway Solar Collector and SDS-1 completely automatic Drain Down System Package. Also available in 66-, 82- and 120-gallon capacities.

Mor-Flo also manufactures a complete line of gas, electric and solar water heaters for residential, commercial, mobile home and recreational vehicle use.

Mor-Flo manufactures the following product brands: Mor-Flo,® Hotstream,® Prestige,® Sands,™ American,® Solarstream,™ Energy Saver® and Energy Saver Plus.®

215. MOTOROLA SEMICONDUCTOR GROUP
P.O. Box 2953
Phoenix, Arizona 85062

Bob Hammond
(602) 244-6511

Motorola manufactures photovoltaic cells and modules and support structures. Applications include: remote communications, navigational aids, offshore equipment, cathodic protection, water pumping. Also have developed a series of solar-powered fluorescent lights utilizing advanced technology solar cells to provide the electricity to power these high-efficiency lights. Also, voltage regulators for photovoltaic power systems.

216. MULTI-DUTI MANUFACTURING, INC.
2736 East Walnut
Pasadena, California 91107

Rod Talley, Sales Manager
(213) 795-2716

Manufacture a line of centrifugal pumps that range from 1/3 to 2 horsepower, have flows of 0-125 and pressure to 95 feet. Available for solar production.

76 SOLAR CENSUS

217. **NATIONAL PRODUCTS, INC.**
900 Baxter Avenue, P.O. Box 4174
Louisville, Kentucky 40204

William K. Ewing
(502) 583-0206

National Products, Inc. manufactures a real glass miror (16 oz. per square foot) with a cotton cloth backing. We score and cut this sheet of mirror (23.5 x 23.5 in. square) into smaller pattern cuts of mirror. Therefore, the sheet of mirror becomes flexible and can be applied convex, concave or flat. It is marketed as wall covering, but also is used in solar energy devices and collectors.

218. **NATIONAL SEMICONDUCTORS LTD.**
331 Cornelia Street
Plattsburgh, New York 12901

Duncan F. Clifton, Marketing Manager
(518) 561-3160

National Semiconductors Ltd., founded in 1958, specializes in the design, manufacture and marketing of photoelectric devices. NSL's current product line includes cadmium sulphide and cadmium selenide photocells, silicon photodiodes and photovoltaic cells, high-voltage optocouplers and customized optoelectronic arrays and assemblies. On-going research and development is conducted in the general field of optoelectronics and solar energy.

National Semiconductors has established facilities in Canada, the United States and the United Kingdom with a worldwide network of representatives and agents to assist the customer in the selection of their optoelectronic components and to provide after-sales support.

219. **NATIONAL SOLAR CORPORATION**
Main Street
Centerbrook, Connecticut 06409

Mark Uhran
(203) 767-2191

National Solar Corporation is involved in the design, manufacture and sale of solar hot water systems for residential and industrial markets, using flat plate hydronic collectors.

220. **NATIONAL SOLAR SUPPLY**
 2331 Adams Drive, N.W.
 Atlanta, Georgia 30318

 Sid Stansell
 (404) 352-3478

 National Solar Supply manufactures flat plate collectors, copper absorber plates and tubes. Also, a wholesale distributor of all related components.

221. **NATIONAL STOVE WORKS**
 Howe Caverns Road
 Cobleskill, New York 12043

 (518) 296-8524

 Manufacturer of solid fuel furnaces and stoves. Also, manufactures collectors and related items. Market area, eastern two-thirds of United States.

222. **NATURAL POWER INC.**
 Francestown Turnpike
 New Boston, New Hampshire 03070

 Stuart Carothers
 (603) 487-5512

 Design, manufacture and marketing of electronic controls, monitors, and site analysis instrumentation for solar collection and wind energy conversion systems. Custom control design, standard products.

 Solar Controls: differential thermostats; Solar Instruments: temperature monitors, Btu meters, insolation (radiation) monitors and recorders; WECS Controls: monitoring protection/voltage regulation systems, dynamic loading switches, alternators, amp/watt-hour meters, frequency sensitive relays; Wind Instruments: data collection, analysis, storage and recording systems, potential energy monitors.

223. NORTH AMERICAN SOLAR DEVELOPMENT CORP.
2800 Juniper Street
Fairfax, Virginia 22030

Albert Nunez, Jr.
(703) 280-2467

Manufacturer's representative dealing with numerous energy conservation products through a network of more than 60 dealers. Also, design and engineering of entire systems.

224. NORTHERN SOLAR POWER CO.
311 S. Elm Street
Moorhead, Minnesota 56560

Bruce Hilde
(218) 233-2515

Manufacturer of the SunBox, a 4 x 8 foot active total system solar air collector for space heating.

225. NORTHRUP INC.
Subsidiary of Atlantic Richfield Company
302 Nichols Drive
Hutchins, Texas 75141

John Borzoni, Marketing Manager/Solar Products
(214) 225-7351

Northrup Inc. manufactures and distributes flat plate liquid collectors and packaged systems for DHW heating and space heating (the DHW82 Freeze-Protected System, and the TS3018 Thermosyphon HA system). We also supply several lines of heat pump products to be used in conjunction with solar equipment.

226. NORTH WIND POWER CO., INC.
Box 315
Warren, Vermont 05674

Phil Tonks/Larry Barber
(802) 496-2955

North Wind is actively engaged in marketing the HRZ, a high-reliability small wind energy conversion system (SWECS) designed and developed under our original DOE contract. We are presently in the design stages of our second contract, which is for a 4-kW line interface SWECS. We have a complete engineering and system design staff and can supply any necessary hardware or services for a complete wind system.

227. **NOVAN ENERGY, INC.**
1630 N. 63rd
Boulder, Colorado 80301

George Anagnost, Vice-President AD/PR
(303) 447-9193

Novan designs, manufactures and markets liquid system DHW, closed and open; combination spa and DHW, closed and open; spa systems, closed and open; space heating modules and pool heating systems.

The Novan OPTIMA collector gives the system designer a great deal of flexibility because of its compliance with building codes and accepted performance standards, and its compatibility with fluids and conventional system components on the market today. For specific applications such as domestic hot water and spa heating, Novan offers pre-engineered system modules to assure consistent performance. Computer-assisted design analysis available for applications of all types to help with proper component selection.

228. **NRG COMPANY**
P.O. Box 306
Ardmore, Pennsylvania 19003

David N. Shoemaker
(215) 896-6850

NRG manufactures humidifiers and spraying systems for agricultural use. We also handle the following accessory items: plastic, nylon and brass valves, 2- and 3-way ball valves; all types and classes of chain, standard and custom;

80 SOLAR CENSUS

fittings—plastic and brass; solenoid valves; plastic tubing; centrifugal and roller pumps; nozzles for sprayers; polyethylene and fiberglass storage tanks; compressors of all sizes. Also, consultation and engineering services.

229. NUTEK
NUCLEAR TECHNOLOGY CORPORATION
P.O. Box 1
Amston, Connecticut 06231

Robert S. Dutton
(203) 537-2387

NUTEK is a manufacturer of coatings for tanks.

230. OEM PRODUCTS, INC.
Route 3, Box 295
Dover, Florida 33527

Donald W. Barlow, Sr.
(813) 752-3121

Manufacturer of heat pumps, thermal storage (phase-change) systems, solar heating and cooling systems, DHW systems. Single source responsibility where required.

231. OLIN CORPORATION/BRASS GROUP
1717 Kettner Boulevard, Suite 250
San Diego, California 92101

John H. Blake, III
(714) 230-1308

Olin Corporation/Brass Group manufactures and markets a complete line of OlinSolar™ collectors. These include liquid collectors for domestic hot water, swimming pool heating, space heating, and other liquid collector applications. Both glazed and unglazed models are available. Absorber plates in the collector are all-copper SOLAR-BOND®.

232. OLIN CORPORATION/BRASS GROUP
East Alton, Illinois 62024

R. J. Miller
(618) 258-2000

Olin Corporation/Brass Group manufactures and markets all-copper SOLAR-BOND® absorber plates. Standard designs include those for general liquid heating and swimming pool heating. Custom designs are available. A variety of sizes are manufactured.

233. OLYMPIC SOLAR CORPORATION
208 15th Street, S.W.
Canton, Ohio 44707

Richard G. Brady
(216) 452-8856

Olympic Solar Corporation, a large electroplating firm, is the leading source for the application of the widely accepted and highly optical selective surface known as Chromonyx Black Chrome. Electroplaters of the BCO No. 91 on solar components—three locations in the United States at present with two other locations being developed. Technology has been licensed to two firms overseas with two more in negotiation.

234. OWENS-ILLINOIS SUNPAK DIVISION
P.O. Box 1035
Toledo, Ohio 43666

Jack Van Hee
(419) 247-9872

Manufacturer of SUNPAK™ solar thermal collectors, which utilize Owens-Illinois evacuated glass tubing technology and vacuum tube insulation; capable of collecting both direct and diffuse light.

The basic component of the SUNPAK solar collector system is a hermetic package consisting of an absorber glass tube and a cover tube. The absorber tube accommodates differential thermal expansion, while a third tube permits the reverse flow path for the heat transfer fluid.

A SUNPAK series solar collector consists of 24 evacuated tubes in a series flow arrangement. The series module is a hydronically filled nondrainable solar unit ideal for those large commercial buildings or industrial process applications where there is a need for maximum solar collection area with minimum piping connections. Once the hydronic circuit is filled, minimum power is required to circulate the solar circuit.

The SUNPAK drainable collector uses the evacuated collector tubes in a parallel flow, one-sided header arrangement—with tubes located above a feed and return manifold.

The SUNPAK air collector combines the many advantages of using air as a heat transfer fluid with the high thermal performance of a selectively coated, low-loss evacuated tubular element. Resembles the modular arrangement of the series collector, with the manifold header located between a group of tubes above and below the unit.

235. PARK ENERGY COMPANY
Star Route Box 9
Jackson, Wyoming 83001

Richard C. Greig
(307) 733-4950

PEC manufactures the Thruflo™ solar heat collector, an air system that uses a simple, efficient, newly developed heat transfer method on its blackened collector sheet. Two new collector models and a domestic water preheater are in the development stage.

236. PHELPS DODGE BRASS COMPANY
Solar Division
2665 Woodland Drive
Anaheim, California 92801

Edward T. Anderson, Manager/Solar Energy Division
(714) 761-4260

The Phelps Dodge all-copper solar absorber plate is effective and economical for use in solar collectors for water heating, space heating, and swimming pool heating. The plates are light in weight, easy to handle, yet use an adequate thickness of copper for effective heat transfer. The riser tubes provide good flow characteristics, 150 psi operating pressure, resistance to blockage, and high surface area for good heat transfer. The fins are wrapped 360 degrees around the riser tubes and permanently secured with continuous roll-formed lock seams that maintain an all-copper thermal path for high efficiency. Headers are provided in a variety of optional sizes. The plates are available in standard and custom sizes and with or without high absorptance coating.

237. **PHIFER WESTERN**
14408 E. Nelson
City of Industry, California 91744

David Ballard
(213) 968-0587

Phifer Western manufactures SunScreen®, woven from vinyl-coated Phiferglas® yarn, which is mounted on the exterior of windows and doors to reduce excess solar heat gain. We are the prime weaver of SunScreen and have fabricators available in most areas.

238. **PHILIPS INDUSTRIES/TWIN PANE DIVISION**
31251 Industrial Road
Livonia, Michigan 48150

(313) 522-8400

Manufacturer of solar collector cover plates.

239. **PIPER HYDRO INC.**
3031 E. Coronado
Anaheim, California 92806

Jim Piper
(714) 630-4040

Piper Hydro manufactures all components for the implementation of Piper Hydro patented combination DHW/space heating community systems.

240. **PITTSBURGH CORNING CORPORATION**
800 Presque Isle Drive
Pittsburgh, Pennsylvania 15239

Robert M. McMarlin
(412) 327-6100

84 SOLAR CENSUS

Manufacturer of FOAMGLAS®, an inorganic cellular glass insulation product having a constant insulating value and zero water vapor permeability.

241. **W. H. PORTER, INC.**
P.O. Box 1112
Holland, Michigan 49423

Sales Desk
(616) 399-1963

Manufacturer of components and total systems for liquid collector/liquid storage systems, as well as all types of insulation.

242. **THE POWER-SONIC CORPORATION**
3106 Spring Street, P.O. Box 5242
Redwood City, California 94063

Mr. Bruno A. Ender
(415) 364-5001

Manufacturer of maintenance-free, sealed lead-acid storage batteries.

243. **PROGRESS INDUSTRIES INC.**
7290 Murdy Circle
Huntington Beach, California 92647

Kenneth Busche, President
(714) 847-7917

Manufacturing and R&D — tracking collectors, tracking drives and controls, passive components.

244. **PSI ENERGY SYSTEMS, INC.**
A Division of Pipe Systems Inc.
1533 Fenpark Drive
Fenton, Missouri 63026

(314) 343-7666

Manufacturer of the Thermol 81™ energy storage rod for passive, hybrid passive/active, forced air solar, and heat pump-coupled systems.

245. RADCO PRODUCTS, INC.
2877 Industrial Parkway
Santa Maria, California 93454

(805) 928-1881

Manufacture solar collectors, glazed and unglazed. Sell as completed collector or components only for customer assembly. Also, distributor for stock DHW and pool system components.

246. RA ENERGY SYSTEMS, INC.
11459-A Woodside Avenue, South
Lakeside, California 92040

Edward W. Schuller
(714) 448-6216

We are a manufacturer of a solar hot water collector using an all-copper absorber, anodized aluminum frame, insulated inside, and glass glazed. We offer part and/or whole solar hot water systems for domestic, pool and spa applications. We also install these systems.

247. LEE RAGSDALE ASSOCIATES
2325 Sutherland Avenue, P.O. Box 10605
Knoxville, Tennessee 37919

Lee Ragsdale, Jr.
(615) 524-2815

Manufacturer's representative, promoting solar space heating and domestic water heating. Sell systems to contractors and wholesalers.

248. RALOS MANUFACTURING CO.
331 Harvard Street
Cambridge, Massachusetts 02139

(617) 661-2081

Manufacturer of The Sun Machine™, tools for quantitative and qualitative assessment of solar site and building potential.

249. RAMADA ENERGY SYSTEMS, INC.
1421 S. McClintock
Tempe, Arizona 85281

Marketing Department
(602) 273-4300

Ramada Energy Systems, Inc. manufactures and markets transparent polymer solar collectors that can be used in active, passive or hybrid applications with both air or liquid transfer media.

250. RAM PRODUCTS INC.
1111 North Centreville Road
Sturgis, Michigan 49091

Paul Sorkoram
(616) 651-9351

Manufacturer of acrylic mirrors for concentrators that are light, virtually unbreakable, and long-lasting. Supply acrylic mirror in formed parabolas as well as flat.

251. RAYPAK INC.
31111 Agoura Road
Westlake Village, California 91359

Allan Boniface
(213) 889-1500

Raypak offers a complete line of solar water heating products including a single-glazed collector panel for domestic hot water, and a double-glazed collector panel for central heating and absorption air conditioning. In addition to the panels, we offer complete package systems for residential use. Raypak also provides all of the components needed for commercial domestic hot water and hydronic heating applications.

Package residential solar hot water systems are available for new construction as well as for retrofit. Other systems for non-freezing climates are available, as are systems incorporating a simplified thermosyphon design.

Commercial components are designed for complete system compatibility. Raypak offers application assistance in the design of commercial hot water systems.

The Raypak collector panel has been developed, tested and refined over more than five years of research. It incorporates advanced design features including a thermally isolated absorber plate with all copper waterways for efficient operation and long life. The collector also features heavy insulation around the complete periphery, break-resistant tempered water white glass with high transmissivity and low reflectivity, rugged baked enamel exterior housing, field replaceable components for simplified service.

Raypak's automatic control concept incorporates a self-draining feature with sensing and control components, automatic freeze protection and an anti-scale feature. When used with direct systems, the automatic control concept provides maximum utilization of solar energy. Raypak's collector panels can be used with either a direct or indirect system. The direct system offers simplicity of design and eliminates potential contamination of domestic hot water. All copper waterways mean that a direct system can be used eliminating the need for a heat exchanger.

252. R C SOLAR CO. INC.
Box 248
Richardton, North Dakota 58652

Ralph Messer
(701) 974-2181

R C Solar manufactures and sells hot air solar furnaces.

253. R & D ENTERPRISE
Route 3, Box 258
Carthage, Texas 75633

Bill Ritmanich, Jr.
(214) 693-5281

Manufacturer of flat plate domestic hot water systems, freon/water heat exchanger systems, and Heat-Stick thermal transfer systems.

254. REACTION RESEARCH LABORATORIES
P.O. Box 356
Groton, Connecticut 06340

Nathan Carver, Chief Engineer
(203) 536-6979

New products in energy gathering systems include power from windmills, compressed air storage batteries, solar panels, thermoelectric generators (metallic), water power systems (may include windmill pumps), compressed air water pumps.

255. REFRIGERATION RESEARCH INC.
Solar Research Division
525 North 5th Street
Brighton, Michigan 48116

Michael W. Ramalia
(313) 227-1151

Manufacturer of DHW components and systems, primarily; also supplemental space heating. Originated refrigerant charged solar systems. A series of developments we have made on refrigerant charged collector systems eliminates many problems associated with hydronic systems and provides greater efficiency. This is an additional link in the inevitable merging of the solar, refrigeration and air conditioning fields.

256. RESEARCH PRODUCTS CORPORATION
1015 E. Washington Avenue, P.O. Box 1467
Madison, Wisconsin 53701

Floyd W. Carlstrom, Advertising Manager
(608) 257-8801

Research Products Corporation manufactures the Suncell® Solar Heating System—a complete air-to-air solar system for domestic hot water for residential use and space heating for residential, commercial and industrial application. Research Products Corporation offers a sophisticated method of determining the feasibility of a Suncell solar heating system for any installation through the f-chart computer program developed at the University of Wisconsin. Regarded as the most reliable method for predicting the thermal performance and economic feasibility of a solar heating system, this method provides a thermal and economic analysis utilizing 41 inputs, including weather data.

257. RESIDENTIAL ENERGY SYSTEMS INC.
5475 East Evans Avenue
Denver, Colorado 80222

Bert Hansen
(303) 753-0440

The Residential Energy Systems store serves as a working model of the products it distributes—solar collectors, photovoltaic cells, glazing materials, storage systems, wind generators, woodburners, etc.—and as a testimony to alternatives. Also provides numerous services, e.g., resource library, seminars, architect/contractor arrangements.

258. REVERE SOLAR AND ARCHITECTURAL PRODUCTS, INC.
P.O. Box 151
Rome, New York 13440

Russell A. Mallard, Vice-President
(315) 337-1505

2107 South Garfield Avenue
Los Angeles, California 90040

William Cole
(213) 724-6982

Revere manufactures and distributes liquid flat plate solar collectors for commercial, industrial, residential and other uses. In addition, we sell complete domestic hot water systems. Swimming pool collectors are also available.

259. RHO SIGMA
11922 Valerio Street
North Hollywood, California 91335

Andrew Davis
(213) 982-6800

Rho Sigma manufactures controls and instrumentation for solar energy systems. The line includes differential temperature controllers, test instruments and sensors. Rho Sigma engineering, with long experience, has developed the most complete line of electronic solid-state control systems available. Rho Sigma has also developed and has available instrumentation for measuring and recording strength of solar radiation as well as Btu meters for measuring the total solar energy collected, energy used and efficiency of a system.

260. RICHDEL, INC.
1851 Oregon Street, P.O. Box A
Carson City, Nevada 89701

Mr. Tom Erwin, Vice-President/Solar Sales
(702) 882-6786

Manufactures complete fluid handling/control component package of valves, pumps, sensors, controls, vents for domestic hot water and swimming pool systems. We have a complete line of industrial, irrigation and solar products.

261. RIGIDIZED METALS CORPORATION
658 Ohio Street
Buffalo, New York 14203

(716) 849-4703

Originators and largest producers of design-strengthened, deep-textured materials. Manufacturer of Solar Rigid-Tex® which increases multi-reflection of incident solar energy. The patterned surface increases the area for solar energy removal, and the undersurface acts as a turbulence inducer, increasing the heat transfer coefficient. The patterning also adds strength to lighter gauges. Rigid-Tex is available in over 70 patterns, in all ferrous and nonferrous metals.

MANUFACTURERS

262. R-M PRODUCTS
5010 Cook Street
Denver, Colorado 80216

Stephen Piro
(303) 825-0203

R-M Products has manufactured and distributed air and liquid collectors since 1974. The solar components have been specified and installed in 225 jobs in 20 states, 115,000 square feet of panel.

Components manufactured by R-M Products include: air blenders, air and hydronic solar collectors, domestic water systems, and solar air handlers.

Through an inhouse engineering staff, R-M Products can aid in the design of a system to meet job requirements around solar components from R-M. Through an inhouse computer terminal providing access to advanced design and simulation programs, R-M can provide the most desirable system the commercial, industrial or residential consumer might require.

263. ROBERTSHAW CONTROLS COMPANY
100 W. Victoria Street
Long Beach, California 90805

C. M. Little/P. S. Johnson
(213) 636-8301

Robertshaw Controls Company manufactures the SD-30 Solar Commander that provides sensitive temperature response to operate the circulating pump of a liquid system or the fan of an air system. Also manufactures Tempstat temperature and pressure relief valves for domestic and commercial water heaters and boilers.

264. ROLL FORMING CORPORATION
Industrial Park
Shelbyville, Kentucky 40065

(502) 633-4435

Linear shape capability—e.g., for panel framing members.

265. ROM-AIRE SOLAR CORPORATION
121 Miller Road
Avon Lake, Ohio 44012

George T. O'Donnell, Vice-President
(216) 933-5000

Manufacturer of ROM-AIRE™ air-type solar heating systems for domestic hot water, space heating, preheating make-up or ventilating air, heat for drying, dehydration, distillation, process heat—residential, commercial, agricultural and industrial applications.

266. RUSTH INDUSTRIES
P.O. Box 1519
Beaverton, Oregon 97075

Manufacture the StratoTherm™ and StratoJet™ heat syphon fan for industrial and residential use. The StratoTherm recirculates warm air over a 2000 square foot area (requires ductwork); the StratoJet covers 300–500 square feet and has a built-in diffuser, eliminating the need for ductwork.

267. RYNIKER STEEL PRODUCTS CO.
P.O. Box 1932, N.P. Industrial Site
Billings, Montana 59103

W. H. Ryniker
(406) 252-3836

Manufacturer of Ryniker/American air-type solar collectors. Dual channeled, polycarbonate glazing; zinc coated steel housing; high R-factor insulation.

268. SANTA CRUZ SOLARWORKS INC.
2715 Porter Street
Soquel, California 95073

Rick McKee
(408) 462-4095

Santa Cruz Solarworks' locally manufactured collectors are all-copper fin-tube plate, 4 x 8 ft; clear heart redwood frame;

3/16-in. tempered glass glazed; 1-in. foil backed foam insulation; designed for easy removal of glass and plate for maintenance; available in a variety of header hook-up configurations for custom installations.

269. **SAV-ON ENERGY PRODUCTS, INC.**
1080 North 11th Street
San Jose, California 95112

Ed Hopper
(408) 293-9471

Manufacture flat plate solar collectors for our own use; however, we plan to have a few other dealers for our Equinox 4000 and 3000 flat plate copper tube and aluminum fin collectors in the near future. Also, a contractor/solar installer service, dealing primarily with swimming pools and hot water systems. We have installed approximately 1200 solar systems over the past 6 years.

270. **SAV SOLAR SYSTEMS, INC.**
550 W. Patrice Place, Suite A
Gardena, California 90248

Dick Isaacs
(213) 327-7210

SAV Solar manufactures high-temperature water heating systems (model LTC 50), which incorporate the solar-powered concentrating track collector with optimal digital temperature control. The SAV 20-gallon water heater (HD 20) utilizes the cylindrical collector and normal water-line pressures (requires no pumps, storage tanks, wiring or control devices).

271. **J. SCOTTS, INC.**
P.O. Box 563
El Toro, California 92630

23011 Moulton Parkway, B-11
Laguna Hills, California 92653

James R. Scott, President
(714) 581-9630

Specialize in single- and double- wall DuPont Tedlar®PVF film glazings for solar collectors and greenhouses, produced to customer specifications. Produce solar greenhouses for retrofit. Distribute Tedlar®PVF film for solar glazing. Consulting service on Tedlar®film fabrication and processing.

272. **SEALED AIR CORPORATION**
Corporate office:
Park 80 Plaza East
Saddle Brook, New Jersey 07662
(201) 791-7600

Selling office:
2015 Saybrook Avenue
Commerce, California 90040

John F. Zaccaro
(213) 921-9617

The Sealed Air Solar Pool Blanket is the number one selling solar cover in the world. It is presently being manufactured in England and France for our European operation and in Toronto for our Canadian market, as well as in Santa Fe Springs and Commerce, CA/Arlington, TX/Sharonville, OH/ West Springfield, MA.

Our entrance into the solar collector field occurred in mid-February 1980 with the acquisition of American HCP of South San Francisco. In the early fall of 1980 we will take occupancy of a 100,000+ square foot facility in Hayward, CA, in which we will manufacture both the Sealed Air Solar Pool Blanket and the Solar Collector.

273. **SEMCOR ENERGY SYSTEMS**
25651 Taladro Circle
Mission Viejo, California 92691

Bernard J. Maxum
(714) 770-9400

"One-Stop Solar Warehouse" provides a single store for the solar professional to obtain components or complete solar systems. In addition to the standard solar components, such as collectors, controllers, pumps, sensors, and storage tanks, Semcor also provides valves, fittings, pipe, tubing, insulation, meters, wire, tape, cement, solder, heat exchangers, expansion tanks, heat pumps, pool and spa products, and photovoltaic products.

Complete packaged systems include pool, spa and integrated pool/spa systems with either solenoid or motorized valving; domestic hot water systems either active or passive, direct or indirect; Heliophase™ Phase-Change heating systems; and space heating and cooling systems.

274. **SERA SOLAR CORPORATION**
3151 Jay Street
Santa Clara, California 95050

Don Barnby
(415) 321-3729

Manufacturing and R&D, photovoltaic cells.

275. **SETSCO**
(SOLAR ENERGY THERMAL SYSTEM COMPANY)
40 Blaine Circle
Moraga, California 94556

Walter Washick
(415) 376-4057

Manufacture liquid collectors for domestic hot water systems.

276. **SHELLEY RADIANT CEILING CO., INC.**
456 West Frontage Road
Northfield, Illinois 60093

William Shelley
(312) 446-2800

Manufacturer of absorber plates.

96 SOLAR CENSUS

277. SILICON SENSORS, INC.
Solar Systems, Inc. division
Highway 18 East
Dodgeville, Wisconsin 53533

Robert L. Bachner
(608) 935-2707

Manufacturer of photovoltaic cells and panels and solar systems/products.

278. SKILLESTAD ENGINEERING, INC.
Old Highway 52 South
P.O. Box 17
Cannon Falls, Minnesota 55009

Craig Coleman
(507) 263-4735

Manufacture and distribute Econ-O-Sun solar systems; complete 8' x 12' collector with duct work and air handler. The system can be installed by the homeowner in 3 to 5 hours.

Distribute waste oil burners which burn used crankcase oil, hydraulic fluid and transmission fluid with no smoke, no odor and less pollution than a regular oil-fired home furnace. Waste oil is cheap, but has more Btu's per gallon than regular heating oil.

Also distribute kerosene portable heaters in southern Minnesota. These efficient petroleum burning devices burn kerosene, no. 1 fuel oil or just fuel with 99.98% efficiency. They are UL approved and they do not need venting except in small areas.

279. A. O. SMITH
P.O. Box 28
Kankakee, Illinois 60901

Bob Jones
(815) 933-8241

Manufacture The Conservationist® solar domestic water heater, tank, and collector system.

280. HAROLD A. SMITH & ASSOCIATES
P.O. Box 42431
Houston, Texas 77042

Harold A. Smith
(713) 493-6207

Manufacturers' agent in the industrial heating and ventilating field.

281. SOLAG
P.O. Box 676
Roseville, Illinois 61473

Dan Stice
(309) 426-2620

Manufacture systems for solar corn drying. Management objective is to market solar products to the agricultural market.

282. SOLAHART CALIFORNIA
3560 Dunhill Street
Sorrento Valley West
San Diego, California 92121

Graham Rose
(714) 452-5805

An Australian manufacturer of total systems selling in San Diego, Fresno and Sacramento through company outlets. Distributor in Phoenix, AZ. Other outlets in Chatsworth, Los Angeles and San Jose, CA, Hawaii and Florida.

283. SOLARA ASSOCIATES INC.
1001 Connecticut Avenue N.W.
Suite 632
Washington, D.C. 20036

Richard L. Hoddinott
(202) 296-7070

Solara Associates Inc. acquired the Solar Products Division of Natural Energy Corporation in June 1979. Solara manufactures and markets flat plate liquid and nontracking concentrating collectors, has a vacuum tube in the development stage and markets complete heating, hot water and air conditioning solar systems of our design. The Design Section is headed by an AIA and offers energy audits as well as solar design.

284. **SOLARADO INC.**
2625 South Santa Fe Drive
Denver, Colorado 80223

Ronald F. Reinisch
(303) 778-0221

Solarado manufactures an active hydronic solar collector panel and serves as the distributor for Denver, CO. Our solar collector is designed to handle heating applications like hot water, space, swimming pools and hot tubs. We install complete systems in metro Denver through direct sales. The pleasing appearance and color choice make the collector panel compatible with both residential and commercial architectural requirements. Exclusive distributorships are available.

285. **THE SOLARAY CORPORATION**
2414 Makiki Heights Drive
Honolulu, Hawaii 96822

Lawrence Judd
(808) 533-6464

Manufactures special thermometers designed for the solar industry. We have also been, for the last five years, a distributor of a complete line of solar products.

286. **SOLAR CITY**
3519 Henderson Boulevard
Tampa, Florida 33609

Ben Bentley
(813) 877-5114

Manufacturer of solar controls, roof flashings, domestic water systems, and home heating systems.

Also, distributor for other major component manufacturers.

287. **SOLAR COLLECTOR SALES CO., INC.**
7766 Auburn Road
Utica, Michigan 48087

Mr. Antonio Trivelloni
(313) 731-8692

Distributor of Solarator collector for use in swimming pool, spa, hot water and heat pump systems. Retail sales of DHW systems, swimming pool systems and energy saving devices.

288. **SOLAR COMFORT SYSTEMS MANUFACTURING CO.**
311 Mount Vernon Place
Rockville, Maryland 20852

Dave DeRiemer
(301) 424-1362

We manufacture and market complete solar systems for heat (space) and hot water, for do-it-yourself installation, for low-cost adaptation to new and/or existing structures.

Sales are: 1. direct to users (primarily do-it-yourselfers); 2. to hardware and woodstove stores for resale to users; 3. to contractors for resale to their customers.

Systems are tested to ASHRAE and NBS standards, comply with HUD requirements, are five-year warranteed with 60-day buy-back warranty upon termination of five-year term, approved for financing through VA, HUD, FHA, FmHA, GNMA and conventional sources.

Company is licensed, bonded, member of U.S. Chamber of Commerce, Better Business Bureau, and Middle Atlantic Solar Association, and has been in business since 1969.

289. **SOLAR CONSTRUCTION AND DESIGN INC.**
6 Lyndhurst Avenue
Wilmington, Delaware 19803

Mr. David Kane
(302) 478-4344

Dealer, installer of Sunwise and SECO systems; insulation; passive builder; general energy consultant.

290. **SOLAR CONTACT SYSTEMS, INC.**
1415 Vernon Street
Anaheim, California 92805

Douglas LaFleur
(714) 991-8120

Manufacture components and total systems for DHW, pool and spa applications.

291. **SOLAR CONTROL CORPORATION**
5721 Arapahoe
Boulder, Colorado 80303

Thomas B. Kent
(303) 449-9180

Solar Control Corporation is a two-year-old Boulder, CO based firm which designs, manufactures and markets solar system components. The company has two primary lines of solar components: solar air movers and solar controllers. for space heating (air and water transport), domestic hot water, swimming pools and spas.

Solar Control also has three subsidiary companies: Solar Environmental Engineering Co., Fort Collins, CO, an R&D company which develops and analyzes new products and concepts in the entire solar energy field; Solar Energy Equipment Corporation, Costa Mesa, CA, which markets swimming pool and domestic water solar heating systems; and E. D. & D. Manufacturing Co., Pacoima, CA, which manufactures electronic parts primarily for the aerospace industry.

292. **SOLAR COOKER PARTS**
3548 Grand Avenue South
Minneapolis, Minnesota 55408

Don Johnson
(612) 825-1678

Solar Cooker Parts manufactures 48-in. precision spun-aluminum parabolic focusing reflectors that can be highly polished by hand. The reflectors have a 5/8-in. rolled edge and carry a 100% money-back guarantee and should last 100 years or more with minimum care.

Also, distributor for Dyna-Tech Wind Chargers.

293. **SOLAR DATA SYSTEMS INTERNATIONAL**
Box 8000
Atlanta, Georgia 30357

(404) 955-0055

Manufactures the Solar Power Calculator.

294. **SOLAR DESIGNS**
470 3rd Street
San Francisco, California 94107

David Jones
(415) 495-7074

Manufacturers' Representative for SUNWORKS and INDEPENDENT ENERGY. We have been in business for three years and will be shortly expanding our line to pool collectors and possibly air-to-air heating and cooling systems.

295. **SOLAR DEVELOPMENT INC. (of Denver)**
11799 East 39th Avenue
Aurora, Colorado 80010

Customer Relations
(303) 343-8154

Manufacturer of SOLSTAR®, SDI designs, manufactures and distributes complete air solar space heating and water heating systems for residential, commercial, industrial and agricultural applications. Products sold through a dealer network of over 80 dealers. Systems have been installed in over 200 homes.

296. SOLAR DEVELOPMENT INC.
3630 Reese Avenue
Riviera Beach, Florida 33404

Ray Jefferson
(305) 842-8935

SDI has been manufacturing and developing flat plate collectors for pools, hot water, space heating and air conditioning for over six years. We also manufacture a circulating pump control for domestic hot water systems.

There are several companies that manufacture our collectors under license. Western SDI in Vacaville, CA, Universal SDI in Orlando, FL, SDI Northwest in Pocatello, ID, and Denersol C.I.A. Ltda. in Quito, Ecuador.

SDI provides solar engineering services and also has an R&D Department. Our latest project is a highly efficient still for the production of alcohol fuel.

297. SOLAR DYNAMICS OF ARIZONA
1100 North Lake Havasu Avenue
Lake Havasu City, Arizona 86403

Floyd B. Hamilton
(602) 855-5051

Solar Dynamics, through an ongoing R&D program, has developed and manufactures flat plate collectors; Sola-Sentry, a drain down valve for open loop systems; Dyna-Stil, a solar water distillation unit; electronic controllers for the Solar Dyne system, and mounting hardware for Solar Dyne collectors.

Research and development continues toward products in the photovoltaic division, and solar production of ethanol.

298. SOLAR ELECTRIC INC.
403 South Maple
West Branch, Iowa 52358

Clarence E. Sewell
(319) 643-2598

Manufacture total self-contained solar-powered auxiliary heater systems. The Delta National auxiliary heater automatically blows heat into the room to give extra warmth that keeps the regular heating system from going on, even with the thermostat turned down to conserve energy.

As the solar-heated air circulates, the cooler air is pulled out of the room and pushed into the solar heat collectors.

299. **SOLAR ENERGIES OF CALIFORNIA**
 11421 Woodside Avenue
 Lakeside, California 92040

 Dwight H. Nicewonger
 (714) 448-4300

 Manufacture SUNMATE™ air- and liquid-type collectors and solar hot water systems for residential, commercial and industrial use.

300. **SOLAR ENERGY OF COLORADO, INC. (SECO)**
 4230 Fox Street
 Denver, Colorado 80216

 D. May
 (303) 433-8761

 SECO manufactures and distributes a sub-ambient solar system (the COOLECTOR) for domestic hot water, space heating, hot tubs, spas and swimming pools. The collector is cooled with a refrigerant to operate in a sub-ambient condition, therefore producing energy (heat) at low-ambient temperatures, even at night in a snowstorm.

301. **SOLAR ENERGY ENGINEERING, INC.**
 31 Maxwell Court
 Santa Rosa, California 95401

 Lynn Selby
 (707) 542-4498

 Solar Energy Engineering, Inc. (S.E.E.) manufactures all-copper absorber plates which are also enclosed and glazed. Besides collectors, S.E.E. manufactures a water-to-water exchanger tank. S.E.E. also designs and installs solar space, hot water and pool heating systems locally.

302. SOLAR ENERGYMASTER
P.O. Box 592
Florham Park, New Jersey 07932

Manufacture flat plate collectors and Solar Energymaster™ DHW systems. DHW systems feature "drainback" design that uses pure water in its "solar loop" piping and prevents freezeup.

303. SOLAR ENERGY RESEARCH CORPORATION
10075 East County Line Road
Longmont, Colorado 80501

Jim Wiegand
(303) 772-8406

SERC manufactures the total system Universal Air/Water Spray Solar Collector for space heating, cooling and DHW application, the solar hot water Heat Recovery Package (HRP), and Heat Exchange Module (HEM).

304. SOLAR ENGINEERING AND MANUFACTURING
East Highway 275 Industrial Area
Fremont, Nebraska 68025

Dean Morrison
(402) 721-9966

Since 1941, Solar Engineering and Manufacturing has produced Solar Hot Tubs and other items for the do-it-yourself market as well as complete systems. Windchargers constitute a major part of the operation. Also design and construct solar alcohol plants with backup electric. Mail order or direct sales.

305. SOLAR EQUIPMENT CORPORATION
P.O. Box 357
Lakeside, California 92040

Jack Hedger, President
(714) 561-4531

Presently tooling up to produce kits of parts (also single components) to be assembled into solar-electric plants by customer. The missing link in the solar industry is the small reciprocating inexpensive engine. We now have this engine in seven varieties.

306. **SOLAREX CORPORATION**
1335 Piccard Drive
Rockville, Maryland 20850

Bob Edgerton
(301) 865-5690

Manufacture photovoltaic cells and panels. Through a DOE grant, are researching and developing semicrystalline silicon, a material which can be poured and cast and then cut into wafers, thus lessening the production costs of photovoltaics.

307. **SOLAR-EYE PRODUCTS INC.**
1300 N.W. McNab Road
Ft. Lauderdale, Florida 33309

Les Botoff, President
Wally Starr, Vice-President
(305) 974-2500

Solar-Eye Products Inc was established in Ft. Lauderdale, FL, and commenced business May 1, 1976, as a result of the merger between Electronic Devices & Controls Company, Inc. and American Solar Heating, Inc., offering 25 years experience in system control engineering and 3 years experience directly in the application and manufacture of solar collector panels, controls and solar systems.

The equipment system offered for domestic, commercial or industrial applications consists of : 1. solar collector panels as required; 2. temperature difference control, temperature sensors and circulating pump; 3. universal field erected mounting frame for collector panels; 4. solar tank (electric hot water heater), optional.

308. SOLAR FAB, INC.
151-23 34 Avenue
Flushing, New York 11354

Chas. Hutcheon
(212) 463-2100

Manufacture air and liquid collectors and systems. Design and consulting services available also.

309. SOLARGIZER INTERNATIONAL INC.
2000 West 98th Street
Bloomington, Minnesota 55431

W. Lee Tallakson
(612) 888-0018

Manufacturer of Solargizer(TM) domestic hot water, space heating and swimming pool systems. Collectors constructed of copper, aluminum and fiberglass.

310. SOLARGY CORPORATION
17914 East Warren Avenue
Detroit, Michigan 48224

Bror Hanson, President
(313) 881-5510

We manufacture the Solargy Tower—a free standing steel tubular column for wind turbine generators. We also distribute wind generation systems.

311. SOLAR HEAT CO.
10347 3rd Avenue
P.O. Box 42
Hesperia, California 92345

Ralph Kipper
(714) 244-5912

We produce liquid collectors with copper waterways using extruded aluminum absorber fins, both unglazed and glazed, with insulation mounted in frames of extruded aluminum.

We obtain design service as required to produce and/or install any part, or all, of a system.

312. SOLAR HEATER MANUFACTURING CO.
1011 Sixth Avenue South
Lake Worth, Florida 33460

Mr. Joseph J. Daetwyler, Owner
(305) 586-3839

Manufacturer of the SHM-10 solar collector. Certified by the Florida Solar Energy Center. Complete design and installation services for all applications of liquid solar systems systems. Retail and wholesale sales of all major components, such as collectors, tanks, pumps, controls, pool covers. Also a distributor for the Convectionaire ductless heat pump system for heating and cooling of residential and commercial buildings. (Manufactured by C.T.I. Inc. of Clearwater, FL).

313. SOLAR HOMES INC./SUNTROL
275 Harborside Boulevard
Providence, Rhode Island 02905

Michael F. Smith, President
(401) 884-0353

Manufacturer and distributor of complete line of Suntrol® solar heating products for residential and commercial application—new or retrofit. Full range of heat pumps, automatic controllers, air collectors, liquid collectors— all components necessary for collecting, storing and distributing solar heat.

314. SOLAR, INCORPORATED
008 Sunburst Lane
Mead, Nebraska 68041

Ron Steinauer
(402) 332-3313

Manufactures Solar-Aire complete residential and commercial systems. Also the Solar-Aire Convertible Yard Unit, a free-standing unit for home retrofit and agricultural applications; and the SI Suntainer eutectic salt heat storage system.

315. **SOLAR INDUSTRIES, INC.**
2525 West 6th Avenue
Denver, Colorado 80204

Larry Stordahl
(303) 534-4133

Solar Industries, Inc. markets, installs and services solar energy systems, energy management systems, and conventional mechanical and electrical systems, primarily in commercial, residential, industrial and institutional applications. We have worked directly with eight solar manufacturers to date, in a variety of applications.

316. **SOLAR INDUSTRIES, INCORPORATED**
Monmouth Airport Industrial Park
Farmingdale, New Jersey 07727

Dwight Macock
(201) 938-7000

Manufactures Slimline™ CMT low- and medium-temperature liquid collectors and controls for DHW and space heating systems, as well as for solar swimming pool/spa heating.

317. **SOLAR INDUSTRIES OF FLORIDA**
(SUNTRAK, INC.)
P.O. Box 9013
Jacksonville, Florida 32208

Gary Laird
(904) 768-4323

Manufactures collectors for water, space, pool and spa heating. Distributes and installs collectors, control equipment, storage systems and complete systems.

Suntrak, Inc., formed in 1978, assumed Solar Industries of Florida, the oldest solar company in northeast Florida.

318. SOLAR KINETICS, INC.
8120 Chancellor Row
Dallas, Texas 75247

John Witt
(214) 630-9328

Manufacture the T-600 and T-700 parabolic trough concentrating collectors. These collectors are primarily used in industrial process heat and heating/air conditioning applications.

319. SOLAR KING INTERNATIONAL, INC.
8577 Canoga Avenue
Canoga Park, California 91304

John Irwin
(213) 998-6400

Solar King International manufactures solar components and energy management controls as well as provides system engineering for total solar system applications. SKI was established in 1976 and has a current contracting license (both General and Solar). All products are warranteed in accordance with California requirements as well as the national SEIA standards.

320. SOLAR MAGNETIC LABS
121 Riley Avenue
Lake Worth, Florida 33461

Richard Dawley, Sr.
(305) 965-0606

Solar Magnetic Labs' goal is not only to produce more practical and efficient solar products, but the economical application of them. We manufacture the "Arctic Solar Heater," (A.S.H.), a convexed-plate liquid solar collector. This patented solar absorber will develop much higher Btu efficiencies per square foot than any conventional flat plate concept.

Also manufacture the "Solar Awning," a basic A.S.H. unit under an aesthetically appealing Colonial, Cape Cod or Salt

Box style awning cover. Classed as a complete system for heating rooms or several units may be used for heating any dwelling using daylight as fuel. It can also directly de-ice driveways or sidewalks.

Presently under testing is the mobile solar still, designed to produce a substitute for fossil fuel, ethanol alcohol, from a mash made from fermented corn or other grains and water. The solar still panel uses only the sun's power to distill the mash, resulting in approximately 38% burnable ethanol from one run through the solar still. Further processing produces a finer, cleaner, nonpolluting fuel.

321. **SOLAR MUSIC INC.**
10615 Chandler Boulevard
North Hollywood, California 91601

Earl Turner
(213) 980-6911/980-6915

Manufactures and distributes solar music boxes and books, and the Solar Energy and Energy Conservation Curriculum and Experiment Kits for grades kindergarten through sixth.

322. **SOLARNETICS CORPORATION**
1654 Pioneer Way
El Cajon, California 92020

Harry Blackstone
(714) 579-7122

Solarnetics Corp. is the exclusive U.S. factory representative for Electra Ltd., manufacture of the "Solarnetics-Electra EC-2" panel—high-efficiency, vertical-finned flat plate collector. Electra also manufactures the "Electra-Solarius" swimming pool system—highly versatile and rugged solar pool heater. We work with developers, distributors and specifying engineers/architects. As more products are developed to a workable level, we will offer other energy conversion systems. Our current emphasis now is on hot water, pool and space conditioning systems.

MANUFACTURERS 111

323. SOLARON CORPORATION
1885 West Dartmouth Avenue
Englewood, Colorado 80110

Lance Löf/Roger Livak
(303) 762-1500

SOLARON is the first publicly held corporation devoted entirely to the development and marketing of solar energy systems. The founders of SOLARON have substantial experience in energy management and solar research. Dr. George Löf, Chairman of the Board and Technical Director, is the past president of the International Solar Energy Society and is Chairman of Colorado State University's Solar Energy Laboratory. The solar heating system that we offer today is based on over 30 years of experimentation and tests conducted by Dr. Löf.

Today, SOLARON is organized as a research and development, systems engineering and marketing firm. We have established a nationwide distributor and manufacturer's representative network. SOLARON works closely with these people in all phases of product application, systems design, training seminars, advertising, sales and promotion.

SOLARON sells both air and liquid flat plate collectors, featuring a variety of domestic hot water systems and complete air space heating systems. This includes air-type collectors, air handlers and automatic system controllers— all the equipment necessary for a solar system installation. SOLARON systems are in operation throughout the U.S., Canada and Europe and are installed in residential, commercial, industrial and agricultural applications.

324. SOLAR PATHWAYS, INC.
3710 Highway 82
Glenwood Springs, Colorado 81601

Robert Clarke/John Ehlers
(303) 945-6503

Manufactures the Solar Pathfinder,™ a site-survey device used to trace the horizon and find energy availability. The use of this instrument is a great timesaver, compared to plotting azimuths and altitudes. It provides the client with a clear depiction of the site's solar accessibility due to (horizon) shading.

325. SOLAR POWER CO. INC.
**556 South Main
P.O. Box 4043
Pocatello, Idaho 83201**

James F. Blake
(208) 232-2320

Active in residential, commercial and agricultural space and water heating, alcohol production and wind generation. We work with mechanical contractors, architects, dealers and individuals to achieve energy efficiency through the most appropriate alternative energy application, and to provide proper system equipment, installation and operating instructions, contractor referral and assistance, as needed.

326. SOLAR POWER CORPORATION
**20 Cabot Road
Woburn, Massachusetts 01801**

William L. Schlosser
(617) 935-4600

Solar Power Corporation, an affiliate of Exxon Enterprises, Incorporated, was established in 1973 as a manufacturer of high-performance, photovoltaic modules and systems. Solar Power Corporation provides rugged and reliable hardware combined with experienced systems and applications engineering to assure the dependability of every Solar Power Corporation solar electric system.

327. SOLAR PRODUCTS, INC.
**2419 20th Street
Rockford, Illinois 61101**

James McGowan/Phil Mosher/James Hartley
(815) 397-0536

In business since 1976, have sold and installed 24,000 square feet of residential space heating solar collectors,

21,000 square feet of residential swimming pool collectors, 4,000 square feet of commercial swimming pool collectors, and portions of many more air, liquid and hybrid systems. We are also active in the construction of atriums, sunspaces and greenhouses and manufacture insulating shutters (modular self-hinging for greenhouses). Also manufacture heat storage phase-change synthetic waxes for air and liquid systems. We distribute our products and those of other manufacturers in Illinois, Iowa, Wisconsin and western Indiana. We have a sales office in St. Charles, Illinois.

328. SOLAR PRODUCTS MANUFACTURING CORP.
1 Alcap Ridge
Cromwell, Connecticut 06416

Alvin L. Trumbull
(203) 635-0266

Solar Products Manufacturing Corp. features the line of SUN-SPOT solar products. We manufacture the six major elements of the SUN-SPOT system: the solar collector—SUN-SPOT, the storage tank—SUN-BANK, the transfer fluid—SUN-TEMP, heat exchangers, differential controls and the SUN-SPOT glass-lined, highly insulated electric hot water heater backup.

329. SOLAR PRODUCTS SUN TANK INC.
4291 N.W. 7th Avenue
Miami, Florida 33127

Mr. Houtkin
(305) 756-7609

We manufacture, distribute and install solar water heating systems for pool and domestic hot water.

330. SOLAR RADIATION INDUSTRIES
2531 Plantside Drive
Louisville, Kentucky 40299

R. W. Rademaker, Sr.
(502) 267-9636

Solar Radiation Industries manufactures heat transfer control modules containing heat exchangers, pumps, control devices and specialties comprising the central control center of the complete solar system.

331. SOLAR RESEARCH SYSTEMS
3001 Red Hill Avenue 1-105
Costa Mesa, California 92626

Joseph Farber, PhD
(714) 545-4941

Manufacture Sol-Heater™ flat plate collectors and solar swimming pool heating systems.

332. SOLAR RESOURCES
a division of United Standard Management Corporation
1735 East Indian School Road
Phoenix, Arizona 85016

Paul Kelley
(602) 263-9447

Solar Resources has been the major distributor of concentrating collectors for domestic use in the Southwest.

333. SOLAR RESOURCES, INC.
Box 1848
Taos, New Mexico 87571

Ms. Leah Alexander
(505) 758-9344

Manufacture the solar greenhouse collector Solar Room,® using a double layer of plastic as glazing. Large single sheets of plastic are stretched over a steel frame and are "zipped" into place along the perimeter of the collector with an aluminum extrustion developed especially for this purpose (patent pending). A low-wattage blower inflates the glazing between the two layers to produce an air tension structure. Because of the minimum of materials used and the simplicity of assembly and installation, Solar Room® has proven to give 4 to 10 times more heat per dollar than other active systems on the market.

Pioneer in the research and development of passive/hybrid solar energy products using thin-film technology. Well-equipped with a prototype R&D shop and thermal testing facility which allows it to comprehensively test the thermal and economic performance of various solar devices. Works under contract for private solar industry on engineering, design and testing projects.

334. SOLAR SEARCH CORPORATION
Rural Route 1, Box 42
Rankin, Illinois 60960

Stuart E. Warnock
(217) 397-2258

Solar Search Corporation, established May 1976, manufactures air-type solar collectors, complete systems (Sun Spotter and Sun Scope); mobile multi-purpose agricultural systems. Continuous research program for product improvement, new uses for solar energy. Also, other energy conserving, heat recovery devices and research: clothes dryers, motor heat shrouds, distillation research, electrical power generation.

335. SOLAR SHELTER, INC.
ENERGY 2000, Marketing Division
800 South Council Street
Muncie, Indiana 47302

Jane Dominique
(800) 428-8644

Solar Shelter manufactures complete air-type domestic hot water systems. Energy 2000, Solar Shelter's marketing division, in addition to marketing and distributing their own products, represent other manufacturers' lines, including hydronic heating systems, unglazed pool and spa collectors, geothermal heat pumps, portable grain dryers, animal confinement heating systems, and control systems.

336. SOLAR SPECTRUM INC.
Seahorse Plastics Corporation
4680 Shoulders Hill Road
Suffolk, Virginia 23435

Jeremy C. Lotz
(804) 488-0472

We manufacture/build solar fiberglass foam roofing systems, tanks, collectors and pans, flat plate trickle collectors, geothermal systems, concentrating collectors, storage systems, insulation systems and dome homes.

337. SOLAR SUPPLY, INC.
6709 Convoy Court
San Diego, California 92111

Donald A. McDonough
(714) 292-7811

We are a manufacturer of pump and control modules for domestic hot water systems.

338. SOLAR TECHNOLOGY CORPORATION
2160 Clay Street
Denver, Colorado 80211

Carl Lehrburger
(303) 455-3309

Solar Technology Corporation is a passive solar company that specializes in the manufacture of the passive solar GreenRoom and the distribution of a variety of products for energy conservation and solar energy. We have four major products as well as approximately 20 energy conservation products for the consumer.

The SOLTEC GreenRoom© is a pre-fabricated assembly, using modular construction. Window-Quilt™ is an innovative movable insulation system, that can be installed on almost any window or sliding glass door, substantially reducing window heat loss. SOLTEC Hot Water© is a complete domestic hot water system that requires no pumps, controls, or reserve tank and uses ordinary tap water as its transfer medium. Acrylite SDP™ double-skinned acrylic sheet is an alternative to thermopane glass. The SDP glazing system is composed of both the sheet itself, and an extrusion system for mounting, that provides great design flexibility.

339. **SOLARTEC INC.**
250 Pennsylvania Avenue
Salem, Ohio 44460

A. T. DeCrow, Jr.
(216) 332-9100

Solartec Inc. manufactures liquid flat plate collectors and packages, and distributes a DHW system.

340. **SOLAR THERMAL SYSTEMS INC.**
90 Cambridge Street
Burlington, Massachusetts 01803

B. Tepper
(617) 272-8460

Manufactures Daystar® collectors and prepackaged systems. Daystar® pre-engineered packages for residential hot water and space heating applications are also expanded to meet the needs of larger structures. Market through nationwide dealers (for residential) and manufacturers' representatives (for commercial).

341. **SOLARTHERM, INC.**
1110 Fidler Lane, Suite 1215
Silver Spring, Maryland 20910

(202) 882-4000

Distributes air and liquid collectors, photovoltaic cells and panels for DHW applications. Distributes the Lordan® Collector, with three-dimensional absorber, light-weight copper tubing and aluminum fin collector, applicable to space, water and pool heating uses.

342. SOLARTHERM MANUFACTURING CORPORATION
768 Vella Road
Palm Springs, California 92264

Reynold Watson, Secretary-Treasurer
(714) 323-2687

Solartherm manufactures a total heating and air conditioning system—the Energy Machine. By means of a three-way valve system and compressor, a heat transfer fluid is pumped between the collector installed on the roof and the valance system installed around the perimeter of your home or business, moving heat into the structure during cold weather, and out of it during warm weather. The direction of the flow is automatically controlled by a thermostat. While the collector is absorbing heat from the sun or surrounding air, the valance on the other end is giving it off.

343. SOLAR-TROL
Division of Trolex Corporation
1078 Route 46
Clifton, New Jersey 07013

Dick Foster
(201) 245-3190

Manufacture motorized TROL-A-TEMP dampers and controls for use with any air heating or cooling system, including solar.

344. SOLARTRONICS INC.
3101 East Reno
Oklahoma City, Oklahoma 73117

C. W. McDonald
(405) 672-0215

Manufactures the Solartronics Solar Heating System which utilizes a digital tracking system with a parabolic collector, a double-walled copper heat exchanger in a stone-lined tank equipped with a thermostatically controlled electric heating element. Each system is individually designed to the customers' needs. Solartronics offers its design and manufacturing capabilities to builders and commercial developers for use in business and industrial construction.

345. **SOLAR UNLIMITED INC.**
37 Traylor Island
Huntsville, Alabama 35801

John W. Crumrine
(205) 534-0661

Manufacturer of Suncatcher® heat exchanger assemblies. Market space heat and hot water systems pre-engineered with mounting gear for quick installation in new or retrofit situations.

346. **SOLAR UNLIMITED, INC.**
240 East Main
P.O. Box 63
Sun Prairie, Wisconsin 53590

Gary A. Brinkman
(608) 837-4788

Manufacturers' representative handling collectors, controls and storage systems.

347. **SOLAR USAGE NOW, INC.**
420 East Tiffin Street
P.O. Box 306
Bascom, Ohio 44809

Joseph Deahl
(419) 937-2226

We manufacture a liquid-type collector. Also, we package and distribute a complete line of solar and alternative energy hardware. We serve primarily the do-it-yourselfers and contractors/installers interested in purchasing components or package kits.

348. SOLAR WATER HEATERS OF N.P.R. INC.
1214 U.S. 19 North
New Port Richey, Florida 33552

Bill Ramsey
(813) 848-2343

Manufacture flat plate, liquid-type, glazed medium-temperature solar collectors—32 square feet. Install total systems for domestic hot water.

349. SOLAR WORLD, INC.
4449 North 12th Street
Suite 7
Phoenix, Arizona 85014

Carl C. Adair, President
(602) 266-1819

Solar World developed and manufactures the electronically controlled and computerized Micro-Thermonic Remote Control Panel, which controls an entire heating, air-conditioning and solar system automatically, and the Solar World Hydronics Assembly, a professionally engineered, automatically controlled, preassembled, one-package hydronics unit, to be used in conjunction with the Micro-Thermonic Panel.

350. SOLATION PRODUCTS INC.
111 West Road
Cortland, New York 13045

(607) 753-9658

Manufacturer of concentrating flat plate collectors and systems for hot water, space heating and cooling, and latent heat storage. Residential, commercial and agricultural applications.

MANUFACTURERS 121

351. SOLEC INTERNATIONAL, INC.
12533 Chadron Avenue
Hawthorne, California 90250

Mr. Ishaq Shahryar
(213) 970-0065

Solec's standard industrial line of products includes 3- and 4-in. diameter solar cells, solar panels to meet a variety of load requirements, voltage regulators, array mounting hardware, and 12-volt batteries. Panels have passed Jet Propulsion Laboratory's qualification tests satisfactorily. Solec has the capability to design panels and arrays to suit any customer requirements.

352. SOLECTRO-THERMO, INC.
1934 Lakeview Avenue
Dracut, Massachusetts 01826

Robert E. Charlton, Vice-President, Marketing
(617) 957-0028

Solectro-Thermo, Inc. has a unique electric/thermal solar energy system (Patent no. 4,080,221), with a simple, in-house, 3-position ($30°E$ of S, $0°S$ and $30°W$ of S) tracking mechanism—designed for onsite or remote generation of electricity and heat energy from one compact system.

Under 1000 wm^2 solar intensity, one square meter of our solar hybrid array will generate 50 watts of electric energy and 350 watts of thermal energy. Tracking generates 35-40% more electricity than does a stationary concentrating collector and 15-18% additional thermal energy.

353. SOLERGY COMPANY
7216 Boone Avenue North
Minneapolis, Minnesota 55428

Lorraine Grimaldi
(612) 535-0305

We manufacture liquid flat plate collectors with selective or nonselective surfaces. We also manufacture a complete line of domestic hot water systems with both aqueous (glycol) and nonaqueous heat transfer fluids. Basic packages and custom sizing with f-chart is done for whole house space heating and solar-assisted heat pump systems.

354. **SOLEX CORPORATION**
     187 Billerica Road
     Chelmsford, Massachusetts 01824

     Charles Edwards
     (617) 256-8724

     Solex manufactures nonmetallic solar collectors and sells solar systems through dealers throughout the U.S. and Canada.

355. **SOLEX SOLAR ENERGY SYSTEMS, INC.**
     444 Bedford Road
     Pleasantville, New York 10570

     Patrick R. Sbarra
     (914) 747-0786

     Manufacture components and total systems for all solar heating applications.

356. **SOL-LECTOR, INC.**
     1529 South Division Avenue
     Grand Rapids, Michigan 49507

     Nancy J. Borr
     (616) 245-7679

     Manufacture solar panels for hot water heating. Distributor for storage tanks and pumps. Provide assistance in design of systems (sizing, installation, etc.)

357. **SOLLOS INCORPORATED**
     2231 South Carmelina Avenue
     Los Angeles, California 90064

     Dr. M. Macha
     (213) 820-5181

Manufacture silicon solar cells, solar electrical modules and 6- and 12-volt battery chargers and solar electric generators.

358. SOLPOWER INDUSTRIES, INC.
10211 C Bubb Road
Cupertino, California 95014

Allan Freeman
(408) 996-3222

Manufacture Solpower all-copper solar collectors, both glazed and unglazed. We also sell hot water heating systems for pool, spa and DHW, which are installed by subcontractors, and sell and install space heating systems.

359. SOL-TEMP INC
1505 42nd Avenue
Suite 4
Capitola, California 95010

J. Pastore
(408) 462-2917

Sol-Temp manufactures a self-contained passive hybrid total energy system that provides hot air, hot water and electrical power, plus heat storage and energy backup.

Sol-Temp has a unique "energy skin" concept, designed to collect, store and supply energy at the exterior walls of a structure, instead of transporting energy to a remote basement area for storage. In effect, the building "skin" provides provides energy and the space needs of the occupant directly, eliminating the usual distribution and storage losses. This highly efficient system also features the flexibility of either see-through or solid wall models, all modular plug-in units fitting into standard construction dimensions. The factory-checked, plug-in Sol-Temp wall module can be easily installed by a semi-skilled technician. Ideal for both retrofit and new construction.

360. SONNEWALD SERVICE
Route 1, Box 1508
Spring Grove, Pennsylvania 17362

H. R. Lefever
(717) 225-3456

Manufacturer's representative—absorbers, heat exchangers, flashings, etc. Also, offer design services relating to use of these products. Consulting services, all solar applications.

361. SOUTHEASTERN SOLAR SYSTEMS, INC.
4705 J Bakers Ferry Road
Atlanta, Georgia 30336

Charles E. Moore
(404) 691-1960

Manufacturer of residential solar systems, including the MINI SYSTEM—a complete, supplementary heating system consisting of solar collectors, a storage tank, pumps, solid-state controls, and fireplace heat extraction grate.

362. SOUTHWEST ENER-TECH, INC.
3020 Valley View Boulevard
Las Vegas, Nevada 89102

Gary Halverson
(702) 876-5444

Manufacturer of air- and liquid-type collectors for DHW and space heating/cooling in residential and commercial applications.

363. SOUTHWEST SOLAR CORPORATION
441 North Oak Street
Inglewood, California 90301

Michael Koberlein
(213) 673-8140

Southwest Solar Corporation's foremost goal is to provide the best possible solar equipment consistent with today's proven technologies. Research, development, testing and quality are all a constant part of our effort to aid those in search of a safe and reliable energy future. All of our collectors and systems, with their accompanying warranties, comply with the tax credit regulations of the California Energy Commission, and therefore qualify for the state solar tax credit.

Southwest Solar Corporation is the manufacturer of the Solar Disc 1000 and 2000 liquid flat plate collectors, as well as spa and pool systems. The Solar Disc 1000 is a 2 x 8 ft collector used in many domestic hot water systems. The Solar Disc 2000 is the 3 x 8 ft collector used in domestic hot water systems as well as spa and pool systems. Both models are manufactured with the same integrity and quality and are similar in terms of appearance and performance. The 1000 and 2000 models incorporate an all-copper absorber plate, selective surface (black chrome on nickel), single glazing low-iron tempered glass, double anodized aluminum frame, closed cell insulation (R11) with asphalt backing, hinge mounts and interlocking lips (for structural integrity), and venting and regenerating dessicant (to prevent condensation). Both models are available in a clear or bronze anodized finish.

Southwest Solar's Disc 2000 has a greater maximum flow rate to enable it to be multipurpose. It incorporates opposing 1/2 in. nipples which allow it to be mounted horizontally as well as vertically. The only major differences between the Solar Disc 1000 and 2000 are in size, flow rate, mounting and increased cost-effectiveness in the 2000 series.

New to our line of products is our 9000 series Solar Spa Heating Systems. These unique systems feature high-performance Solar Disc 2000 collectors, high-accuracy temperature limiting dial, a modular design with guaranteed flow rate, low energy usage pump and a fail-safe antiscald system with backup. The do-it-yourself simplicity allows the use of inexpensive polybutylene piping and simple sizing affords 75% -plus solar fraction.

Southwest Solar Corporation incorporates these same standards of quality and performance in each phase of its operations. Our R&D department is continuously striving to meet evolving energy needs, i.e., by providing sizing of residential, commercial, and industrial projects by computer analysis. We have a west coast distribution network from Oregon to Hawaii. These are all supported by our large warehousing facilities for immediate product availability, engineering support for special projects, creative promotional materials and cooperative advertising and lead programs.

364. **SPECTROLAB**
A Subsidiary of Hughes Aircraft
12500 Gladstone Avenue
Sylmar, California 91342

James A. Albeck, Marketing Manager/Aerospace Products
(213) 365-4611

Spectrolab, one of the nation's leading manufacturers of solar cells, solar cell arrays, solar simulation systems and related instrumentation, was first organized in 1956 in response to the needs of the Aerospace industry for a means of supplying satellite electrical energy by solar photoelectric conversion techniques.

Spectrolab's interdisciplinary approach, combining many complementary academic fields, has resulted in significant advances in photovoltaic technology designed to provide Spectrolab customers with the most advanced and reliable products available within the state-of-the-art.

365. **STATE INDUSTRIES, INC.**
Ashland, Tennessee 89015

(615) 792-4371

Manufacturer of Solarcraft water heaters—totally self-contained, automatic, all-climate solar water heating systems for domestic use.

366. **STRUCTURAL COMPOSITES INDUSTRIES INC. (SCI)**
6344 North Irwindale Avenue
Azusa, California 91702

Vicki L. Morris, Marketing Engineer
(213) 334-8221

SCI develops and manufactures high-performance composite structures. Primary product areas are commercial pressure vessels, large energy and transportation structures, and aerospace products. Applications in the areas of ocean thermal gradient systems and wind generation systems.

367. **SUNBURST SOLAR ENERGY, INC.**
P.O. Box 2799
Menlo Park, California 94025

(415) 327-8022

Manufacturer of single-glazed flat plate collectors. Copper tube/aluminum fin absorber; aluminum frames.

368. **SUNDEX SOLAR EQUIPMENT SALES**
677 Valley View Road
Talent, Oregon 97540

D. Clayton Gangnes
(503) 535-4402

Manufacturer of solar equipment for swimming pools and domestic hot water supply. Dealerships will be available in the near future.

369. **SUNDUIT, INC.**
281 E. Jackson
Virden, Illinois 62690

Jerry Harlin
(217) 965-3352

Primarily in Agri-Solar using hot water collectors for in-floor heat, air systems for ventilating air and grain drying.

Sunduit manufactures all-copper absorber plates (water) for solar roof construction and all-copper absorber plates enclosed in all-aluminum frames for use in existing facilities. Supplier of other components. Also, system design and engineering as well as technical assistance in construction.

370. SUNEARTH OF CALIFORNIA
P.O. Box 1729
Rohnert Park, California 94928

Art Brooks
(707) 584-7414

Sunearth of California manufactures high-quality solar panels (Sunshine energy liquid collector systems) for use in heating swimming pools, spas and domestic hot water systems. We have systems throughout the United States.

371. SUNEARTH SOLAR PRODUCTS CORPORATION
352 Godshall Drive
Harleysville, Pennsylvania 19438

RD 1, Box 337
Green Lane, Pennsylvania 18054

Howard S. Katz
(215) 256-6648

Sunearth manufactures liquid collectors, including a high-performance series, and complete DHW systems. The collectors feature a variety of options for roof or frame mounting, and double glazing consisting of heat, scratch and craze resistant acrylic and heat-proof Teflon®(duPont) film. The absorber plate is formed and coated for optimal heat transfer. All waterways are 100% copper with brazed connections. The collector is designed for easy installation and service with features including integral supply and return headers, removable glazing frames, no solder brass interconnections, and self-sealing roof brackets.

The Sunearth domestic hot water heater is a totally automatic package. The system comes complete with collectors, tank, controls, fittings, pump, nontoxic antifreeze, and instructions. This system employs a heat exchanger/antifreeze loop to provide freeze protection.

372. SUNFLOWER ENERGY WORKS
101 Main Street
Lehigh, Kansas 67073

Waldo Voth
(316) 523-3621

Sunflower Energy Works manufactures the V-12, an airflow collector suitable for array installation, capable of heating both air and domestic hot water. Designer of the unit is Dr. Nelson Kilmer, Professor of Physics at Hesston College, Hesston, KS. SEW was formed in 1977 by several Kansas businessmen/educators interested in decreasing the rate at which natural resources are being depleted. Testing of various prototype collectors throughout 1977-78 culminated with the V-12 design. Manufacturing began in late 1978. Over 600 V-12 solar collectors have been installed on homes, workshops, commercial buildings, bowling alleys, and will be used on a 45,000 square foot industrial warehouse.

373. SUN INDUSTRIES, INC.
P.O. Box 1547
Carlsbad, California 92008

Prescott Spalding/John Van Tuyl
(714) 565-9451

Design and manufacture solar/geothermal electric plants for use in homes, small businesses, agriculture, buoys, islands, and any other location without electric power. Also build heating units for use in the above locations.

374. SUN LIFE SYSTEMS INC.
6210 N.E. 92nd Drive, Suite 101
Portland, Oregon 97220

Glenn Howk
(503) 257-9183

Sun Life Systems is a fully integrated solar company engaging in five major activities: (1) panel manufacturing — SLS panels have the largest market share of all panels sold in Oregon and are a major entrant in the west coast market; (2) wholesalers of complete domestic solar hot water systems — SLS packages and sells five complete systems tailored to the needs of builders and individual homeowners; (3) wholesale distributor of solar and solar-related products — SLS is the exclusive regional distributor for A.O. Smith solar tanks and a full line of other solar products; (4) installation — SLS provides complete installation of domestic hot water solar systems on a turn-key basis; (5) commercial spec work — SLS provides complete engineering and systems analysis for commercial clients. Jobs completed include the solar heating system at the Odessa Brown Clinic in Seattle, WA.

375. SUNMASTER CORPORATION
12 Spruce Street
Corning, New York 14830

Gary Fradin
(607) 937-5441

Manufacturers of the Sunmaster solar hot water and space heating systems featuring evacuated tube collectors. Systems also feature the Winston parabolic reflector and a fully drainable manifold.

376. SUN OF MAN SOLAR SYSTEMS
Box 1066
Guerneville, California 95446

Bob Tillotson
(707) 869-3331

Manufacturer of the Solarsyphon Diode™, a valve system that eliminates the back-syphoning problems of thermosyphon type systems and the need for a pump or sensor.

377. SUN-PAC, INC.
P.O. Box 8169
Alexandria, Louisiana 71306

Jim Nordin
(318) 445-6751

Sun-Pac is a manufacturer of hydronic flat plate collectors. Our collectors are constructed of a fiberglass housing with a copper absorber plate, selectively coated with black onyx and chrome and glazed with water white glass. We are active in the design, architecture and engineering of complete active solar systems for residential, commercial and industrial applications.

378. **SUNPOWER SYSTEMS CORP.**
510 S. 52nd Street
Tempe, Arizona 85281

John Gamero
(602) 894-2331

Manufacturer of solar concentrating collectors used for water heating, space heating, and space cooling applications.

379. **SUN RAY OF CALIFORNIA**
P.O. Box 46033
Los Angeles, California 90046

Irwin Weingarten
(213) 654-3416

Manufacturer of collectors and systems. Total design, procurement and installation of solar systems; heat recovery systems. Computer-assisted design, consultation, and energy audits.

380. **SUN-RAY SOLAR EQUIPMENT CO., INC.**
4 Pines Bridge Road
Beacon Falls, Connecticut 06403

(203) 888-0534

Manufacturer of Sol-Arc patented panels and Sol-Arc Solariums.

381. **SUN-RAY SOLAR HEATERS**
4898 Ronson Court
San Diego, California 92111

M. N. Waiche
(714) 278-7720

Manufacturer of components and total heating systems. Active in residential, commercial and industrial design.

382. **SUNSHINE DEVELOPMENT CORPORATION, INC.**
A Subsidiary of RISE Inc.
1641 East Main Street
Salem, Virginia 24153

John Lensch/Ed Papworth
(703) 563-5800

Presently manufacture a cost-effective hot air flat plate collector and accessories (manifolds, mounting brackets, etc.). All-aluminum collector with baked enamel flat black absorber, using ASG's Solatex glazing.

383. **SUNSOURCE OF ARIZONA**
3441 N. 29th Avenue
Phoenix, Arizona 85017

William Seginski
(602) 258-1549

Innovator/packager of the Problem Solver Solar Control Kit (nonelectric solar system) which includes a self-powered control valve, solar pump, balance valve and solar pump timer. Sunsource of Arizona is also a wholesale/retail outlet for Hydronic Solar Specialties.

384. **SUNSPOOL CORPORATION**
439 Tasso Street
Palo Alto, California 94301

Dr. Harry T. Whitehouse
(415) 324-2022

Manufacturer of the SUNSPOOL drain-down valve for freeze protection for solar arrays used in direct water heating systems. The drain-down technique avoids use of heat exchangers and elaborate tanks, while providing high thermal performance through direct heating of potable water.

MANUFACTURERS 133

385. **SUN STONE COMPANY, INC.**
 P.O. Box 138
 Baraboo, Wisconsin 53913

 C. Anderson
 (608) 356-7744

 Sun Stone Company manufactures solar energy equipment for space heating and domestic hot water heating. Sun Stone provides the collectors, blowers, controls, storage media, heat exchangers, pumps, storage tanks, dampers, and engineering for a complete installation. Dealers will size, design, install and service systems.

386. **SUNTEK SOLAR PRODUCTS LTD.**
 2285 Meyers Avenue
 Escondido, California 92025

 (714) 743-0446

 Manufacturer of solar water heating equipment—domestic and industrial applications. Also, pool, spa and hot tub heating systems.

387. **SUNTIME INC.**
 101 High Street
 Belfast, Maine 04915

 (207) 338-4038

 Manufactures a passive, self-pumping, self-regulating solar system.

388. **SUNTREE SOLAR COMPANY**
 20 Austin Avenue
 Greenville, Rhode Island 02828

 Walter H. Barrett
 (401) 949-2972

 Suntree Solar Company is a manufacturer of solar heating systems for home, domestic hot water, and industrial heating. We are also working on an installation seminar.

Suntree solar collectors come in two sizes: 36 x 77 x 5-3/4 in. and 48 x 96 x 5-3/4 in. Each has an all-copper absorber plate, Filon or Tuffak Twinwall glazing and aluminum over wood housing. We have installed many systems using this collector. We have also developed a water-to-water heat pump—soon to be on the market. Suntree designs, engineers, installs and services solar heating systems for all types of applications.

389. SUNTRON
 A Division of K&G Manufacturing Inc.
 2302–10 Pennsylvania Street
 Fort Wayne, Indiana 46802

Richard Barnes, Manager
(219) 424-6270

Involved in the design and manufacturing of consumer solar and energy-saving products. Also design for commercial and industrial application.

390. SUNWORKS DIVISION OF SUN SOLECTOR CORP.
 P.O. Box 3900
 Somerville, New Jersey 08876

Ed Jones
(201) 469-0399

We are one of the largest manufacturers of flat plate solar collectors in the United States and a major supplier of solar water heaters. Manufacturer of the Solector® solar energy collectors for water heating, space heating, and process heating. Solector®Pak—domestic hot water systems. SunSym™ computer programs.

391. SYenergy METHODS, INC.
 1367 Elmwood Avenue
 Cranston, Rhode Island 02910

Douglas C. Creed
(401) 467-2600

Wall and roof systems for new and retrofit. Insulation placed on the exterior side of wall and roof construction.

392. TECHNITREK CORPORATION
1999 Pike Avenue
San Leandro, California 94577

David L. Andersen
(415) 352-0535

TechniTrek Corporation manufactures a complete line of all-copper absorber collectors. TechniTrek also manufactures the TechniTrol Domestic Hot Water Control Module. The module features a pump, controller, and draindown valve, plumbed, wired and mounted.

TechniTrek also manufactures electromagnets for science and industry, and insect control devices.

393. TELEDYNE MONO-THANE
1460 Industrial Parkway
Akron, Ohio 44310

Michele Weber, Sales Coordinator
(216) 633-6100

Manufacturer of Foamedge Pipe Cover and Foamedge Insulation. Foamedge has a soft, flexible polyurethane foam core, clad in a tough, resilient vinyl cover. Urethane foam compresses under pressure to conform to the shape of any gap or opening, providing a permanent, tight seal. When pressure is relieved, it bounces back to its original shape.

The vinyl cover, formed simultaneously with the core, provides an attractive, wear-resistant outer surface. The vinyl cover extends to form a flap. Foamedge is provided with or without self-adhesive flaps, depending on requirements.

394. TERRA-LIGHT INC.
P.O. Box 493, 30 Manning Road
Billerica, Massachusetts 01821

(617) 663-2075

Manufacturer of a patented copper solar heating system for pools with 23- and 29-square foot absorber plates available. No secondary headers are used in the design, as it is engineered to be lightweight and offer uniform efficiency. The

136 SOLAR CENSUS

systems utilize round manifolds that accommodate the solar heated water flow through the system and absorber plate. Also, manufactures a patented high-pressure copper absorber plate for OEMs and suppliers of solar panels and systems.

395. TEXXOR CORPORATION
9910 N. 48th Street, P.O. Box 14337
Omaha, Nebraska 68124

Stan Schlorholtz
(402) 453-7558

The TEXXOR Heat Cell™ is a practical, safe and simple system with no moving parts. Each cell is a thermal storage module that stores or releases heat as it undergoes a phase change (utilizing BISOL II© phase-change compound) to meet the requirements of heating system. Due to its unique size and packaging—it is cylindrically shaped with a high thermal conductivity and a surface-to-volume ratio carefully selected to effect a thermally responsive system—it is easily adaptable to hundreds of different uses ranging from direct gain passive solar applications to forced air solar heating systems and large-scale waste heat conservation systems.

396. THERMACO
6828 7th Avenue
Rio Linda, California 95673

Clint Walker
(916) 991-3579

Thermaco manufactures all sizes of insulated and noninsulated tanks as our emphasis. Panels and heat exchangers constitute a lesser percentage of production.

397. THERMAL ENERGY STORAGE, INC.
10637 Roselle Street
San Diego, California 92121

Jim Hitchin
(714) 453-1395

Manufacturer of thermal energy storage devices using phase-change salt hydrates. Phase-change temperature is 115F with approximately 9300 Btu/cubic foot storage. The units are 3 feet in diameter by 4 feet high and contain two heat exchangers: one for heat input (from solar panels), the other for heat output from storage (for water heating or space heating). Total latent heat storage capacity is 250,000 Btu. Smaller or larger units are available.

398. **THERMAL ENGINEERING & DESIGN CO.**
 44 N. Summit Street
 Akron, Ohio 44308

 R. M. Wells
 (216) 535-5761

 Thermal Engineering & Design Co. is a manufacturer of controls and instruments.

399. **THERMATOOL CORPORATION**
 Solar Products Division
 280 Fairfield Avenue
 Stamford, Connecticut 06902

 R. E. Russell
 (203) 357-1555

 Manufacturer of Thermafin®, a liquid solar energy all-copper absorber plate fabricated from high-frequency welded fin-tube elements. Also provide fintubes to other absorber manufacturers.

400. **THERMEX SOLAR CORPORATION**
 1050 N. Kraemer Place
 Anaheim, California 92806

 Robin Scott
 (714) 630-5882

 Thermex Solar Corporation manufactures Kopper King solar panels—all copper from case to absorber plate, waterways to rivets. Also feature no-iron tempered glass. Thermex manufactures all-copper solar energy systems for domestic, commercial and industrial heating.

401. THERMODULAR DESIGNS, INC.
5095 Paris Street
Denver, Colorado 80239

Bob Seserman
(303) 371-4111

Manufacturer of greenhouse windows, solar windows, garden walls and skylights.

402. THERMOGRATE, INC.
2785 N. Fairview Avenue, P.O. Box 43566
St. Paul, Minnesota 55164

John Traynor, Director of Marketing and Sales
(612) 636-7033

Manufacturer of the Thermograte fire control insert and total control insert stainless steel heat exchangers for fireplaces, and freestanding fireplace stoves with stainless steel heat exchanger tubes and glass door enclosures.

403. THOMASON SOLAR HOMES, INC.
609 Cedar Avenue
Fort Washington, Maryland (Washington, D.C.) 20022

(301) 839-1738

Manufacturer of the Solaris solar space and water heating and air conditioning systems, featuring the open-flow flat plate Solaris collector. The system provides 50—90% of space heating and DHW needs; 100% of air conditioning needs. Invented and patented by Dr. Harry E. Thomason, who has been active in the solar field since 1956, Solaris systems are distributed by a nationwide network of Thomason licensees.

404. SVEN TJERNAGEL SOLAR SYSTEMS
477 Woodcrest Drive
Mechanicsburg, Pennsylvania 17055

Sven L. Tjernagel
(717) 761-5838

Sven Tjernagel Solar Systems designs, manufactures and distributes to its authorized dealers full-house solar heating systems (domestic hot water included), designed to provide 50% of the heating and domestic hot water load on an annual basis. The system includes pumping and controls for a coal/wood-burning boiler—fireplace insert or free-standing unit—designed to provide the balance of the annual energy load.

This system integrates automatically with a hot air furnace, heat pump, or hot water baseboard conventional fired heating system through our solid-state, programmable electronic control package.

405. TONA-MIDSOUTH, LTD.
P.O. Box 1160
Poplar Bluff, Missouri 63901

O.P. Wren
(314) 785-0161

Distributor for total solar heating and cooling systems.

406. TOTAL ENERGY
55 Knickerbocker Avenue
Bohemia, New York 11716

(516) 567-1064

Manufacturer of solar collectors and storage tanks. Electronic control equipment.

407. TRI-EX TOWER CORPORATION
7182 Rasmussen Avenue
Visalia, California 92377

(209) 625-9400

Manufactures towers for wind generation systems, engineered to customer's application needs.

408. TURBONICS INC.
11200 Madison
Cleveland, Ohio 44102

Jack Jordan
(216) 228-9663

Turbonics Inc. is the manufacturer of the CHILL Chaser™, supplemental heating system that utilizes heat from any clean, economical hot water source—domestic hot water heaters, boilers, solar storage or tempering tanks, and waste heat recovery systems.

409. **TURNER GREENHOUSES**
Highway 117 South
Goldsboro, North Carolina 27530

(919) 734-8345

Manufactures kits for constructing greenhouses of single-glazed fiberglass, single or inflated-double polyethylene, and a zinc- and aluminum-coated steel frame.

410. **UNARCO-ROHN**
6718 West Plank Road, P.O. Box 2000
Peoria, Illinois 61601

Philip W. Metcalfe
(309) 697-4400

Unarco-Rohn designs and manufactures a full line of support structures for horizontal and vertical axis wind turbines.

411. **UNITED ENERGY CORPORATION**
666 Mapunapuna Street
Honolulu, Hawaii 96819

Mikuel Jackson
(808) 836-1593

Manufacturer of solar components. R&D in the area of photovoltaic cells.

412. **UNITED STATES SOLAR INDUSTRIES**
5600 Roswell Road N.E., Suite 350
Prado East
Atlanta, Georgia 30342

Regional Office:
25 W. Fifth Street
London, Ohio 43140

David A. Higerd, Operations Manager
(614) 852-0032

United States Solar Industries manufactures domestic/commercial hot water and space heating systems; also, fireplace heat extraction systems. Regional office (London, Ohio) — wholesale and retail, complete installation services by factory trained technicians. Custom design services.

413. UNIVERSAL SOLAR DEVELOPMENT, INC.
1505 Sligh Boulevard
Orlando, Florida 32806

Russell Flanigan
(305) 423-8727

Manufacturer of solar panels.

414. URETHANE MOLDING
RFD 3, Box 190
Laconia, New Hampshire 03246

Randy H. Annis, Vice-President
(603) 524-7577

Urethane Molding produces INSULJAC®, a patented procedure. Insuljac is urethane insulation with a rigid PVC exterior. Urethane Molding produces a solar insulation with the supply and return line within the same rigid PVC exterior.

415. U. S. SOLAR CORPORATION
P.O. Drawer K
Hampton, Florida 32044

(904) 468-1517

Manufacturer of Eagle Sun Collector Systems. Fourteen models available. All have double glazing, copper tubing, brass pipe nipples, and rigid anodized aluminum frames.

416. VALLEY PUMP COMPANY
Division of Valley Industries, Inc.
Aermotor Plant
P.O. Box 1364, Commerce & Exchange Streets
Conway, Arkansas 72032

Andy Dunlap
(501) 329-9811

Manufacturer of pumps for the solar industry and of wind generation systems.

VALOUR FIBERGLASS MANUFACTURING See (430B)

417. VANGUARD ENERGY SYSTEMS
9133 Chesapeake Drive
San Diego, California 92123

Ted C. Reirr
(714) 292-1433

Manufacturer of water source heat pumps. Applications of the Vanguard water-to-air heat pump include: direct heating and cooling—residential, commercial and industrial—utilizing an external source of water; zoned heating and cooling within a hydronic system; direct heating and cooling of residences (single- and multi-family) utilizing water from a solar-heated swimming pool. The Vanguard water-to-water heat pump is applicable to: backup heating plus cooling in a two-tank solar system: offpeak heating and cooling utilizing a hydronic storage system; heating and cooling in commercial/industrial structures using a hydronic system.

418. VAUGHN CORPORATION
386 Elm Street
Salisbury, Massachusetts 01950

James F. Vaughn III
(617) 462-6683

Vaughn Corporation is a manufacturer of solar water heaters.

419. VEGETABLE FACTORY, INC.
100 Court Street
Copiague, New York 11726

Fred Schwartz
(516) 842-9300
1-800-221-2550

Manufacturer and distributor of solar panel greenhouses, lean-to and free-standing. Glazings all double-wall (fiberglass or Lexan®) bonded to aluminum I-beam grid core. Complete structure is a volume air collector that may be retrofit with storage and distribution systems to qualify for energy tax credits. Five-year warranty (panels only available).

420. VIRACON/DIAL
1315 N. North Branch
Chicago, Illinois 60622

L. Sarno
(312) 943-4200

We bend the glass used to collect the sun's rays.

421. VULCAN SOLAR INDUSTRIES, INC.
6 Industrial Drive
Smithfield, Rhode Island 02917

J. Michael Levesque
(401) 231-4422

Vulcan Solar Industries, incorporated in 1977, is currently involved in solar residential, industrial and commercial applications for both new and retrofit installations.

In 1979 Vulcan was awarded what is reported to be the second largest flat plate collector system in the country, a $1.2 million DOE contract. Vulcan provided 600 flat plate liquid collectors for hot water and space heating to the Ballston Spa Wood Road School Complex, with 300 more anticipated. The complex was heated primarily by electricity.

In 1978 Vulcan was chosen to provide to the State of Rhode Island 17 hot water systems, in a competitive bid through the RI Department of Community Affairs. The systems were installed in low-income houses throughout the state. After strict monitoring of these systems, it was estimated that savings were 4525 gallons of oil or equivalent in a year, which averages to 266 gallons of oil saved per house. The systems provided 60% of the homes' annual hot water needs.

A manufacturer of light weight flat plate collectors and complete DHW, space and pool heating systems, Vulcan is currently involved in all aspects of manufacturing and installation. The company is now in the process of formulating a national dealer network to market Vulcan products.

422. WATERMEISTER
P.O. Box 87433
Chicago, Illinois 60680

James Keogh
(312) 664-6303

Manufacture and install hot tubs. Have used hot tub as heat sink for solar system.

423. WEATHER ENERGY SYSTEMS, INC.
39 Barlows Landing Road, P.O. Box 968
Pocasset, Massachusetts 02559

James Oliver Brown, Jr.
(617) 563-9337

Weather Energy Systems designs and builds air control systems for use in passive solar settings ranging from retrofit greenhouses and sunspaces to entire solar homes. Our Plexus Symmetrical Controllers give computer-based regulation and monitoring for effective heat distribution in simple or complex structures. Solar heat sources can be effectively integrated and prioritized with other sources such as wood stoves or fossil fuel furnaces. W.E.S. offers a variety of sizes of blowers and fans to best suit the requirements of the job.

Sun Haus™ is not a kit-type greenhouse. It adds a permanent room on a south-facing wall. Because they are designed with thermal efficiency in mind, they do not lose heat the way most other greenhouses/solaria do. In addition, the plans for this design are sold nationally to people interested in having a passive solar structure built onto their home.

W.E.S. provides experienced design services complete with a computer to model important parameters of solar design: direct gain, thermal mass, insulation, etc.

424. WESTERN SOLAR DEVELOPMENT, INC.
1236 Callen Street
Vacaville, California   95688

Charles Crouse
(707) 446-4411

Western Solar is a manufacturer of all-copper flat plate solar collectors (5 models). Collectors are certified by IAPMO, California Energy Commission, Florida Solar Energy Center, and are listed by the City of Los Angeles. The collectors have all-copper absorber plates, aluminum frames, closed cell insulation. Glazing can be either solar glass or Filon fiberglass-reinforced plastic. Collector absorber plates can be either black chrome selective coated — black painted.

425. LAWRENCE WITTMAN & CO. INC.
1395 Marconi Boulevard
Copiague, New York   11726

Charles Wittman
(516) 842-4770

Custom molder of fiberglass reinforced plastics. Manufacturer of fiberglass solar collector shells and translucent swimming pool enclosures.

426. WOLVERINE SOLAR INDUSTRIES INC.
13450 Northland Drive
Big Rapids, Michigan   49307

Howard W. Beatty, Jr., P.E.
(616) 796-5501

Manufacture the Zodiac brand solar systems, including the Aquarius domestic water heater (using thermal collection and storage means) and the Zodiac space heating solar systems for forced air, hydronic and electrical space heating interfaces. Also produce the Zodiac Taurus systems for commercial and industrial use.

427. **YING MANUFACTURING CORP.**
 1957 West 144th Street
 Gardena, California 90249

 Frank Kraus
 (213) 770-1756

Involved in the manufacture of solar collectors (concentrating and liquid) and storage tanks. Design and engineering of total active systems, including hot water systems, space heating systems, pool and spa heating systems.

428. **R. M. YOUNG COMPANY**
 2801 Aero-Park Drive
 Traverse City, Michigan 49684

 David Kipley
 (616) 946-3980

Manufacturer of sensitive wind and temperature instruments for wind power generation site studies, insulation and heat loss studies, measurement of air movement for heating and ventilating studies.

429. **ZEPHYR WIND DYNAMO COMPANY**
 P.O. Box 241
 Brunswick, Maine 04011

 Richard Vaglia
 (207) 725-6534

Zephyr Wind Dynamo Company manufactures direct drive alternators for use in wind powered systems. They are very low speed permanent magnet alternators providing a three-phase alternating current of a frequency and voltage corresponding to rpm. This may be used directly for resistance heating, rectified to dc for battery charging, or inverted to interface with a utility grid.

The Zephyr VLS-PM alternator was designed specifically for wind systems. Their low operating speeds make them ideal for direct drive applications, thereby avoiding the efficiency losses and maintenance problems associated with gearing.

Zephyr manufactures three models of the VLS-PM alternator: Series 311-B, rated 1.5 kW at 450 rpm; Series 432, rated 3.5 kW at 250 rpm; and Series 647, rated 8 kW at 220 rpm. All models are available with rectifier/regulator and motorizing circuitry (for use with non self-starting turbines).

430. ZIA ASSOCIATES, INC.
1830 N. 55th Street
Boulder, Colorado 80302

Joanne Brown
(303) 449-9170

Design and manufacture of solar controls.

430. INTERNATIONAL SOLAR TECHNOLOGIES, INC.
(A)
Route 2, Box 321
Plainfield, Indiana 46168

Henry H. Leek
(317) 272-2996

ISTI manufactures the Sunkit™, a supplemental solar heating system for residential retrofit applications. The Sunkit™ package consists of a flat plate collector, aluminum reflector shield, rock storage chamber, air handler, motors and dampers, and controls. Connects easily with existing ductwork.

430. VALOUR FIBERGLASS MANUFACTURING, INC.
(B)
3703 East Melville Way
Anaheim, California 92806

Mr. Burt Buser
(714) 630-2500

Valour manufactures the Suntank for Dalton Tank & Supply Co. Engineered to be free-standing, it needs only footing foundation. The Suntank is constructed of isophthalic polyester resin and can withstand 212F temperatures and is impervious to alkali. The Suntank carries the International Association of Plumbing and Mechanics Official Approval (IAPMO), the only tank manufactured for solar applications with the seal of approval by its independent research committee. Approximately 3000 Suntanks are now in use.

430. MASS ENERGY SYSTEMS, INC.
(C) P.O. Box 1311
Rapid City, South Dakota 57709

Dale N. Skillman
(605) 342-1787

Market portable roll-forming machines which produce a flat plate, air-type collector to any custom length. Construction: two roll-formed aluminum pans (top pan has pre-applied selective absorber coating), double cover has Tedlar®(outer) and Teflon®(inner). EPDM extrusions for glazing attachment. Panels for retrofit and new construction.

* * *

Also see the following entries—

433. ACORN STRUCTURES
499. CROFT & COMPANY
502. DAS/SOLAR SYSTEMS
503. DAVIS ENGINEERING, INC.
523. EKOSE'A, INC.
532. ENERGY ASSOCIATES
547. ENVIRONMENTAL INSTRUMENTATION
572. GOODWIN CONTROLS CO.
581. GREEN HORIZON
616. JACOBS-DEL SOLAR SYSTEMS, INC.
640. LE BLEU ENTERPRISES
664. MILLER & SUN ENTERPRISES, INC.
666. ROBERT MITCHELL, SOLAR SYSTEMS DESIGN INC.
743. SOLAR DESIGN ASSOCIATES INC.

748. SOLAR ENERGY DESIGN CORPORATION OF AMERICA
773. S S SOLAR INC.
828. WESTINGHOUSE ELECTRIC CORPORATION
880. CRYSTAL SYSTEMS, INC.
890. ECOTRONIC LABORATORIES, INC.
911. KAMAN AEROSPACE CORPORATION
956. SCIENTIFIC BUILDING AND ENERGY CONSULTANTS
982. UCE INC.
990. WINDEPENDENCE ELECTRIC POWER CO.

DESIGN Architecture and Engineering

431. ABACUS GROUP INCORPORATED
P.O. Box 8006
Boulder, Colorado 80302

Don Huffman
(303) 444-3111

Computer service for solar industry. We have put together a package of commonly used solar information computer systems. We offer large complex programs to companies on a time-shared basis. One package that we offer is the f-chart, the computer tool for estimating the long-term thermal performance and life cycle of active solar heating systems.

432. ABERNATHY SOLAR CONSULTANT
1250 North Walden Lane
Anaheim, California 92807

Bill Abernathy
(714) 970-1446

System design—specializing in air systems, including retrofit of air systems.

433. ACORN STRUCTURES
Box 250
Concord, Massachusetts 01742

M. E. Kelley
(617) 369-4111

Designers and manufacturers of energy-efficient homes and light commercial buildings. A complete array of active and passive techniques are employed. We manufacture our own solar collectors, storage tanks, and space heating/DHW systems under the tradename Sunwave® As a solar component manufacturer, we sell our systems with our homes or to qualified dealer/installers, rather than directly.

434. ADOBE SOLAR LTD.
930 Matador S.E.
Albuquerque, New Mexico 87123

Martin Selinfreund
(505) 299-5901/255-9184

Designer, engineer and builder of passive solar homes.

435. ADVANCE AIR CONTROL
1506 Terry Andrae Court
Sheboygan, Wisconsin 53081

Ted Kuck, P.E.
(414) 452-4476

Have designed a model home in Sheboygan, complete with stone storage, plus eutectic salt storage trays, split system heat pump, liquid fireplace, electric furnace, two step coils, and passive solar heating.

436. ADVANCED ENERGY SYSTEMS
3744 W. 63rd Street
Chicago, Illinois 60629

Daniel J. Sullivan
(312) 581-6809

Solar engineering—residential, commercial and industrial. Advanced Energy Systems have available various system designs covering all aspects of alternative energy usage. Standard designed systems can be used with new structures or retrofitted to existing structures. Firm also manufactures HDPP storage tanks.

437. AFFILIATED ENGINEERS INC.
FUAD & ASSOCIATES
P.O. Box 5039, 625 North Segol Road, Suite C
Madison, Wisconsin 53705

John Nelson
(608) 238-2616

Engineering affiliate of largest architectural/engineering firm in Wisconsin. Have offices in Madison, Wisconsin, and Gainesville, Florida. Have designed several operating solar systems. Specialize in commercial/industrial projects.

438. AIDCO MAINE CORPORATION
Orr's Island, Maine 04066

R. Multer
(207) 833-6700

Consulting engineers specializing in renewable energy and waste resource recovery and pioneering commercial development of new technology, AIDCO helps industrial, commercial and residential clients solve a wide variety of thermal, chemical, biochemical, electrical, and mechanical problems. Operating worldwide from headquarters on the Maine seacoast near Portland, AIDCO offers a complete range of engineering services from initial consulting, through design and construction supervision, to operator training and performance monitoring. AIDCO's renewable energy systems are operating in Maine and elsewhere for satisfied owners of new and old residential and commercial facilities; several designs were selected for funding by the U.S. government. Under its StrataSOL trade name, AIDCO is perhaps the first firm anywhere to offer full engineering services for the versatile and well documented nonconvective solar pond energy system. Design review and troubleshooting services are offered in relation to energy systems designed by others.

439. ALKAR ENGINEERING & MANUFACTURING CO.
25520 Ingleside
Southfield, Michigan 48034

Alexander Kargilis
(313) 353-0696

Activities include: energy audit services, heat transfer equipment (design and build), engineering services for heating and cooling, energy conversion systems (design and build), heat exchanger manufacture.

440. ALLEN & SHERIFF
3020 South Robertson, BL S4
Los Angeles, California 90034

Garth Sheriff, AIA
(213) 837-5552

Architectural design—a wide range of active and passive solar energy applications in residential, commercial and industrial buildings.

441. ALTENBURG AND COMPANY
587 Spring Street
Westbrook, Maine 04092

William M. Altenburg
(207) 856-6348

Design/engineering of wind generation systems.

442. ALTERNATE ENERGY INDUSTRIES CORPORATION (AEIC)
420 Lexington Avenue, Suite 1628
New York, New York 10017

Hector Guevara
(212) 682-8220

AEIC is a diversified engineering, marketing and consulting firm dedicated to appropriate technology and the promulgation of alternative energy sources in general, and solar energy systems in particular.

While our primary activity is marketing solar hot water heating systems for homes, businesses and institutions, we are also involved in other areas. These include solar swimming pool heaters, thermal shades for windows, insulation, custom engineered solar space heating/cooling projects, passive/active solar consultation, and other interests in wind power, micro-hydropower, and heating controls for space heating systems.

We were installation contractor for the solar domestic hot water heating system for The White House in Washington. We were involved with the demonstration installation at the New England Solar Energy Center in Cambridge, MA. On a larger scale, we were involved with the 102 panel solar hot water system for the Joliet, IL Housing authority, and acted as solar consultant for the 110 panel system to be built on the 127th Street basketball arena in New York City. Of course, we have installed numerous systems in residential and commercial applications.

Together with associate companies, we can supply a wide range of consulting, design and installation services for all types of solar applications from residential hot water to industrial process heat.

443. ALTERNATIVE ENERGY CONCEPTS OF CALIFORNIA INC.
P.O. Box 564
Palm Springs, California 92060

David Ames/Mark Mainquist
(714) 324-5880

Engineer, design and install solar, geothermal and heat recovery systems, specializing in high ambient geographic locations throughout the West.

Primary emphasis: new construction, residential, commercial and industrial. Secondary emphasis: reworking system failures. Staff includes licensed engineers and hydronic specialist.

444. ALTERNATIVE ENERGY WORKS, INC.
310 West Fourth Street, P.O. Box 271
Newton, Kansas 67114

Gilbert Buller
(316) 283-5993

AEW has the capability to design and install solar hot water systems, solar space heating systems, both active and passive, and passive solar greenhouses. We have installed four solar hot water heaters, one solar space heater, and are currently involved in the design stage of a greenhouse. Three of the DHW systems were installed in new residences, and the other projects were retrofit applications. We also have extensive experience in energy conservation in the residential, commercial, institutional, and industrial sectors.

445. ALTERNATIVE RESOURCES, INC.
4 East LaVerne
Mill Valley, California 94941

Roan Browne/Denise Cousineau
(415) 388-1466

We are a contracting firm in solar heating of homes, water, pools, spas. We also install windmills and generators and photovoltaic cell panels. We have also been active in various forms of passive design since 1976.

446. ANCO ENGINEERS INC.
1701 Colorado Avenue
Santa Monica, California 90404

Mike Anderson/Douglass Taber
(213) 829-9721

We provide a complete range of passive, active, and earth-sheltered architectural and solar design. In addition, we do structural design, seismic evaluation, and microprocessor research and development (as applied to solar systems). ANCO's energy/design group also conducts full-scale energy audits of hospitals, schools and industrial plants.

ANCO energy management/design projects for the government include: DOE (Richland, WA)—comprehensive energy survey of 10 sites and 145 buildings; Argonne National Laboratory—West energy survey; U.S. Navy Air Conditioning Survey (Pearl Harbor); DOE Hanford Reservation—conceptual design cogeneration; feasibility study for a Hanford energy monitoring and control system.

Industrial/commercial projects include: energy audit for ABC Entertainment Center (LA and NYC); alternative drying process for Rochelle Laboratories (Long Beach, CA); heat recovery opportunity investigations for Garrett AiResearch Co. (Torrance, CA); electricity use analysis for Pacific States Steel Corp. (Union City, CA); energy audit of Altec Sound Corp. (Anaheim, CA).

447. JOHN ANDERSON ASSOCIATES ARCHITECTS
1522 Blake Street
Denver, Colorado 80202

Alan G. Gass, AIA, Vice-President
(303) 534-5566

Although the firm has an extensive background in a variety of building types ranging from office and municipal buildings to recreational projects, the focus of the practice through its entirety has been on educational and institutional facilities.

For many years, energy conservation and solar energy applications have been of central concern to the firm. Once a structure is initially planned as energy conservative, maximum consideration is given to the use of passive solar energy for both heating and natural light. Only when the firm has carefully considered energy conservation and passive solar energy use, does active solar system design receive consideration. If it can then be demonstrated that a major contribution can be made at a reasonable additional cost, active solar becomes part of the project.

Over the years, the firm has won a number of awards for its work. Most recently, its awards have been for both design excellence and energy conservation.

448. LEROY ANDREWS ARCHITECTS, INC.
2284 South Victoria Avenue, Suite 2A
Ventura, California 93003

LeRoy M. Andrews
(805) 642-3288

We have been directly involved in the design of active solar systems since 1974 and passive solar since 1960. Our design work includes the first solar heated fire station in the United States (operating since June 1977 in Ventura, CA). Mr. Andrews lectures extensively on solar energy and his own residence is included in the California Department of Energy book of 20 properly designed houses using passive solar energy. This house was built in 1961.

449. ANGEL, MULL & ASSOCIATES, INC.
3049 Sylvania Avenue
Toledo, Ohio 43613

Joseph M. Angel, AIA
(419) 474-5496

Active and passive design, specializing in multi-family residential and commercial structures—new and retrofit.

450. THE ARCHITECTS PARTNERSHIP, INC.
717 Cherry Street
Columbia, Missouri 65201

K. Renner, AIA
(314) 449-3971

Architectural design—passive systems for residential (single- and multi-family) and commercial application.

451. ARCHITECTURAL ALLIANCE
400 Clifton Avenue South
Minneapolis, Minnesota 55403

Peter Pfister, AIA
(612) 871-5703

Architectural Alliance/Energy is committed to integrating energy concerns in the team design process. The result

DESIGN—ARCHITECTURE & ENGINEERING 159

of this approach is buildings that meet the client's functional needs, are energy efficient, and are aesthetically pleasing.

Commercial and institutional projects include: long-range development plan for Carlton Park Research Facility, 3M Company, Lake Elmo, MN — a planning study to set the framework for the development of an 11 million square foot research facility over a 20-year construction period; Land O' Lakes headquarters and laboratory facility — a 290,000 square foot corporate office and laboratory facility; University of Minnesota, St. Paul, vocational technical education facility — renovation of a 30,000 square foot building and addition of 70,000 square feet of new space to house offices, classrooms and labs; Minneapolis—St. Paul International Airport — perform an energy audit on the 750,000 square foot main terminal complex.

Housing and residential projects include: Creek Terrace Apartments, Minneapolis — an 82-unit apartment building for elderly residents with provisions for low income and handicapped residents. A HUD grant financed a 6,000 square foot flat plate solar heating system. The system provides for domestic hot water heating and ventilation air heating. MASEC Solar 80 Homes, Plymouth, MN —The design of two tract-type passive solar homes. Both homes incorporate a direct gain passive solar system with thermal storage in concrete floors. Movable night window insulation covers major glazed areas to minimize heat loss. The passive systems will provide about 45—50% of the space heating needs and an active solar domestic hot water system will provide 80% of hot water needs.

Pfister Residence/Passive Solar Retrofit, Minneapolis — the weatherization and integration of a passive solar heating system in a two-story house built in 1920. The direct gain passive system was incorporated by enlarging south facing windows and by the addition of a two-story solarium to the kitchen. Thermal storage in the solarium is provided in a concrete structural floor. Phase change energy rods are used for thermal storage in an enlarged window bay. Exterior reflectors augment radiation. Nearly all windows are covered with movable insulation. A simple fan and duct system augments natural convection. The residence has been instrumented and monitored under a DOE grant. The design also received a HUD Passive Solar Design Award in 1979.

452. ARCHITECTURAL BUREAU
1900 E. Broad Street
Columbus, Ohio 43209

W. B. Ireland, FAIA
(614) 252-5209

Contractor with the Ohio Department of Energy passive solar mentally retarded workshop under construction for the state. Solar school for the mentally retarded in production for the state. Several commercial and residential solar applications in production for the private sector.

453. ARCHITECTURAL COLLECTIVE ARCHITRONICS, INC.
4615 W. Streetsboro Road
Richfield, Ohio 44286

Chas. Green
(216) 659-3151/633-0697

Architectural firm working with solar whenever possible—residential, commercial and industrial projects.

454. ARCHITECTURAL CONCEPTS
3140 O Street
Lincoln, Nebraska 68510

Kenneth Stadler
(402) 477-4944

Architectural Concepts provides building design services including preliminary design, design development, construction drawings and documentation, bid negotiation and construction supervision. Our firm provides services for models, perspectives and all interior design aspects. Principal of firm registered in Nebraska and Colorado. Solar systems are developed as needed for each custom project, including active and passive solar energy systems for space heating and cooling, domestic water heating, and swimming pool heating. We have design solutions for earth-sheltered structures and greenhouses.

DESIGN—ARCHITECTURE & ENGINEERING 161

455. ARCHITECTURAL & INDUSTRIAL MARKETING, INC.
1203 Salzburg, P.O. Box 308
Bay City, Michigan 48706

Sandra Proctor/Russell Campbell
(517) 686-1312

Design, engineering, sales — industrial, commercial and residential applications of solar systems.

456. ARCHITECTURE ONE
8 First Federal Center
Brainerd, Minnesota 56401

(218) 829-0525

Architectural design (new and retrofit) for residential, commercial and industrial application. Active and passive system design.

457. ARCHITEKTON, INC.
700 Walnut Street
Cincinnati, Ohio 45202

William J. Brown, AIA/John M. Purdy, AIA
(513) 421-5430

Architekton, Inc., a Cincinnati-based architectural, engineering and planning firm, has been actively involved in designing energy-efficient buildings for a number of years. We provide onsite energy surveys, identify energy saving opportunities, calculate the estimated cost savings, and assist in evaluation and design of alternate energy sources. Some recent experience includes a manufacturing facility with a solar-assisted thermal roof, and a church-owned recreational facility providing a passive solar system design.

458. ARIES CONSULTING ENGINEERS, INC.
2021 S.W. 43rd Avenue
Gainesville, Florida 32608

William J. Fielder, P.E.
(904) 372-6687

Solar design for more than 30 installations, including: health clinics, indoor sports facilities, office complexes, apartment complexes. Design of solar heating system for *Mother Earth News* offices, Hendersonville, NC. Design of world's largest solar-assisted heat pump, Gainesville, FL. Solar applications studies for The General Electric Co., Alachua, FL; Coca-Cola USA, Atlanta, GA; Grand Turk Island, BWI. Solar instrumentation design for HUD demonstration projects.

459. **ARMSTRONG, TORSETH, SKOLD & RYDEEN**
4901 Olson Highway
Minneapolis, Minnesota 55422

Roy Palmquist
(612) 545-3731

Architects and engineers concentrating on solar system application to multi-family residential, commercial and industrial structures.

460. **ATKINSON/KARIUS/ARCHITECTS**
1738 Wynkoop Street
Denver, Colorado 80202

Lawrence C. Atkinson
(303) 571-0096

Full-service architectural design for any climate. Consultation on energy applications. Prototype building design development: modular, mobile, dismountable, etc.

461. **AYRES, LEWIS, NORRIS & MAY, INC.**
3983 Research Park Drive
Ann Arbor, Michigan 48104

John L. King III, P.E.
(313) 761-1010

Energy-related work includes: generation systems (hydroelectric, etc.), energy recovery systems, energy auditing and technical assistance programs, retrofit design, and contract administration. Residential, commercial and industrial projects.

DESIGN—ARCHITECTURE & ENGINEERING

462. BAKEWELL CORPORATION
8820 Ladue Road
St. Louis, Missouri 63124

Ted Bakewell III
(314) 862-5555

Specialize in real estate development and commercial and industrial design. Superinsulated structures. Designer/builder of The Autonomous Dwelling Vehicle—prototype energy independent mobile home (100%) equipped with solar photovoltaic array for the structure's electrical needs. (Ref. *Popular Science* feature article, April 1980)

463. BALANCE ASSOCIATES
201 Summit Avenue East
Seattle, Washington 98102

(206) 325-4500

Computer programs—heat loss.

464. MARK BECK ASSOCIATES
762 Fairmount Avenue
Towson, Maryland 21204

Research and development—solar energy system design.

465. BECKMAN, BLYDENBURGH & ASSOCIATES
P.O. Box 100
Providence, Rhode Island 02901

Jeff Blydenburgh
(401) 274-3690

Specializing in passive solar applications in the commercial and industrial areas, new and retrofit, Beckman, Blydenburgh & Associates is an organization whose purpose is to promote a systematic and creative effort to bring appropriate technology into active existence and make it generally visible and available. BB&A is committed to the design and development of environments and products which are supportive of the users' needs. Services provided include planning, design and engineering of energy systems.

466. HOWARD BELL ENTERPRISES, INC.
5931 East Low Road, P.O. Box 413
Valley City, Ohio 44280

Howard Bell, President
(216) 225-3184

Designer/builder of single-family solar homes.

467. ROBERT A. BELL ARCHITECTS LTD.
115 N. Marion
Oak Park, Illinois 60301

Bob Bell/Cliff Nordling
(312) 383-0890

Robert Bell Architects Ltd. provides full architectural services, including design, engineering and consultation. Our office does solar-related research and feasibility studies for both active and passive solar projects. One project already constructed is a cemetary field operations center containing offices, lunchroom facilities, locker rooms, and an 18,000-square foot equipment maintenance garage. It uses an active solar air heating system using a flat plate collector. Architectural design included the field-constructed collector, rock heat storage, and air controls system. Robert Bell Architects Ltd. deals with the investigation of emerging solar technologies while providing a wide range of architectural services.

468. BERG & ASSOCIATES DESIGN/BUILD INC.
3140 Harbor Lane
Plymouth, Maine 55441

L. W. Berg
(612) 559-0272

Berg & Associates is an architectural/engineering firm specializing in environmental technology and energy efficiency. The firm's creative nucleus is composed of design professionals in architecture and mechanical/structural engineering.

Berg & Associates specializes in the practical utilization of solar energy for both new and existing projects. We can design and specify either passive or active space and water heating systems. We can provide state-of-the-art passive heating and cooling design using techniques such as: direct, indirect and isolated solar gain; natural and induced ventilation; dessication and evaporative cooling; night sky radiation; and diurnal and underground temperature systems.

Our firm utilizes and recommends earth-sheltered underground construction technology as an energy saving technique whenever feasible. We can also provide innovative yet practical design of wind technology as an alternative energy source.

Awards and grants include: 1979 HUD Cycle 5 Residential Solar Demonstration Design/Build Grant; 1978 HUD Passive Solar Residential Design/Build Grant; 1978 Twin City Metropolitan Council Modest Cost Housing Design Award.

469. BERKELEY SOLAR GROUP
3026 Shattuck Avenue
Berkeley, California 94705

Jan Phillips
(415) 843-7600

Consulting services for design of energy conserving buildings, specializing in multi-family housing, commercial and industrial structures. We have extensive experience in solar applications, thermal analysis, computer science, mechanical engineering, economic analysis, architecture and education.

Computer services/software development. Computer simulation for evaluating design strategies and energy conservation methods. Life-cycle cost calculations and feasibility studies. Climate analysis. Building performance monitoring. Solar system design and design review. Specialized training programs and seminars. Time-sharing service on inhouse Data General Eclipse computer; dial-in access to public domain and BSG proprietary software.

470. BERNHEIM, KAHN & LOZANO
One North Wacker Drive
Suite 205
Chicago, Illinois 60606

Offers architectural and engineering services to the solar community; engaged in earth-covered shelter research.

471. W. J. BLACK, AIA/ARCHITECT
1122 Westwood Avenue
Columbus, Ohio 43212

W. Jay Black
(614) 488-7343

Design of solar residences (single- and multi-family), and commercial buildings. Active and passive systems, new and retrofit.

472. BLACK—HUETTENRAUCH—SCHUYLER ASSOCIATES, ARCHITECTS, INC.
330 West Silver Spring Drive
Milwaukee, Wisconsin 53217

Peter Schuyler
(414) 332-9340

Member of State of Wisconsin energy audit team; listed as an energy auditor by the state. Passive energy design projects. Energy conservation studies and retrofit projects for commercial, industrial and multi-family clients. Active solar study for commercial clients.

473. J. A. BOCKMAN SOLAR ENGINEERING
1912 Northfield Court
Naperville, Illinois 60540

Jeffrey A. Bockman
(312) 355-4669

Residential and commercial solar design. A recent active system (Green Lake, WI) exhibits 18 vertical collectors and a phase-change storage room (350 square feet). Design sys-

DESIGN—ARCHITECTURE & ENGINEERING 167

tems using phase-change storage for wood and solar combinations and for a low-temperature heat source/heat pump.

474. PAUL F. BOGEN, ARCHITECT
2350 Columbia Street
Eugene, Oregon 97403

Paul F. Bogen
(503) 344-2368

Comprehensive solar design service integrated with all other design factors. Residential, commercial and industrial projects.

BRATTLEBORO DESIGN GROUP, *See (842).*

475. ROBERT J. BREGAR ASSOCIATES, ARCHITECTS
22700 Shore Center Drive, Suite 303
Cleveland, Ohio 44123

Schematic design, residential, with heat-saver fireplaces, for each of two adjoining rooms, solar collectors to rock storage, windmill and pond. Provisions for future solar heat applications in commercial buildings.

476. BROWN & BROWN
726 North Country Club Road
Tucson, Arizona 85716

Involved in passive solar design.

477. BUCKMASTER INDUSTRIES, INC.
P.O. Box 730, 23846 Sunnymead Boulevard
Sunnymead, California 92388

Warren D. Buckmaster
(714) 653-8461

Design/construction of energy-efficient homes that include active solar systems for domestic hot water and space heating. Also, sell, install and maintain active solar systems for space heating, domestic hot water, pools and spas.

478. BURT HILL KOSAR RITTELMANN ASSOCIATES
400 Morgan Center
Butler, Pennsylvania 16001

Michael W. Adsit/Jean B. Purvis
(412) 285-4761

Burt Hill Kosar Rittelmann Associates is an architectural/engineering firm with more than eight years of experience in energy conservation/solar energy research, development and applied design. In addition to comprehensive architectural services, we offer: solar energy architectural design/engineering (both active and passive); research and development in active and passive solar; photovoltaic, including systems design, module design, code analysis, other research; fuel cell and battery storage research; solar pond design; community energy systems; and co-generation. In addition, Burt Hill has served as consultant to manufacturers for solar component design.

479. ROBERT H. BUSHNELL CONSULTING ENGINEER
502 Ord Drive
Boulder, Colorado 80303

Robert H. Bushnell
(303) 494-7421

Robert H. Bushnell, meteorologist and consulting engineer working exclusively in solar energy since 1974, has consulted on numerous solar heating projects in Colorado. He has consulted on active air-type space heat, hybrid space heat with rock storage, collector design, swimming pool heat and heat loss of buildings. He reported on the observed heat use of buildings at the 1978 Denver conference of the International Solar Energy Society. He has developed a method of finding climatic temperature distributions, published in the *Monthly Weather Review,* and developed the Climatic Design Method, published in *Solar Energy,* which is particularly useful for highly insulated buildings and combined passive and active solar heat. His buildings have appeared in several publications. He is a member of ISES, ASHRAE, AMS and IEEE. He is a registered engineer in Colorado.

DESIGN—ARCHITECTURE & ENGINEERING

480. CALDWELL CONSTRUCTION
240 Commercial Street
Nevada City, California 95959

Richard Caldwell
(916) 265-6828

General building contractor. Solar installer of water and space heating systems. Design of buildings and solar heating systems. Performance estimates by computer simulation.

481. CAL ENERGY CONSULTANTS, INC.
711 E. Walnut, Suite 205
Pasadena, California 91101

L.O. Ford, President/L.E. Ford, Office Administrator
(213) 440-1490

We are an energy consulting firm, specializing in computerized energy analysis. We use the DOE computer/energy programs as well as TRACE, BLAST, TVENT, F-CHART, SOLCOST, etc. We research/optimize buildings and size equipment and life-cycle costing. For solar, we help size equipment and lay it out for optimum utilization. We also do energy audits for retrofits, plan checks for various cities, and Title 24 compliance documentation.

482. CALIFORNIA SOLAR DESIGNS
383 Union Street
Encinitas, California 92024

Greg Meyer
(714) 753-3330

Residential design (single- and multi-family), new and retrofit. Active and passive systems. Consulting service; evaluation; troubleshooting. Computer simulation.

483. CANDREX PACIFIC
693 Veterans Boulevard
Redwood City, California 94063

Solar energy engineering; computer-based applications.

170 SOLAR CENSUS

484. CAPITOL CONSULTANTS INC.
1627 Lake Lansing Road
Lansing, Michigan 48912

W. T. Harvey/R. E. Tadgerson
(517) 371-1200

Consulting engineers in active and passive solar technology.

485. CENTRAL STATES ENERGY RESEARCH CORP.
128-1/2 E. Washington Street, P.O. Box 2623
Iowa City, Iowa 52240

James L. Schoenfelder
(319) 351-2441

Solar energy consulting and design of residential, commercial and industrial structures. Active and passive systems incorporated in new and existing buildings.

486. CIDER BLUFF INC.
5445 Holmes
Kansas City, Missouri 64110

Kenneth W. Spere
(816) 333-8862

We design systems for single-family residential application, new or retrofit. We also distribute and install Suncatcher® solar systems (Solar Unlimited of Huntsville, AL), which have various applications from domestic hot water, zone heating, and multi-room heating to full central heating and wood furnace backup systems.

487. CLARK & WALTER ARCHITECTS
513 S. Union
Traverse City, Michigan 49684

E. Terry Clark
(616) 946-3627

Design of earth-sheltered structures utilizing passive solar heating. Numerous residential projects; some commercial and retrofit projects.

DESIGN—ARCHITECTURE & ENGINEERING 171

488. CLOSE ASSOCIATES, INC.
3101 E. Franklin Avenue
Minneapolis, Minnesota 55406

Winston A. Close
(612) 339-0979

Active and passive design for residential (single- and multi-family) applications. On two recent projects (12 row houses, 1 detached house), the University of Minnesota Underground Space Center will operate monitoring equipment for 2 years.

489. E. C. COLLINS II ASSOCIATES
40 Lowell Road
Concord, Massachusetts 01742

Keith B. Gross, Architect
(617) 369-1833

Architects in solar and energy conservation projects—single- and multi-family residential.

490. C. PHILLIP COLVER AND ASSOCIATES, INC.
0855 Mountain Laurel Drive
Aspen, Colorado 81611

C. Phillip Colver, Ph.D., P.E.
(303) 925-4088

We are a consulting firm primarily involved in the teaching of one- to five-day short courses on solar technology, energy management and conservation. Some direct consulting engineering is involved.

491. COMMERCIAL PLUMBING CORP.
732 Washington Street
Weymouth, Massachusetts 02188

S. Junta
(617) 337-8550

Design and install active solar systems; commercial structure emphasis.

492. COMMUNITY ACTION COMMISSION FOR THE CITY OF MADISON AND THE COUNTY OF DANE, INC.
1045 East Dayton Street
Madison, Wisconsin 53703

William Benisch
(608) 266-9731

CAC is an anti-poverty community action agency involved in a demonstration project of alternate energy usage based on renewable energy sources of wood and solar in Madison nieghborhoods. The project demonstrates production, use and benefits of renewable energy systems relative to savings in the overall costs of energy bills, and provides a comprehensive profile of energy patterns for low-income residents.

493. COMMUNITY BUILDERS
Shaker Road
Canterbury, New Hampshire 03224

Designer/builder of passive solar homes. Authority on double-shell, buffering air envelope. Consultant on passive and double shell.

494. COMSTOCK CONSTRUCTION CO., INC.
5848 W. Higgins Road
Chicago, Illinois 60630

Edgar Garcia Smoot
(312) 545-9200

Design and construction of both experimental and working solar systems for multi-family residential complexes, commercial and industrial structures. Active systems, new and retrofit.

495. CONKLIN & ROSSANT
251 Park Avenue South
New York, New York 10010

Active in site planning, building design, solar collector research.

DESIGN—ARCHITECTURE & ENGINEERING

496. CONTEXTUS CORPORATION
110 Orange Street
Chico, California 95927

Peter Straus
(916) 891-0848

Contextus Corporation offers a full line of solar services, from active solar sales and installation to passive solar design and construction. Our design team has drawn up plans for a number of passive solar homes using the envelope concept. In 1979, we were awarded a first place and special commendation for applying the envelope concept to mobile homes, thus creating a solar mobile home. We are currently planning a passive solar branch bank.

The construction crews are made up of highly skilled carpenters and craftsmen. They are currently involved in three solar retrofit projects.

497. COORDINATED SYSTEMS, INC.
1007 Farmington Avenue
West Hartford, Connecticut 06107

Ed Curley, Director of Operations
(203) 233-6248

Coordinated Systems, Inc. is an energy consulting firm providing energy management services to large commercial, institutional, and industrial clients. It offers its customers partial or complete turn-key packages for energy conservation, including study, design and general contracting services. Special HVAC and alternate energy study and design services are also offered.

498. CREATIVE ALTERNATIVES
Route 1
Long Prairie, Minnesota 56347

John Weber
(612) 732-6304

Design and installation of active and passive systems. Provide both a low-cost, site-built collector system and modules.

Design of sunspaces and greenhouses. Heat exchangers for confinement barns.

499. **CROFT & COMPANY**
ENGINEERING DIVISION
708 W. 10th Street
Austin, Texas 78701

James Logan
(512) 474-1985

Solar engineering design primarily for the commercial/industrial sector. Energy management, energy audits, and educational activities.

OEM—microprocessor controllers. Integration, installation and maintenance of digitally based building automation systems. Programmable microprocessors network transmissions to/from mechanical and electrical subsystems to simulate and optimize building operations (services); monitor, indicate and assign process control functions; automate pointset/position reset of remote transducers, controllers and actuators; and support query, log and management reporting. Modular design (vertically integrative logic and plug-in peripherals) is compatible with a full range of scale and control requirements; permissive/distributive electrical load optimization; maintenance scheduling and detection; space conditioning, security, fire protection, and lighting control.

500. **CROWTHER SOLAR GROUP**
310 Steele Street
Denver, Colorado 80206

Rich Heinemeyer
(303) 355-2301

Architecture, energy design consultation, energy audits and evaluation of existing structures. Passive and active solar system design.

501. DANIEL ENTERPRISES, INCORPORATED
SOLAR TECHNOLOGY DIVISION
P.O. Box 2370, 1291 S. Brass Lantern Drive
La Habra, California 90631

G. Appleman, P.E.
(213) 943-8883

Computing services. System performance simulation. Solar consulting. We also provide conventional engineering and consulting services to the solar, HVAC and process industries. These include residential heat loss and heat gain analysis (in California, Title 24 reports), solar collector design analysis, solar system analysis, feasibility studies.

502. DAS/SOLAR SYSTEMS
188 Flatbush Avenue, Ext.
Brooklyn, New York 11201

William Ross
(212) 522-0400

DAS/Solar Systems designs, merchandises and installs solar systems for residential, commercial, industrial, and institutional applications. The company operates its own comparative test facility to gauge actual performance of various solar components before they are integrated into pre-engineered or custom designed solar systems for its customers. DAS/Solar Systems purchases industry-standard components to assemble its own working solar systems.

Since 1976 the company has installed more than 10,000 square feet of solar collectors for its clients. Applications for solar energy to date have included: service hot water for residences and apartment buildings; space heating—both direct and heat pump assisted; pool heating; and industrial process water.

DAS/Solar Systems provides design consulting, system engineering, project management, and turn-key solar system delivery services to architects, builders, and government agencies. Some of our many corporate clients include Phelps Dodge, Citibank, the State of New Jersey, and the General Magnaplate Corporation.

503. DAVIS ENGINEERING, INC.
20976 Currier Road
Walnut, California 91789

Greg Newton
(714) 594-1681/594-1896

Solar R&D and design engineering for residential, commercial and industrial applications. Active systems for new structures and retrofit.

504. DAVIS, JACOUBOWSKY, HAWKINS ASSOCIATES, INC.
299 Cannery Row
Monterey, California 93940

Bob Jacoubowsky
(408) 649-1701

Design and construction documents for active and passive solar heated homes and condominiums.

505. ROBERT E. DES LAURIERS, ARCHITECT
9349 El Cajon Boulevard
La Mesa, California 92041

Robert DesLauriers, AIA
(714) 469-0135

We are primarily an architectural/engineering firm, providing design for solar installations of all types of uses and functions. Residential, commercial and industrial design and R&D.

506. DETROIT ENVIRONMENTAL CONTROL ENGINEERS INC.
39108 Charbeneau Street
Mt. Clemens, Michigan 48043

(313) 468-3225

Detroit Environmental Control Engineers Inc. is involved with conventional heating and air conditioning system design and installation, adaptation to solar capabilities as well as passive and active residential and commercial design and construction. Also, consulting, energy management and energy auditing capabilities, including industrial. Also see listing for SOLAR AGE DESIGNERS, R&D section.

507. dh2W, INC.
705 Franklin Square
Michigan City, Indiana 46360

Raymond C. Warren
(219) 872-9406

Design of residential and light commercial structures. Active systems include closed liquid systems of collectors, piping, storage tank, differential temperature controllers, pumps, insulation, boiler backup, air handling units with coils, chillers, economizers, etc.

508. DICKEY/KODET/ARCHITECTS/INC.
4930 France Avenue South
Minneapolis, Minnesota 55410

Ed Kodet
(612) 920-3993

The firm of Dickey/Kodet/Architects is involved in both passive and active solar design. The firm currently has a residence in southwest Minneapolis that is an exhibit of its passive solar design. Furthermore, we have been involved in commercial projects involving passive solar and the retrofitting of active solar.

As an architectural firm involved with the establishment of the criteria for the national energy code, the firm has been active in a variety of projects regarding energy conservation. It furthermore seeks such projects and develops a program with each client to establish what energy considerations are applicable and how they become economically viable.

509. DIMETRODON
Route 1, Box 160
Warren, Vermont 05674

William Maclay
(802) 496-2907/496-2787

Dimetrodon is a cluster of private and common living and working spaces employing integrated wood, wind and solar energy systems and agricultural use of the land.

A central heating system of the whole building consists of a large flat plate solar collector and a central wood furnace. Domestic hot water for the building is supplied by the wood furnace and solar collector in winter and by the solar collector during the summer. Heat exchangers in solar and wood storage tanks isolate potable and heating system water. Electricity for all pumps, blowers and controls for the mechanical system is supplied by two 32-volt wind generators.

Dimetrodon Research Station is a nonprofit research center investigating the interrelationship of energy, lifestyle and culture.

510. DIXON/CARTER ARCHITECTS
P.O. Box 797
Granby, Colorado 80446

Dan Dixon
(303) 887-2200

We design residential and commercial structures with both active and passive space heating. Projects include Solar Plaza in Granby, CO (which houses Dixon/Carter Associates and several other firms). The solar system is an active air system using rocks as the storage medium. It was designed to provide 80% of the total space heating and domestic hot water energy requirements.

511. DONOVAN ENTERPRISES
3642 Eagle Street
San Diego, California 92103

T. W. Donovan
(714) 297-5220

Consulting and engineering for solar hot water, pool, spa and space conditioning systems. Also, marketing and installation.

512. DOOLING & SIEGEL ARCHITECTS
84 Bowers Street
Newtonville, Massachusetts 02160

Mark Dooling
(617) 332-1694

Serving a wide range of institutional, commercial and residential clients, we provide comprehensive architectural services, including programming, site selection assistance, design, project budgeting, work drawings, specifications, and construction administration.

Passive design emphasis is applied to the design of new buildings, additions, alterations to existing buildings, space planning.

513. DOWNING LEACH ARCHITECTS & ENGINEERS
3985 Wonderland Hill Avenue
Boulder, Colorado 80302

Jim Leach
(303) 443-7533

Downing/Leach and Associates has chosen to respond strongly to the energy challenge by offering complete solar design services. Since 1975 and the creation of a solar program for Wonderland Hill Development Co., Downing/Leach has had extensive solar experience, and has developed expertise in many aspects of solar design—passive, hybrid, land planning for solar development. Downing/Leach solar experience includes assembly and management of successful solar grant applications totaling more than $600.000. These federal grants have been awarded for HUD Cycle II, III, IV and IVA solar projects, involving more than 150 housing units.

514. DRESSLER ENERGY CONSULTING AND DESIGN CORPORATION
4550 West 109th, Suite 170
Overland Park, Kansas 66211

Wm. E. Dressler
(913) 341-1743

Dressler provides consultation services in the areas of energy efficiency and solar energy feasibility and solar energy applications. We offer complete building energy studies with the aid of the most advanced computer technology. We have engineered tremendous energy saving devices and reclaim methods in several commercial and industrial facilities. We are highly experienced in solar energy systems design and operation.

Our personnel have been responsible for the five-story Meadowlark Hills Highrise located in Manhattan, KS, a design utilizing a large scale solar-assisted heat pump system. We solicited and received a grant for $198,000 from the HUD Solar Heating and Cooling Demonstration Program. With a grant from ERDA, we designed a solar system for the Kaw Valley Bank (Topeka, KS), which will furnish 70% cooling, 55% heating and 95% hot water, using concentrating collectors.

515. DSS ENGINEERS, INCORPORATED
1850 N.W. 69th Avenue
Fort Lauderdale, Florida 33313

Mr. O. J. Morin
(305) 792-6660

DSS Engineers, Inc. was founded in 1968 under the name of Desalting Systems and Services. As this name implies, the company was initially organized to fill a need for an independent consulting engineering firm which could provide a full range of engineering services essential to the design, construction and successful operation of sea water desalination facilities and dual-purpose desalination in power plants. It was soon recognized that the technology and experience obtained as a result of intensive research, design and operation efforts in desalination could be applied to related engineering projects. The name was then changed to DSS Engineers and the firm has since expanded to provide engineering and design services in a number of related advanced technology fields. These are as follows:

Mineral extraction from sea water and brines; Pollution control and environmental impacts; Water quality improvement; Ocean thermal energy systems; Utilization of geothermal fluids; Solar energy systems.

During its 12-year history, DSS has executed over 90 separate contracts, some of which are continuing on a permanent consulting basis. These contracts have included the following services:

Technical, economic and feasibility studies; Research and development; Master planning; Site investigations; Conceptual designs; Systems optimization and engineering studies; Plant and process design; Preparation of detailed construction plans and specifications; Estimating; Construction management; Negotiation of contracts; Procurement and expediting of materials and equipment; Training of the client's engineering and operating personnel; Initial plant startup; Plant operations and maintenance.

516. DUBIN-BLOOME ASSOCIATES, P.C.
42 West 39th Street
New York, New York 10018

G. M. McDougal
(212) 840-6700

Dubin-Bloome Associates is an organization of registered professional engineers engaged in design of mechanical and electrical systems for all types of buildings, and construction administration; master planning and engineering for utility systems to serve campuses, cities and industrial complexes; solar energy, energy conservation and energy management; basic and applied research for universities, institutions, government and industry.

517. DULANEY & ASSOCIATES
P.O. Box 346
Ames, Iowa 50010

Dave Dulaney
(515) 232-1178

Design of active and passive residential and commercial structures. Special interest in earth sheltering. Lecturing on general energy conserving design, passive solar design, earth-sheltered design, earth-sheltered structural considerations.

518. EARTH SHELTERED HOUSING SYSTEMS
Route 1
Marshalltown, Iowa 50158

Jacob Kvinlaug
(515) 752-2489

We do custom design and construction of passive solar earth-sheltered residential and commercial buildings. All of our buildings are 80—90% solar heated and 100% cooled. We use poured reinforced concrete walls with pre-cast concrete roof and approximately 2 feet of earth cover. Our passive systems are almost always of the direct-gain type utilizing either primary or secondary storage, or both.

One of our homes, recently featured by the press, operated through this last Iowa winter on less than 3/4 of one cord of firewood as its only auxiliary heat. We estimate that the home is 92—95% solar heated and it will be 100% passively cooled.

We also incorporate interior insulated window shutters into each of our designs. We also employ various electricity and water conservation devices to further lower the total energy consumption of our buildings.

519. EARTH—SUN—DESIGN
P.O. Box 2473
Springfield, Illinois 62705

Robert L. Litvan
(217) 528-5355

We are a subsidiary design firm of Maslauski/Litvan Architects and Planners, Inc., 901 South Second Street, Springfield, Illinois 62704. Earth—Sun—Design specializes in the custom design of residential, commercial and institutional buildings utilizing earth-sheltered and solar technology.

520. ECOTOPE GROUP
2332 East Madison
Seattle, Washington 98112

Belinda Boulter, Internal Administrator
(206) 322-3753

Ecotope's design services emphasize the use of solar, energy-efficient designs for residential and commercial buildings in the Northwest. We consult with home owners, builders, architects, engineers and designers, and often serve as design team managers to bring these people together.

We also use two computer models to assist evaluation of design options, and design consultation is available for solar water heater systems and residential wood heating systems as well.

We provide comprehensive passive solar design services for residential and commercial buildings from the conception of projects through construction and monitoring the performance of finished buildings. Our services have been used for owner-built, contractor-built, and community workship projects.

We also provide analysis of existing structures and develop design recommendations to combine a passive solar system with energy-efficient remodeling.

521. ECS—ENVIRONMENTAL CONTROL SYSTEMS
525 East 10th Street
Jeffersonville, Indiana 47130

B. Van Cleave
(812) 282-7894

Retrofit of two solar systems (residential), 2400 square feet each, and one solar domestic hot water system.

184 SOLAR CENSUS

522. EDGE RESEARCH
RD 1, Box 394B
Kingston, New York 12401

(914) 336-5597

Offers a kit for do-it-yourself construction of an attached greenhouse of beadwall or double-glazed plastic, with a treated wood frame.

523. EKOSE'A, INC.
573 Mission Street
San Francisco, California 94105

Bill Pearson/Robert Acher
(415) 543-5010

Ekose'a is presently engaged in design construction documentation, construction supervision and monitoring of a constantly growing number of projects which maintain comfortable interior conditions without supportive systems using electricity or fossil fuels. These projects include: single-family dwellings, multi-family dwellings, commercial buildings and remodeling/renovation.

Ekose'a is the creator and developer of the geothermal convection loop envelope concepts. Thirty houses have been completed (100% passive heating and cooling), and 100 houses are now in various planning and construction phases.

524. ELECTRICAL CONSTRUCTION COMPANY
P.O. Box 1028
Palmdale, California 93550

Dudley Foster
(805) 273-2556

Deal primarily in heating and cooling, and calculations and evaluation of structures for California Title 24 energy standards. Also do HVAC design and analysis. Single-family and multi-family residential, commercial and industrial structures.

525. ELECTROSTATIC CONSULTING ASSOCIATES
P.O. Box 552
Groton, Massachusetts 01450

Paul Malinaric
(617) 448-5485

Design and retrofit of heat pump systems for hot water heating. Specifically, heat pump control and defrost systems and controls. Also upgrading efficiency of existing heat pump systems.

526. CHARLES ELEY ASSOCIATES
342 Green Street
San Francisco, California 94133

Charles Eley
(415) 398-6535

Design/consulting services—residential, commercial and industrial, passive and active systems. Projects include: solar consultation services for a 64-unit apartment project in Vacaville, CA, combining active solar systems for domestic hot water and pool heating with passive solar greenhouses. Design of an energy-efficient office building of 240,000 square feet in Sacramento that incorporates passive solar systems to heat and cool all circulation and lobby spaces. Energy impact assessment for a 140-acre industrial/office development in Menlo Park. Design of four solar duplex units in Davis, CA for low and moderate income families. Preparation of design guidelines and an energy element for the codes, covenants and restrictions of a planned unit development of about 1500 dwelling units near Kennewick, WA. Design consultation and solar feasibility study for the 600-acre ATO Desert Project, Palm Springs. Energy design consultation for a 106-unit multi-family housing project in Sacramento. Energy code compliance assessment for San Francisco International Airport. Design, development and testing of a new line of passive solar single-family homes for a major Sacramento builder.

186 SOLAR CENSUS

527. ELLMORE/TITUS/ARCHITECTS/INC.
736 Chestnut Street
Santa Cruz, California 95060

S. A. Titus
(408) 427-1010

Architectural firm specializing in passive system design for residential, commercial and industrial application.

528. ELSWOOD–SMITH–CARLSON, ARCHITECTS
5700 Broadmoor, Suite 703
Mission, Kansas 66202

Dean R. Smith
(913) 432-5655

We designed one house for an active system with rock storage; also one earth contact/passive system combined with central fireplace and wood furnace.

We have done preliminary drawings for a custom house with a swimming pool in an atrium; also six preliminaries for speculative houses and one custom house incorporating greenhouses. Construction drawings for one of the speculative houses are currently being drafted.

Also, currently working toward design and construction of a two-story underground home. Active in workshops on the subject of passive solar and earth contact construction to remain cognizant of new developments and technology.

529. ENERCOM, INC.
2323 South Hardy Drive
Tempe, Arizona 85282

A. R. Colen
(602) 894-2279

Residential, commercial, industrial and institutional system design, new and existing applications, as a part of the HVAC. Applications as a part of an energy audit. Audit program design utilizing solar.

530. ENERCON, LTD.
500 Davis Street
Evanston, Illinois 60201

A. S. Butkus, Jr.
(312) 328-3555

Enercon, Ltd. is a consulting engineering firm specializing in energy management and alternative energy systems. Services include: energy conservation studies for existing and new facilities; solar energy system feasibility, analysis and design; building automation system design, specification and evaluation; energy audits; life cycle cost studies; architectural energy impact reviews; computerized economic and financial analyses; fuel source selection; proposal preparation for federal grants; preventative maintenance programs.

531. ENERGY APPLICATIONS
Long Reach Village Center, Suite 227
Columbia, Maryland 21045

James L. Coggins
(301) 730-0663

Specializing in solar energy applications and energy conservation. Offer consulting services to industry, institutions, and government for the reduction of fuel and utility costs. Recent projects include documentation of construction costs and as-built system design for federally funded commercial solar demonstration projects; a brief feasibility study of a solar system for a hotel in Peking, China; development of a photovoltaic laboratory session for a seminar; an investigation of industrial process utilization of solar energy; and an investigation of the solar cooling of buildings.

532. ENERGY ASSOCIATES
Box 6602
Greenville, South Carolina 29606

Roger R. Varin, President
(803) 277-7884/277-6221

Involved in commercial/industrial applications of solar. Two major areas: low-cost active collection system being designed;

low-cost insulating structures being evaluated for commercial exploitation.

533. **ENERGY CENTER, INC.**
1726 W. Mission, P.O. Box 2086
Escondido, California 92025

George Phillips
(714) 743-3119

We are a small super-active solar heating company. Five years experience, specializing in hot water systems, pools and spas, and space heating. Solar heating is our only business.

534. **ENERGY DESIGN CONSULTANTS AND BUILDERS, INC.**
118 Brook Bridge Drive
Cary, Illinois 60013

John H. Stitt
(312) 639-3343

We design passive solar homes, build solar homes, and consult on existing structures.

535. **ENERGY DESIGNS/ARCHITECTS**
201 Woodrow Street
Columbia, South Carolina 29205

Dick Lamar AIA
(803) 799-7495

Architectural design of energy conscious and solar buildings —residential, commercial and industrial.

536. **ENERGY ENGINEERING GROUP, INC.**
P.O. Box 130, 1115 Washington Avenue
Golden, Colorado 80401

Richard L. Casperson
(303) 279-1851

Consulting, Research and Development, A/E Design Services. Areas of Specialization: active and passive solar energy systems, heating and cooling, residential and commercial; building energy conservation and optimal design techniques; thermal, structural and economics analysis.

Computer services: provide time-sharing services for solar design and analysis, structural analysis, economic feasibility analysis; provide software and hardware (turnkey) systems for builders, contractors, architects and engineering firms; implementation of software on certain minicomputer or microcomputer systems.

537. ENERGY MANAGEMENT CONSULTANTS, INC.
1180 South Beverly Drive, Suite 315
Los Angeles, California 90035

Douglas S. Stenhouse, President
(213) 553-8725

EMC is a multidisciplinary firm organized for the express purpose of performing work in the energy conservation field. The firm offers a wide variety of services which include: building energy audits; energy conservation planning; energy conservation planchecking for compliance with Title 24; energy-conserving analysis and design for new buildings; energy conservation retrofit studies and designs; analysis of building energy conservation legislation and regulations; energy system and component equipment analysis and design; solar energy (passive and active) system design; system economic (life-cycle cost) analyses; energy conservation program management support; and energy conservation training programs and seminars.

EMC has completed energy surveys and analyses of over 12 million square feet of buildings, representing various nonresidential types and sizes. The work involves development of a work plan, field surveys, interviews with maintenance and management personnel, data collection and analysis, and the development of a comprehensive list of energy conservation opportunities prioritized for payback of investment. Each recommendation is carefully explained and includes detailed calculations of energy savings and estimates of initial costs. The items recommended may include not only ways to save energy for lighting, heating and cooling buildings, for reduction of energy to power equipment and heat water, but also practical ideas about how to manage energy resources better and how to encourage wise use of energy by the people who use buildings.

The firm has worked with architects, engineers, builders, and building owners in the initial design of various types and sizes of new buildings. We have completed numerous energy audits and system redesign proposals for existing buildings.

538. **ENERGY MANAGEMENT AND CONTROL CORPORATION (EMC2)**
634 Harrison, Suite B
Topeka, Kansas 66603

Lee V. McQueen
(913) 233-0289

EMC2 provides services during all phases of project development, from initial concept (including computer simulations) to final testing and balancing. Complete energy analysis of existing or proposed buildings is included. EMC2 can also provide load reduction studies for buildings or utilities.

539. **ENERGY PLANNING & INVESTMENT CORPORATION (EPIC)**
833 N. 4th Avenue
Tucson, Arizona 85705

Larry Medlin
(602) 623-6406

Design and development of projects and solar communities focused upon energy-conscious design including solar utilization, lightweight and industrialized construction techniques, and adaptable living.

DESIGN—ARCHITECTURE & ENGINEERING

540. ENERGY SPECIALTIES
9312 Greenback Lane
Orangevale, California 95662

Philip Leen
(916) 988-1208

Design, engineering, installation and repair services for several types of energy-producing equipment. Also, contracting services for residential and commercial projects.

541. ENER-TECH, INC.
1924 Burlewood Drive
St. Louis, Missouri 63141

James A. Ray
(314) 878-1586

Ener-Tech has as its primary emphasis energy conservation and alternative energy design. Solar design is integrated into the toal design with specific design toward passive. In addition to design, Ener-Tech pursues a research and development program to test new design and products prior to incorporating them into company designs.

542. ENGINEERING CONSULTING SERVICES
8 South Montgomery Street, P.O. Box 1809
San Jose, California 95109

Dr. Joseph C. Olsen, P.E.
(408) 297-8778

Engineering design, research and development. Certified testing laboratory—ICBO and IAPMO verification testing.

543. ENGINEERS—ARCHITECTS, P.C. (EAPC)
408 First Avenue Building
Minot, North Dakota 58701

Gordon L. Rosby
(701) 838-3618

Engineering/architectural applications of solar energy— residential, commercial and industrial. Because of feasibility at present, several buildings have/are being designed for future solar systems. A military project now in design is receiving favorable consideration for solar hot water and space heating.

544. ENTEC
1900 Point West Way, Suite 171
Sacramento, California 95815

Robert Sieglitz
(916) 920-1441

ENTEC's management engineering services include: physical survey, measurement and testing, analysis and reporting, engineering, implementation, and monitoring. Residential, commercial and industrial projects.

545. ENVIRONMENTAL DESIGN ALTERNATIVES, ARCHITECTS
1951 Brookview Drive
Kent, Ohio 44240

Douglas G. Fuller
(216) 673-2158

We are a small architectural firm involved in many aspects of energy conservation in building design. We have been involved in many and varied applications of alternative energy. Other services include: community energy conservation planning, lectures and workshops, energy audits.

546. ENVIRONMENTAL DESIGN GROUP
269 West Second Street
Claremont, California 91711

Mark von Wodtke
(714) 621-1440

Site planning, architecture and landscape architecture for housing, commercial/office and institutional projects. Projects include: Helios solar townhouses, Mammoth Lakes, CA, a 15-unit complex with passive solar space heating, solar domestic hot water, and spa heating; double envelope houses with solariums. Claremont Estates (Claremont, CA), a 40-acre development, housing with optimal solar orientation. We have also designed a number of single-family residential retrofits.

547. ENVIRONMENTAL INSTRUMENTATION
922 S. Barrington Avenue, Suite 202
Los Angeles, California 90049

Winfield B. Heinz
(213) 820-6111

Environmental Instrumentation was started by Winfield B. Heinz after his retirement from teaching Systems Engineering at UCLA. Mr. Heinz designed a nucleus line of flat plate solar collectors for domestic water heating, for swimming pools and for air heating. Four panels have been heating domestic water in a large residence outside Malibu for more than two years, with storage temperatures which can be 200F or above. The collectors and their advanced control system are of our own design and manufacture. They are high in performance and low in manufacturing cost. Collectors of lower cost are available for swimming pools and for heating air.

548. ENVIRONMENTAL POWER CORPORATION
12750-146 Centralia Street
Lakewood, California 90715

R. J. Zwerling
(213) 865-3409

EPC provides engineering, consulting, and a full service management program for solar and energy conserving projects. EPC management services include integrating our engineering staff with solar energy architects, installers, and tax consultants to produce a complete design, engineering, installation and financial package. Our engineering services include energy conserving design and/or analysis on new or existing structures, system and component design and/or analysis on proposed or existing systems, solar design and economic analysis. All services extend to retrofit and new construction efforts on residential or commercial structures for: pool heating, hot water systems, space heating and cooling, energy conserving projects.

549. ENVIRONMENT ASSOCIATES ARCHITECTS/PLANNERS
2777 Allen Parkway, Suite 207
Houston, Texas 77019

LaVerne A. Williams
(713) 528-0000

Operating under the name LaVerne A. Williams, AIA, Architect/Energy Conservation Consultant from 1973 to 1978 until which time Environment Associates was formed, the firm specializes in the passive design of new buildings and retrofit/revitalization applications of appropriate alternative energy technologies. Projects include a passively designed office building, several earth-sheltered passive solar homes, several above-ground passive solar homes, and numerous residential and commercial building appropriate alternative energy retrofit design and consultation commisions.

The firm is also the architect for the Southern Solar Energy Center's "Interactive Passive Design" demonstration home for Houston, TX.

550. ESR – ENVIRONMENTAL SYSTEMS RESEARCH
1102 Coloma Way
Roseville, California 95678

Mr. Martin Dowanoe
(916) 782-8349

P.O. Box 372
Running Springs, California 92382

Mr. Dar Grossman
(714) 867-7874

In 1977 we established a formal dual partnership business for the purpose of assisting others in saving our energy resources. Since 1977 we have been actively engaged in all aspects of energy conservation specializing in the domestic and commercial sectors.

The past year has involved us in heavy active and passive activity with emphasis on passive greenhouse augmentations with active solar backup systems. Next year we anticipate an increase in consulting to include commitments in wind generation systems.

551. JOHN M. EVANOFF, ARCHITECT
3405 River Road
Toledo, Ohio 43614

John M. Evanoff
(419) 382-5376

Research and design of residential structures with emphasis on heliothermic planning. Since heliothermic is an expanded version of what is now called passive solar and earth-sheltered structures, my work incorporates these concepts.

I also have researched active solar systems, but feel that a heliothermic structure with storage integrated within the structure, with a possible domestic hot water collection system (hybrid), is a more cost-effective design.

My interest in solar actively began in 1969 with the design of my own home (a heliothermic), which serves as a prototype for future designs in solar.

552. THE EVJEN ASSOCIATES, INC.
Box 152
Hudson, Wisconsin 54016

Richard Evjen
(715) 386-2658

Full-service architectural/engineering firm involved in the design of active and passive systems—residential, commercial and industrial.

553. PAUL FELLERS, ARCHITECT
244-C Commercial Street
Nevada City, California 95959

Paul Fellers
(916) 265-2745

The primary purpose of the designs that our office produces is to integrate sensitive and functional spatial designs that are based on user's needs and desires, with efficient, low-cost passive space heating and cooling systems. Primarily residential and small-scale office structures.

554. **FEUER CORPORATION**
2601 Ocean Park Boulevard
Santa Monica, California 90405

Stanley Feuer
(213) 450-0363

Design of mechanical systems for multi-family, commercial and industrial buildings. This includes heating, ventilating, air conditioning, plumbing, and solar systems.

555. **FISHER/ROBERTS**
220 Pier Avenue
Santa Monica, California 90405

Thane Roberts
(213) 392-3086

Fisher/Roberts is an architectural firm specializing in the use of passive energy systems for the heating and/or cooling of buildings. Design services include the conceptualization, engineering, design and construction supervision of all systems. Consultation is also offered for other architects, engineers, developers, or owners wishing to incorporate energy saving measures and/or systems. Special emphasis is placed on simple, economical and workable designs which do not require sophisticated equipment or exorbitant expense. Fisher/Roberts has experience with all sizes and types of projects.

556. **W. S. FLEMING AND ASSOCIATES, INC.**
840 James Street
Syracuse, New York 13203

William Fleming
(315) 472-4405

The firm of W. S. Fleming and Associates, Inc. provides services to building owners, architects, institutions, city, county, state and federal government. The following services are handled by local offices in Syracuse, NY, Albany, NY, and Landover, MD:

Professional design of active and passive solar HVAC and domestic hot water systems; professional design of active and passive solar/heat reclamation HVAC and domestic hot water systems; contributors and authors of two energy analysis computer programs which simulate active and passive solar and/or heat reclamation systems (e.g., DOE 2.1 and SEE); professional energy and economic analysis of active and passive solar systems to determine operating cost, capital expenditure, payback and present worth cost/benefit ratio; professional advice to correct current solar energy system problems which may be a result of system design, equipment and/or operation; professional advice pertaining to solar HVAC system experience as related to design, operation, equipment, system, technical and economic feasibility.

557. W. S. FLEMING AND ASSOCIATES, INC.
3 Computer Drive
Albany, New York 12205

Dr. Dennis R. Landsberg
(518) 458-2249

Passive and active solar design for multi-family residential and commercial structures. Energy conservation solar retrofits as part of building energy audits where practical. Consultant to state, federal and municipal governments.

558. FLINN SAITO ANDERSEN ARCHITECTS
604 Mulberry
Waterloo, Iowa 50703

Tom Gardner
(319) 233-1163

Specialize in earth-sheltered housing, passive solar housing, passive solar commercial offices.

559. JAMES FOLLENSBEE & ASSOCIATES LTD.
Architects—Planners—Engineers
311 West Hubbard Street
Chicago, Illinois 60610

James Follensbee, AIA, President
(312) 467-4767

Architects and planners for solar-equipped projects from feasibility studies through site planning—construction document and construction administration phases of work.

560. MARIO FONDA-BONARDI
3111 3rd Street, No. 19
Santa Monica, California 90405

Mario Fonda-Bonardi
(213) 392-3273

Architect emphasizing the use of passive, superinsulated, high-mass thermal strategies for heating and cooling residences in the relatively benign Southern California climate.

561. FOREST CITY DILLON, INC.
a subsidiary of Forest City Enterprises, Inc.
10800 Brookpark Road
Cleveland, Ohio 44130

Kenneth M. Yarus
(216) 267-1200

Developer/general contractor specializing in energy-efficient multi-family dwellings, senior citizen high-rises.

562. FORSTER-MORRELL ENGINEERING ASSOCIATES, INC.
P.O. Box 9881
Colorado Springs, Colorado 80932

T. J. Forster
(303) 574-2127

We perform solar feasibility studies, design active and passive solar systems for commercial and industrial application, perform computer simulations of building energy systems to include solar systems, and perform energy code compliance checks. We provide consulting services in these same areas, and survey and inspect solar systems.

563. **FREEBERG COMPANY**
 8624 Pine Hill Road
 Bloomington, Minnesota 55438

 Roger Freeberg
 (612) 941-7092

 Design and construction of passive solar residences (single- and multi-family). Also, residential solar system retrofit.

564. **RONNIE FREEMAN**
 Solar Energy Consultant
 Box 215
 Trent, Texas 79561

 Ronnie Freeman
 (915) 862-3292

 Have designed and installed three solar hot water and air systems using (R11) regrigerant in a passive loop of the collectors and six solar hot water systems of the same passive design.

565. **FRIEDMAN SAGAR McCARTHY MILLER AND ASSOC.**
 353 Folsom Street
 San Francisco, California 94105

 John McCarthy
 (415) 495-0550

 We are interested in exploring passive solar design. Our work at present includes: clothing factories in Texas, North Carolina, Shanghai, Mexico and Sweden; research laboratories; hospital renovations; space planning; corporate offices; elderly housing; nursing homes; warehouses.

566. RANDOLPH STEVEN GADE
Energy Consultant
P.O. Box 248
Riverside, California 92501

Mr. Gade
(714) 682-9788

Design and architectural services. Building envelope analysis. Sizing and selecting HVAC and domestic water systems. Complete system design.

567. G. E. ASSOCIATES, INC., ARCHITECTS
G-6235 Corunna Road
Flint, Michigan 48504

Mr. Clifford E. Hull, R. A.
(313) 732-2010

Architectural firm specializing in commercial application of solar systems.

568. GENERAL SOLAR CORP.
P.O. Box 15835
Tulsa, Oklahoma 74112

J. W. Johnson
(918) 664-1677

Design of active solar systems for heating and cooling residential, commercial and industrial structures. Also, manufacturer's representative for solar hardware.

569. GENESIS ARCHITECTURE
417-A West Litchfield Avenue
Box 107
Willmar, Minnesota 56201

Phil Anderson
(612) 235-8663

Primary architectural emphasis on low- to medium-cost commercial, residential, educational buildings. Presently involved in several multi-family residential projects. Our involvement in solar has been only passive, single-family residential.

570. **GARY GLENN, AIA — ARCHITECTS**
10 South Euclid
St. Louis, Missouri 63108

Gary Glenn
(314) 367-1234

Gary Glenn, AIA — Architects is a professional design firm established to provide complete architectural services. These include needs analysis, site planning, space planning, architectural design, construction drawings and specifications, bidding assistance, construction supervision and project administration.

Many of our projects involve renovation, expansion or adaptive re-use of existing structures. Many involve the functional and aesthetic integration of new facilities with existing structures and the environment.

We are active proponents of energy conservation and solar energy, and have been one of the few architectural firms in the St. Louis area to utilize passive solar heating and cooling techniques.

571. **GLUMAC & ASSOCIATES, INC.**
1 Embarcadero Center
San Francisco, California 94111

(415) 398-7667

Design of multi-family residential and commercial structures, new and retrofit.

572. **GOODWIN CONTROLS CO.**
1430 Ranier Lane
Plymouth, Minnesota 55447

R. C. Goodwin
(612) 473-7888

202 SOLAR CENSUS

Design solar systems for residential, commercial and industrial applications. Also specialize in the engineering of control systems.

573. **GO SOLAR INC.**
835 Grand Avenue
Grover City, California 93433

Jeff Kennemer/S. Lane
(805) 481-3444/481-2132/489-3304

Designer/installer of active systems for residential and commercial application.

574. **GOVE ASSOCIATES INC.**
1601 Portage Street
Kalamazoo, Michigan 49001

Bill McDonough P.E./Adrian Noordhoek AIA
(616) 385-0011

Residential, commercial and industrial application of solar technology. Professional services include: community planning, municipal engineering, development engineering, industrial engineering, land surveying, structural design, mechanical engineering, electrical engineering, land development, sanitary engineering.

575. **RONDAL GOWER ASSOCIATES**
2404 Windsor Place
Champaign, Illinois 61820

Ron Gower/JoLynn Gower
(217) 352-2448

Design of active and passive systems for application to new and existing residential and commercial structures.

576. **GRAHECK, BELL, KLINE & BROWN**
Architects & Engineers, Inc.
220 West Washington Street, Suite 110
P.O. Box 789
Marquette, Michigan 49855

Robert C. Stow, AIA
(906) 228-7720

Interested in developing more efficient buildings through passive solar concepts, tuning structure to work with environmental factors rather than against them. Projects include: passive residences and studies for active domestic hot water system for an apartment complex, Marquette, MI.

577. CARLETON GRANBERY, ARCHITECT, FAIA
111 Old Quarry
Guilford, Connecticut 06437

C. Granbery
(203) 453-2449

Design services and R&D — residential and industrial applications of solar systems.

578. GRAYSON ASSOCIATES, INC.
68 Leonard Street
Belmont, Massachusetts 02178

Paul John Grayson, President
(617) 484-8820

Grayson Associates is an architectural and planning firm working in liaison with affiliated solar firm to develop solar building projects. The principal has been a member of ISPS and NESEA.

The firm has designed a large-scale "S" dome hot water heater for a large institution and is currently designing a solar community in Rhode Island.

579. PETER H. GREEN & ASSOCIATES
Architecture—Planning—Urban Design
25 South Bemiston Avenue
St. Louis, Missouri 63105

Peter H. Green, Principal
(314) 727-9591

Full architectural services, marketing/feasibility studies, site and landscape planning, building engineering, passive and active solar design, restoration, renovation and retrofit— for commercial, institutional, multi-family, business park projects.

580. WM. E. GREEN ASSOCIATES
Consulting Engineers
P.O. Box 38674
Sacramento, California 95838

(916) 922-1326

We are registered Mechanical and Safety Engineers specializing in commercial and industrial design.

581. GREEN HORIZON
Route 7 Box 124 MS
Santa Fe, New Mexico 87501

Valerie Walsh
(505) 982-3961

Green Horizon specializes in the design and building of solar greenhouses, and builds direct-gain skylights as well. Stress is on quality, function and beauty. The two basic models are (1) curved laminated mahogany, and (2) pine shed style (see *Solar Age* feature, June 1980). Both are with double Exolite for overhead glazing, and insulating glass for the vertical faces. Custom orders designed as well. Although most work is local, nationwide shipment of components is planned.

582. GEORGE GREER DESIGN & CONSTRUCTION
5411 Colodny Drive
Agoura, California 91301

George Greer
(213) 991-5959

Design and construction of energy-efficient structures, including planning and conscientious land development.

DESIGN—ARCHITECTURE & ENGINEERING

583. GLENN F. GROTH ARCHITECT AIA
2619 Camelot Boulevard
Sheboygan, Wisconsin 53081

Glenn Groth
(414) 452-3602

Professional planning and architectural services. Over twenty years of educational, religious, commercial, residential and recreational building design combined with extensive work in renewable energy resources such as solar, wind and biomass are used to solve each individual problem.

584. HABITEC: ARCHITECTURE & PLANNING
445 Washington Street
Santa Clara, California 93110

Michael O'Hearn
(408) 984-8515

Designers of nation's first solar industrial buildings, in Santa Clara, CA.

585. HAINES TATARIAN IPSEN & ASSOCIATES
121 Second Street — 4th Floor
San Francisco, California 94105

Earle Ipsen
(415) 392-8731

Architects specializing in the design of educational facilities and other public facilities with special concern for natural energy responsive planning.

586. HAMMEL GREEN & ABRAHAMSON INC.
2675 University Avenue
St. Paul, Minnesota 55114

Perry R. Bovin, Vice-President
(612) 646-7500/378-3833

Design firm specializing in multi-family residential, commercial and industrial areas. Apply active and passive system technology to new and existing structures.

587. HAMMER, SILER, GEORGE ASSOCIATES
1140 Connecticut Avenue, NW
Washington, D.C. 20036

Bruce Kin Huie
(202) 223-1100

Involved in the design/engineering, R&D and commercial application of solar energy, HSGA is an economic consulting firm with a specific focus on public and private decisionmaking. Founded in 1954, it serves a worldwide clientele composed of businessmen and public officials having a need for objective economic research.

Most of HSGA's work is concerned with the economics of physical development. In the private sector, its consulting assignments range from feasibility analyses of projects and investments to evaluations of sites, markets, manpower, competition and development strategies. In the public sector, its activities cover the economics of growth and management policies, fiscal accountability, public service delivery, and the deployment of public investments.

588. HARRIS ARCHITECTS, INC.
3821 Wales Road NW
Massillon, Ohio 44646

Gary G. Olp
(216) 837-4248

The design, innovation and incorporation of passive heating, cooling and lighting concepts into all design projects. Currently acquiring 40% efficiency out of systems constructed. We now are attempting to provide 80 to 100% passive heating and cooling efficiency and 20 to 50% passive lighting efficiency into all of our design problems.

589. HAVERSTICK & ASSOCIATES
Architects—Planners—Designers
Lake Tekakwitha
Pacific, Missouri 63069

Charles D. Haverstick, AIA
(314) 257-4798

DESIGN—ARCHITECTURE & ENGINEERING

Involved in residential and commercial projects utilizing solar (passive, active, combined and some subterranean) designing. Involved in new structures, additions and retrofit along such lines. Principal of firm is certified as an energy auditor by the State of Missouri Department of Natural Resources, listed on the roster of the National Solar Heating & Cooling Information Center, a licensed architect, a member of the Missouri Council of Architects, The National and St. Louis Chapter of the American Institute of Architects, as well as the Missouri Solar Resource Advisory Panel.

590. HAWKINS & ASSOCIATES
375 S. Main Street, Box 161
Lakeport, California 95453

Scott S. Bennett
(707) 263-8888

We are an architecture firm specializing in the design of passively heated and cooled structures—residential, commercial and industrial. We also work closely with several local solar equipment dealers and installers, engineering active solar systems for optimum performance.

591. HAWKWEED GROUP, LTD.
4643 North Clark
Chicago, Illinois 60640

Robert Selby
(312) 784-5025

The Hawkweed Group, Ltd. concentrates on architectural and planning projects utilizing solar space heating, energy conservation and other alternate energy sources.

Our first solar building, a passive solar house, was designed and built in 1960. Since 1973, we have accepted only solar projects and have worked on over 100 buildings, as well as doing site planning and related studies.

592. HEMPHILL, VIERK & DAWSON
908 Terminal Building
Lincoln, Nebraska 68508

Robert C. Dawson
(402) 476-2709

Architectural firm applying solar technology to residences and commercial/industrial facilities—new and retrofit.

593. HGF/CENTRUM ARCHITECTS INC.
6311 Wayzata Boulevard
Minneapolis, Minnesota 55416

Leslie Formell
(612) 545-5678

Solar design for residential, commercial and industrial application—new and retrofit.

594. HOLT + FATTER + SCOTT INC.
2525 Wallingwood, No. 501
Austin, Texas 78746

Joseph Holt
(512) 327-0454

We have designed numerous passive single-family dwellings including the design and building of Austin's first speculative solar house equipped with a passive direct gain system and with a solar-assisted dual source heat pump. We were also the architects for Austin's only solar air-conditioned house. We have two passive-oriented commercial structures in design stages now.

595. HOOKER/DE JONG ASSOCIATES ARCHITECTS
409 Frauenthal Building
Muskegon, Michigan 49440

Edgar A. De Jong
(616) 722-3407

We have been involved in passive solar design for new residences, commercial (new and retrofit) and religious buildings.

596. HOSTER'S HVAC ENGINEERING CO.
1102 Brickyard Road
Seaford, Delaware 19973

Bruce L. Hoster
(302) 629-3871

We are a complete design and construction firm providing commercial/residential design.
We experiment with all types of heat pump/solar and heat pump/windmill combinations.

597. **HOYEM-BASSO ASSOCIATES, INC.**
25 West Long Lake Road
Bloomfield Hills, Michigan 48013

Ralph Steele, Executive Vice-President
(313) 645-0400

1. Designing a Passive Solar Reception Center for Detroit Edison at Greenwood Energy Center, Port Huron, MI.

2. Designed active solar system for the private residence of Ralph Steele, Oxford, MI, and several others.

3. Designed Energy Management and Control Systems for the majority of the State of Michigan's Institutes for Higher Education. Of particular interest is the monitoring system for General Motors Institute, Flint, MI, which includes an active solar system and roof-mounted panels.

4. Studies for solar adaptation are included with the majority of our design programs. Of particular interest— the use of passive solar for Michigan State University's Plants & Soil Science Building in the extensive greenhouse area.

598. **HSR ASSOCIATES, INC.**
Architecture Engineering Planning
100 Milwaukee Street
La Crosse, Wisconsin 54601

James F. Michaels
(608) 784-1830

Architectural/engineering firm specializing in design of commercial, industrial and multi-family residential projects. Design active and passive systems for new and existing structures.

599. HUNTER HUNTER ASSOCIATES ARCHITECTS
8990 Manchester Road
St. Louis, Missouri 63144

Vincil F. Hunter Jr.
(314) 962-6298

Emphasis on passive solar design. Received a design award and grant from the Department of Housing & Urban Development for the design of an underground 87% passive solar residence.

We are also involved with energy conservation as illustrated in our design for an underground pistol range for the State of Missouri Highway Patrol.

600. JOE HYLTON & ASSOCIATES
566 Buchanan Street
Norman, Oklahoma 73069

Joe Hylton
(405) 364-4865

Architectural design and retrofit—single-family residential and commercial applications of active and passive systems. Also specialize in subterranean buildings.

601. IBE ENERGY
656 West Main Street
Palmyra, Pennsylvania 17078

Robert J. Carlson, President/W. Reece Newton,
W. Reece Newton, Technical Supervisor
(717) 838-9016

IBE Energy researches and develops solar/heat pump energy systems for space and domestic water heating. We both retrofit these systems and install them in new structures in conjunction with IBE Construction and Carlson Homes, Inc. (same address).

DESIGN—ARCHITECTURE & ENGINEERING

602. i e associates, inc.
3704 11th Avenue South
Minneapolis, Minnesota 55407

(612) 825-9451

Design of active and passive systems applicable to residential, commercial and industrial structures, new and retrofit. R&D in the areas of anaerobic digestion (methane), combustion, fermentation, gasification, alcohol distillation.

603. UNIVERSITY OF ILLINOIS
Small Homes Council/Building Research Council
1 East St. Mary's Road
Champaign, Illinois 61820

Karen Ouzts
(217) 333-1912

PASSIVE: Development of Illinois Lo-Cal House design through computer simulation of thermal performance, also have houseplans available and a related publication, *Technical Note 14: Details & Engineering Analysis of the Illinois Lo-Cal House.*

In the near future, will be monitoring the performance of a Lo-Cal house now built.

ACTIVE: Information clearinghouse; possible work on D-I-Y DHW systems in near future.

604. IMAGINEERING SOLAR
296 Vista Conejo
Newbury Park, California 91320

Al Ottum
(805) 498-4075

Engineering firm involved in system design and application of solar heating and cooling to residential and commercial structures, new and retrofit. Design and research of all solar related systems, e.g., collector design, package DHW systems, crop drying, geothermal ground storage.

605. M. R. IMMORMINO
23950 Hazelmere Road
Cleveland, Ohio 44122

M. R. Immormino
(216) 751-1848

We are an architectural firm that applies active solar system design to residential, commercial and industrial structures.

606. INATOME & ASSOCIATES, INC.
10140 West Nine Mile Road
Oak Park, Michigan 48237

Joseph Inatome
(313) 542-4862

Designer of the largest (8000-square-foot) solar heating and cooling system in Michigan (3-year-old, 3-story office building) which features: the utilization of clear water and glycol storage tanks (dual); the integration of flat plate collectors into the wall of the building; and high output.

607. INDIANA SOLAR DESIGNS
R. R. No. 2, Box 59B
Paoli, Indiana 47454

Mike Bruner
(812) 723-3881

Designer and building contractor. Primary work areas include design and actual building of small residential and commercial buildings; also earth-sheltered and underground homes with active solar systems.

608. INNOVATIVE ENERGY CORPORATION
216 Conrad Street
New Orleans, Louisiana 70124

Don Wicks
(504) 283-1045

Consulting services, energy audits—residential, commercial and industrial applications—primarily in New Orleans area.

DESIGN—ARCHITECTURE & ENGINEERING

609. INTEGRAL DESIGN
3825 Sebastopol Road
Santa Rosa, California 95401

John Burton
(707) 528-0616

Integral Design provides design, consulting, education and installation services for low-cost passive solar measures.

The primary focus is on the passive integral solar water heaters (commonly known as Breadbox). A complete set of plans to build your own is available. Workshops provide an educational experience and a completed installation. Contracted installations are also available. R&D on passive integral solar water heaters is also underway.

610. INTEGRATED ENERGY SYSTEMS
301 North Columbia Street
Chapel Hill, North Carolina 27514

Daniel R. Koenigshofer
(919) 942-2007

Integrated Energy Systems, P.A., has specialized in architecture, energy engineering, energy and environmental planning since 1976. The design team consists of architects, energy, mechanical, electrical and structural engineers. Services include: engineering design of energy efficient systems, including wood and solar; construction management; renewable energy sources management (biomass); air quality management; energy audits; community energy planning; environmental impact analysis; and energy forecasting.

Our solar work has included: active heating, cooling, hot water; air and liquid (2 DOE awards); passive heating and cooling Trombe, water and tilt-up concrete walls (2 HUD awards); solar- and groundwater-assisted heat pump systems systems; construction management of solar homes.

IES also performs feasibility studies and engineering designs for wind generation system integration, and is involved in R&D of methane digestion systems.

611. INTERACTIVE RESOURCES, INC.
Comprehensive Professional Services Group
117 Park Place
Point Richmond, California 94801

Dale A. Sartor
(415) 236-7435

Energy conservation design and engineering services include: energy audits and conservation studies, energy building code compliance, solar feasibility studies, active and passive solar applications design (new and retrofit—residential, commercial and industrial), and construction management.

612. IONIC SOLAR INC.
8934 J Street
Omaha, Nebraska 68127

Garry D. Harley
(402) 339-2420

Branch Office:
148 8th Avenue North
Nashville, Tennessee 37203

(615) 242-1017

Active and passive system design through turn-key installation—residential, commercial, industrial applications. Feasibility studies, grant applications, construction management, and solar land planning. Recipient of HUD passive research grant.

Nashville branch is an "Approved TVA" installer for the "Solar Nashville" project, using solar DHW systems.

613. IRONWOOD, INC.
115 North First Street
Minneapolis, Minnesota 55401

Dave Bryan
(612) 339-2520

Services offered: architectural design; solar engineering—active and passive; computer simulation—building heat loss and gain, f-chart, economic investment analysis.

Distributor of solar components, solar systems, energy conservation items, window installation, wood- and coal-fired heating equipment.

614. IRVINE ENGINEERING
P.O. Box 246
Suffern, New York 10901

R. Gerald Irvine, P.E.
(914) 357-6156

Consulting engineering activities in energy conservation, electric space conditioning, electric water heating, and other electric engineering matters.

615. KAMAL S. ISKANDER AND ASSOCIATES
Consulting Engineers
350 North Acaso Drive
Walnut, California 91789

Kamal Iskander
(714) 595-8566/(213) 481-5843

Consulting engineers specializing in mechanical and electrical solar system design, and energy management.

616. JACOBS-DEL SOLAR SYSTEMS, INC.
A Member of the Jacobs Engineering Group
251 South Lake Avenue
Pasadena, California 91101

(213) 449-2171

Jacobs Engineering Group, Inc. is an international engineering construction firm which has served the process industries on a worldwide basis since 1947. Jacobs is responding to the current energy crisis by optimizing the utilization of energy on all company projects, though Jacobs has been pioneering the use of solar energy for many years. We were among the first to employ evaporation ponds for industry. In 1974 Del Manufacturing Company initiated an intensive research and development program on solar collector systems. In May 1977 Jacobs-Del Solar Systems, Inc. was formed as a subsidiary of the Jacobs Engineering Group to centralize our solar activities. Our specialty is industrial applications.

617. J & D SOLAR CONTRACTING
11400 Bacon Road
Plainwell, Michigan 49080

Hank James
(616) 664-5282

Building and design of passive solar homes and retrofit for solar adaptations to existing homes. First house finished in Spring 1980. Main living area (downstairs)—800 sq. ft. (earth-sheltered); utility area (upstairs)—680 sq. ft.; heating systems—50% solar, 50% wood (2 cords last winter), gas backup not used; collecting areas—cement floors and walls, sand-filled block walls and chimney column. Attached garage and stairwell—1250 sq. ft. Average utility bill—$5.50/month; average LP gas—$4.00/month; average fuel bill—$0.-$5.00/month for chain saw.

618. BRION S. JEANNETTE & ASSOCIATES
Architects & Planners
470 Old Newport Boulevard
Newport Beach, California 92663

Brion S. Jeannette
(714) 645-5854

We are designing many custom homes with total hybrid systems. We presently have completed five projects, and 19 more are on the board—passive and active. We are also applying solar to commercial projects.

619. J.F.A. SERVICES, INC.
7340 Kingsgate Way, Suite 230
West Chester, Ohio 45069

Ronald E. Buckley
(513) 777-4600

J.F.A. Services is a consulting engineering company. Our primary function is to design, layout and specify HVAC, plumbing, electrical and fire protection systems for residential, commercial and industrial buildings. We have designed both active and passive systems.

Mr. Buckley serves on the Board of Directors of both the Ohio Solar Energy Association and the Southwest Ohio Alternate Energy Association.

620. JIRA HEATING & COOLING, INC.
SOLAR-AIR BY JIRA
Route 5, Box 400
Columbia, Missouri 65201

Charles J. Jira
(314) 445-4187

Jira, Inc. serves the eastern two-thirds of Missouri with solar systems and other energy-conserving equipment. Our growing dealership network will make available products that conserve water, fuel and money.

621. JOHNSON ENGINEERING
340 North Riverside Avenue
Rialto, California 92376

G. Johnson
(714) 874-5555

Engineering design, inspection and construction management for residential, commercial and industrial projects—new and retrofit.

622. M. DEAN JONES, ARCHITECT
Architecture—Community Planning
45 Sycamore
Mill Valley, California 94941

M. Dean Jones
(415) 435-4515

Our main emphasis, from an architectural direction, is to design self-sufficient buildings. We design residential and commercial buildings that can be heated and cooled passively, while not looking like solar collectors—to do that in an artful expression is our goal in architecture.

218 SOLAR CENSUS

623. ARLAN KAY & ASSOCIATES
5685 Lincoln Road
Oregon, Wisconsin 53575

Arlan Kay
(608) 835-5747

We are an architectural firm, having designed a number of facilities, primarily passive, since 1972.

624. KEANE ASSOCIATES
22 Monument Square
Portland, Maine 04101

John Keane
(207) 773-0577

The majority of our work deals with the design and construction of passive and active solar residential buildings and commercial buildings. This includes both the design of new structures and retrofits of existing structures. We are currently working toward use of wind generation systems.

625. KELBAUGH & LEE ARCHITECTS
240 Nassau Street
Princeton, New Jersey 08540

Doug Kelbaugh
(609) 924-9576

Five-person architectural firm specializing in passive solar heating and cooling. Winner of six energy design awards. Completed over 10 passive solar buildings including the Kelbaugh House, published in over 50 books and periodicals. Design and consulting services on an hourly or contract basis.

626. STANLEY KENISTON, AIA
666 State Street
San Diego, California 92101

Stanley Keniston/Jon Mehnert
(714) 231-1312

DESIGN—ARCHITECTURE & ENGINEERING

Consultation to architects, developers, public agencies in conservation, design, solar access and policy, etc.

Design of site-specific structures.

Analysis and design of site plans responsive to energy conservation.

627. C. KESSEL CO.
17940 Laramie Lane
Twain Harte, California 95383

Sheri Hoffman/Cooper Kessel
(209) 586-5458

Passive system design, residential and commercial, for new and existing structures.

628. KESSEL INSOLAR DESIGNS
19550 Cordelia Avenue
Sonora, California 95370

Ken Kessel
(209) 532-2996

Design, research and development on passive solar utilizing earth heat, glass wall, indoor pool storage and convective loop.

629. THE KIENE & BRADLEY PARTNERSHIP
1st National Bank, Suite 925
Topeka, Kansas 66603

Don Curry
(913) 234-6615

Design of residential, commercial and industrial structures— new and retrofit, passive and active systems.

630. KINGSCOTT ASSOCIATES, INC.
Architects & Engineers
511 Monroe Street
P.O. Box 671
Kalamazoo, Michigan 49005

Nelson B. Nave
(616) 343-2657

We design and supervise the construction of schools, hospitals, industrial and commercial structures, institutions, apartments, churches, with emphasis on energy conserving systems so far as our clients permit. In the planning stages are an active trombe wall highrise office building with rock, ice and water storage with a VAV internal system, a passive (and active) elementary school in northern Indiana, and various other buildings with a touch of passive—and all with high insulation qualities.

631. KIRKHAM, MICHAEL & ASSOCIATES
Architects, Engineers, Planners
7601 Kentucky Avenue North
Minneapolis, Minnesota 55410

Charles E. Sullivan, AIA/Kent O. Lande
(612) 425-5777

Kirkham-Michael architects and engineers, aided by our in-house computer, can simulate the life cycle effects of various building designs and mechanical-electrical systems. Guided by extensive professional experience, KM's design teams are well-equipped to devise individualized energy solutions for a wide range of projects and clients.

Kirkham-Michael design teams orient new buildings to derive the maximum advantages from the angle of the sun, cooling breezes and shade. KM engineers specify environmental control systems (heating, cooling and ventilation) that respond to the occupied/unoccupied cycles of the building. KM electrical engineers plan maximum lighting for task and reading areas, but design lower levels of illumination for "non-seeing" general areas. KM mechanical engineers adapt renewable sources of energy—solar, water gradation, geothermal—to supplement or replace conventional sources.

KM's complete total energy profile interrelates three basic energy categories: (1) those which consume energy (HVAC-heating, ventilating, air conditioning); (2) those which influence how much energy will be consumed (wall construction, insulation, glazing); and (3) individuals who operate, use or maintain the systems. Kirkham-Michael then assists the client in developing a long-term total energy management plan. KM energy management services often include: architectural evaluation of the building envelope (exterior walls and roof) and estimated costs of building modifications to curb heat loss; engineering evaluation of existing HVAC systems to determine the cost and practicability of converting to alternate sources of energy and/or adding supplemental heat recovery systems. Evaluation of the operating procedures and schedules of mechanical and electrical systems; development of necessary plans and sketches for proposed architectural and engineering modifications.

632. **FREDERICK H. KOHLOSS & ASSOCIATES, INC.**
    345 Queen Street, Suite 401
    Honolulu, Hawaii 96813

    Frederick Kohloss
    (808) 536-1737

    We are consulting mechanical and electrical engineers involved in active solar system application to residential, commercial and industrial structures. Other offices of the firm are:

    2301 East Broadway   Mr. Delbert R. Hartzer
    Tucson, Arizona 85719

    500 Sansoma Street, No. 202   Mr. Richard R. Hughes
    San Francisco, California 94111

    132 Albert Road
    South Melbourne, Vic.   Mr. David V. Bibby
    Australia 3205

633. DR. JAN F. KREIDER, P.E.
Consulting Engineers
1455 Oak Circle
Boulder, Colorado 80302

Dr. Jan F. Kreider
(303) 447-2218

Solar system design—active and passive, feasibility studies, seminar series, critical design review.

634. KRUMBHAAR & HOLT ARCHITECTS
66 Main Street
Ellsworth, Maine 04605

Rick Malm
(207) 667-5575/8661

Won the Maine Office of Energy Resources passive solar design award for the Carrabassett Valley recreation center. Designed the United College environmental science building using passive solar, active solar and wood space heating. Designed Medway rest center using passive solar. Promotes use of passive solar in schools and residences.

635. JAMES LAMBETH, ARCHITECT
1591 Clark
Fayetteville, Arkansas 72701

James Lambeth
(501) 521-1304

Passive solar design (new and retrofit) of residential (single- and multi-family), commercial and institutional structures. Consideration is given to siting, materials, construction and climatic data to design structures which make the best use of solar gains, while suiting the needs of the occupants.

636. LANCASTER COUNTY COMMUNITY ACTION PROGRAM
Solar Project Program
630 Rockland Street
Lancaster, Pennsylvania 17602

Steve Steinbacher
(717) 291-1051

Solar Project's main objective concerns the reduction of heating bills for low-income persons. Our principal installations to date have been solar domestic hot water systems, solar hot air collectors, solar attached greenhouses, glazed-in porches and large sunspaces. Our installations in the upcoming year will be mainly passive solar types on a large scale. Our retrofits will be attached greenhouses, Trombe walls, direct gain systems, sunspaces with mass storage and possibly glazed roof systems with water storage. Part of the installations will concern the addition of optimal weatherization techniques wherever it is feasible. Each installation will reduce the recipient's heating bill a minimum of 40%. Preliminary calculations indicate that a 40% reduction, in many cases, is a conservative figure.

637. PAUL LARKIN, P.E.
Solar/Mechanical Engineer
7202 Bodega Avenue
Sebastopol, California 95472

Paul Larkin
(707) 823-1168

Design of mechanical systems for active and hybrid solar heating systems: air and water. Consulting in solar heating design and passive solar building design. Design of conventional/backup heating systems. Rock bed heat storage design. Energy audits and conservation. Product design.

638. GENE LA TOUR, GENERAL CONTRACTOR
1414 Ashland Avenue
Santa Monica, California 90405

Gene La Tour
(213) 391-8522

Specialize in passive solar room additions.

639. K. T. LEAR ASSOCIATES, INC.
53 Lyness Street
Manchester, Connecticut 06040

Alfred Eggen
(203) 647-9795

1. Design and construction of passive and earth-sheltered buildings, mainly residential, some nonresidential.

2. Design, development, construction of site-built active air systems and water heating systems.

3. R&D alcohol systems.

4. Design and construction of greenhouses and other passive retrofits.

640. LE BLEU ENTERPRISES
844 17th Avenue
Santa Cruz, California 95062

Robert Le Bleu
(408) 476-9497

Design heating and air conditioning systems for industrial, commercial and residential applications.

Design and engineer passive/active solar systems for new and retrofit.

Energy budgets for building permits to meet state regulations.

Build our own controllers (5 types), autovalves and accessories.

641. LEE & ASSOCIATES
499 Van Buren
Monterey, California 93940

Richard Lee
(408) 649-8000

Active and passive system design for residential, commercial and industrial applications—new and retrofit.

DESIGN—ARCHITECTURE & ENGINEERING

642. STEVEN M. LESLIE, CONSULTANT
2561 Gondar Avenue
Long Beach, California 90815

Steve Leslie
(213) 598-9349

Engineering and design—active systems for residential, commercial and industrial application. Specification on turn-key operation.

643. LEVI & CO. — CONCEPT BUILDERS, INC.
P.O. Box 2254
Evansville, Indiana 47714

Don L. Wilson, Jr.
(812) 424-1014

Designer/builder of passive solar residences and light commercial structures.

644. JOHN R. LEWIS, P.E.
Mechanical Engineer
9 First Street, Suite 819
San Francisco, California 94105

John Lewis
(415) 495-5082

Feasibility study for 3000 gpd laundry hot water system utilizing solar collectors (flat and concentrating).

Feasibility study for hospital domestic hot water heating by solar-assisted heat pump.

645. MALCOLM LEWIS ASSOCIATES
220 Park Avenue
Laguna Beach, California 92651

M. Lewis
(714) 768-6911

Energy conservation engineering and design—active and passive system application to residential, commercial and industrial structures, new and retrofit.

646. LISS ENGINEERING, INC.
2862-A Walnut Avenue
Tustin, California 92680

Bonnie Stice/Sheldon Liss
(714) 730-0222

Liss Engineering provides electrical engineering services in the fields of lighting, power distribution and controls, for residential, commercial, industrial and recreational buildings.

647. LIVING SYSTEMS
Route 1, Box 170
Winters, California 95694

Mary Oak
(916) 795-2111

Living Systems is a multidisciplinary research and design group, specializing in passive solar buildings and self-sufficient community design, as well as R&D of these areas. We are architects, ecologists, landscape architects, mathematicians, systems analysts, artists and craftspeople.

Our research programs emphasize energy conservation and passive heating and cooling. We have developed thermal storage techniques, inexpensive movable insulation, accurate methods of predicting the thermal performance of buildings, an efficient fireplace/wood heater and the nation's first building codes (for Davis, Indio and Sacramento County, CA). Our ongoing work is designed to promote energy conservative land use through planning which includes solar access, solar rights and aesthetic, functional landscaping. Currently planning a solar heated and cooled community in the San Joaquin Valley. Also preparing a design manual, *Planning Solar Neighborhoods*, for the American Society of Planning Officials and HUD.

We have designed solar heated/cooled homes in a variety of climates. Current work includes design of low-cost solar housing, moderate-cost frame houses, ferro-cement "Living Structure."

Our own office is 100% naturally cooled and 95% naturally heated, and has 100% daylighting.

Awards: AIA Sunset Homes Special Award 1977-78; American Institute of Planners: Meritorious Program Award 1978, California Chapter Certificate of Distinction 1978; *Progressive Architecture* 26th Awards Program Citation 1979.

Current Projects: Design of a passive solar heated/cooled Community Center in Winters, CA; 13 housing units currently under construction using a variety of passive concepts; DOE-funded R&D of the Cool Pool, a powerful passive cooling system invented at Living Systems. Designing a 20-acre energy-efficient development at Dinuba, CA, with 120 passive solar heated/cooled units.

648. RUDOLPH B. LOBATO ASSOCIATES
10075 East County Line Road
Longmont, Colorado 80501

Rudolph B. Lobato
(303) 651-0670

Rudolph B. Lobato Associates is a professional architectural design firm specializing in energy-efficient architecture. We are committed to bringing solar utilization to the marketplace in ways that are economical and effective, and enjoy a track record with a diversity of solar design applications in commercial and residential projects. Besides the designing of residences using both passive and active solar technology, our latest project is a fast-track medical clinic in Delta, CO, utilizing passive design technology and a wood furnace as a heat backup source, and an active hydronic collector system to meet therapeutic whirlpool and domestic hot water loads.

649. LONDE-PARKER-MICHELS, INC.
7438 Forsyth, Suite 202
St. Louis, Missouri 63105

Timothy Michels
(314) 725-5501

Londe-Parker-Michels, Inc. is a professional consulting firm which has as its purpose to advance the knowledge and to implement the use of energy-saving methods, solar energy and energy-integrated farming. The firm's areas of interest and activity are: passive solar building design; development of computer programs and hardware; energy auditing; design reviews; consulting to industry, building owners and professionals. R&D of community energy independence; alternative fuel production.

650. **EDWARD J. LONG CONSULTING ENGINEERS**
516 East Monroe Street—8th Floor
Springfield, Illinois 62701

Edward J. Long
(217) 525-1161

Engineering of active systems for new and existing structures—residential, commercial and industrial.

651. **LSW ENGINEERS**
9455 Ridgehaven Court
San Diego, California 92123

John Littrell
(714) 268-3224

Mechanical and electrical engineering design of active and passive solar systems for residential, commercial, industrial and military installations.

652. **LUNDQUIST, WILMAR, SCHULTZ & MARTIN, INC.**
614 Endicott-on-4th Building
St. Paul, Minnesota 55101

Len Lundquist
(612) 291-1293

Design of passive and active solar systems for space heating, cooling and DHW. Have three active systems on-line and numerous passive systems.

653. **PAUL H. LUTTON**
50 St. James Place
Piedmont, California 94611

Paul H. Lutton
(415) 482-3660

DESIGN—ARCHITECTURE & ENGINEERING 229

All architectural work is passive and/or energy-conserving. Firm engaged in extensive R&D for thermal shutters and operable insulation.

654. MADLIN'S ENTERPRISES
P.O. Box 1443
Palm Springs, California 92263

Cathy
(714) 324-1731

We assist architects and engineers in designing solar systems both passive and active. We have a computer program that calculates the heat loss and gain in any given structure and assists us in space heating solar system load requirements.

655. MAINEFORM ARCHITECTURE
295 Water Street
Augusta, Maine 04330

Jim Harley
(207) 622-7171

Residential and commercial design—new and retrofit. Involved in giving seminars concerning passive solar design; site concerns—wind, sun, earth; Russian fireplaces; thermal siphon envelopes; convective loop concepts.

656. MARSHALL'S DRAFTING/CALIFORNIA BUILDERS
P.O. Box 471
Hesperia, California 92345

Bruce Wood
(714) 244-6421

Design and installation of appropriate solar systems according to client/user needs—residential, commercial, industrial. Application of technology to retrofit existing structures. Design of greenhouses, swimming pool heaters and passive control systems for houses.

657. JAMES A MARTIS, JR. ARCHITECTS
28790 Chagrin Boulevard, No. 250
Cleveland, Ohio 44122

James A. Martis, Jr.
(216) 831-0757

Firm is a design/building/construction management organization. We utilize both active and passive solar design in residential, commercial and industrial projects.

658. MASSDESIGN ARCHITECTS & PLANNERS, INC.
138 Mt. Auburn Street
Cambridge, Massachusetts 02138

Gordon F. Tully, President
(617) 491-0961

Architects specializing in solar buildings of all kinds and solar system design. Consultants to government, industry, private clients. Research and development—passive design; photovoltaics, subcontractor to GE and MIT. Author of design text. Author of hand-held solar design program. Teaching solar course at Harvard University.

659. MATH/TEC INC.
118 S. Catalina
Redondo Beach, California 90277

Dan Young
(213) 374-8959

We provide passive building performance calculations using a computer modeling program for residential and small commercial buildings. Architectural design, active and passive systems, new and existing structures. Computer aid for architects and engineers.

660. McCONNELL, STEVELEY, ANDERSON ARCHITECTS & PLANNERS
860 17th Street, SE
Cedar Rapids, Iowa 52403

Develop and design solar power systems, especially passive solar constructions.

DESIGN—ARCHITECTURE & ENGINEERING

661. McCRACKEN SOLAR CO.
329 W. Carlos
Alturas, California 96101

Horace McCracken
(916) 233-3175

Over the past 22 years, have designed, built and installed 250 solar devices and systems: approximately 200 solar stills, 10 DHW systems, 15 pool heaters, 10 pool blankets, 10 space heating/DHW systems; 5 passive cooling systems. Residential and commercial applications, new and retrofit.

662. STEPHEN MERDLER ASSOCIATES
300 Calle Sierpe
Santa Fe, New Mexico 87501

Stephen Merdler
(505) 988-4137

Involved in entire gamut of energy-efficient and self-sufficient single- and multi-unit residential structures. We design in many vernaculars, but particularly are oriented toward organic and sculptural architecture with adobe/plaster materials. Energy-saving concepts revolve around life cycle costing and complete passive systems involving collection, storage, retention and backup systems.

663. MERRYMEETING ARCHITECTS
Lincoln Building
Brunswick, Maine 04011

Douglas Richmond
(209) 729-0989

Residential and commercial design. Have designed two residences with active solar systems, a greenhouse, and employ principles of passive solar energy in all of our design work.

664. MILLER & SUN ENTERPRISES, INC.
P.O. Box 19151
Portland, Oregon 97219

William R. Miller
(503) 246-2175

Specialist in solar engineering equipment, products and design. Residential and commercial applications, new and retrofit.

665. **MILTON INTERNATIONAL, INC.**
808 River Acres Drive
Tecumseh, Michigan 49286

Milton J. Appel, President
(517) 423-4988

Provides comprehensive environmental service which includes engineering, design, material specifications, project management and site supervision including feasibility and energy analysis on new or existing structures—industrial, commercial and residential.

666. **ROBERT MITCHELL SOLAR SYSTEMS DESIGN INC.**
RD 3, Box 239
Selkirk, New York 12158

Robert Mitchell
(518) 767-3100

Design of passive solar structures; design of passive and active retrofits; large and small scale energy analysis; education programs. Also, new product design and development; manufacturer of controllers, window shutters, glazing details.

667. **MITCHELL & JENSEN ARCHITECT & ENGINEER**
4247 Philadelphia Drive
Dayton, Ohio 45405

Harry Jensen
(513) 277-9338

In 1978 our firm entered actively into the alternate energy advocacy field. We built a new office building for our firm which has an alternate energy solar and hydrothermal system that provides us with energy cost reduction of 90% over conventional electrical energy systems. This engineering and design activity has grown rapidly. Our system has been personally inspected by the Chairman of the President's White House Energy Task Force, federal and state government congressional representatives, state DOE and EPA representatives, engineers, scientists, business concerns, Ohio and out-of-state universities, visitors from three foreign nations, and about 500 other visitors. We have testified before the Federal Department of Energy and U.S. Internal Revenue Service in Washington concerning the application of alternate energy systems, grants and tax credit proposals. We have currently and in the past year designed alternate energy systems for approximately 250,000 square feet of retrofit and new construction, including two government buildings. We can provide integrated architectural and engineering design for alternate energy systems which will provide at least a 25-year life cycle on the equipment and use of nondepleting hydrothermal, geothermal, solar and enthalpy systems to provide up to 90% energy cost reduction. The applications are tested and cost-competitive. We have been judged by highly qualified experts in this relative new field to be at least two years ahead in the state-of-the-art, experience and credentials to provide the needed professional services.

Services: architectural design and development; programming; mechanical, electrical and sanitary engineering; interior design; fire protection design; graphic design; special use space planning; parks and landscape design; design-build projects; fast-track design; product research; cost estimating; field observation; site surveys and design; neighborhood priority planning; grant assistance; historic restoration and adaptive use; investigation of existing facilities; value and life cycle cost studies; energy resource surveys; energy audits; hydrothermal/geothermal, solar and passive design systems; resource recovery; computer monitoring systems.

668. **MOGAVERO & UNRUH**
ARCHITECTURE AND DEVELOPMENT
811 J Street
Sacramento, California 95814

David J. Mogavero
(916) 443-1033

Architectural firm—all solar, both passive and active systems—throughout California. Some projects in the Midwest and East. Single-family and multi-family residential and commercial projects, new and retrofit. Presently developing a multi-family housing project incorporating active/passive solar. Award-winning work.

669. **JAMES A. MONSUL & ASSOCIATES**
642 Brooksedge Boulevard
Westerville, Ohio 43081

Jim Monsul
(614) 890-3600

Provide energy analysis for retrofit or new construction. Will provide design/build services for small commercial or residential needs. Experience includes design/build of a 5000 square foot active/passive office building, and active/passive residences.

670. **MOORE GROVER HARPER, P.C.
ARCHITECTS AND PLANNERS**
P.O. Box 235
Essex, Connecticut 06426

William H. Grover
(203) 767-0101

Moore Grover Harper's most ambitious solar-heated project is the Armed Forces Reserve Center in Norwich, CT. This is the nation's first solar energy armory. The system cost $88,849 and conserved 17,250 gallons of heating oil in the first year. Savings are expected to pay back the initial investment in 12 years.

In 1977 the firm designed a medical office building in which, through the combined strategies of solar heating and other energy conservation devices, heating costs have been reduced by 75% compared with conventional construction. The solar heating system includes 57 liquid-type flat plate collectors, a thermal storage tank, and a gas-fired backup boiler.

In the early 1970s, the firm assisted in the design of an experimental prototype house for NASA at Langley Research Center in Virginia. A contemporary family residence designed to be resource-efficient and cost-effective, it included solar space heating, DHW and nocturnal heat dissipation through solar collector/radiation systems with dual function thermal storage, optimum insulating materials, thermal shutters and a variety of other energy saving devices.

During the past five years the firm has designed and built some 20 single-family residences which incorporate some form of energy conservation into their design. Both passive and active solar systems have been used. Computer programming determines the type and extent of the heating system required and the degree to which it will be cost-effective. The firm believes that the important consideration in developing successful solar heating systems is to develop concepts that can be adapted to fit the requirements of each different site or of the particular interests of the client, rather than to produce a prototypical design.

Most of the buildings are in the Northeast where, in spite of less bountiful sun than elsewhere, they require only minimal backup heating systems. Three houses have active solar systems, others are designed in anticipation of future conversion. Other buildings combine heat sources which differ for the occupied or unoccupied periods.

671. **MOORE/WEINRICH ARCHITECTS**
    49 Pleasant Street
    Brunswick, Maine 04011

    S. Moore
    (207) 729-0451

    Design of solar residences—single- and multi-family—incorporating air collectors and trombe wall design.

672. **JIM D. MORELAN**
    ARCHITECT & ASSOCIATES
    2242 Camden Avenue, Suite 6
    San Jose, California 95124

    Jim Morelan, AIA
    (408) 371-1572

General practive of architecture including both active and passive design. Primary experience with single-family custom homes in California. Award: Second Place, State of California Passive Solar Design Competition, Multi-Family Category 1979.

673. **MORELAND ASSOCIATES**
904 Boland
Fort Worth, Texas 76107

Frank L. Moreland
(817) 335-2883

Moreland Associates is an architectural firm specializing in the design of earth-covered buildings. A full range of architectural services are available, with engineering services provided by the nation's best consultants. Besides projects in California, Texas and Minnesota, Moreland Associates is also active in research for both the private and institutional sectors.

674. **MORMEC ENGINEERING, INC.**
115 West Main Street, C
Visalia, California 93277

Leland Morgan, P.E.
(209) 625-5273

Our firm provides solar engineering from preliminary feasibility studies to construction review on all types of projects. We are consulting mechanical engineers and we do consulting, design engineering, and research and development in solar.

675. **GLEN H. MORTENSEN, INC.**
1036 W. Robinhood Drive, No. 201
Stockton, California 95207

Glen H. Mortensen, AIA
(209) 478-2670

We provide basic design and are involved in research and development centered around residential and commercial construction. Active and passive systems for new and retrofit applications.

676. C. F. MURPHY ASSOCIATION
224 South Michigan Avenue
Chicago, Illinois 60604

Helmut Jahn/Susan Froelich
(312) 427-7300

Design of the Program Support Facility, Argonne National Laboratories, for the U.S. Department of Energy, an office building incorporating the latest passive and active solar and energy conservation techniques. Design included the use of concentrating collectors, thermal storage, passive solar heat collection, utilization of natural lighting, specialized building configuration, and building orientation.

677. RONALD NABOZNY
710 Randolph
Jackson, Michigan 49203

Ronald Nabozny
(517) 783-1176

Design and size systems for residential use. Passive and active systems, new construction and retrofit applications. Also teach course on alternative energy for Jackson Community College and University of Michigan Extension Service.

678. NATURAL ENERGY WORKSHOP, INC.
Route 1, Box 130
North Freedom, Wisconsin 53951

Edwin K. Doerr
(608) 544-3013

Natural Energy Workshop, Inc. concerns itself with the research, development, education of alternative energy systems and environmental designs. We deal with these concerns by means of design, drafting, and special services. These services are done on the basis of existing projects and their operations with proven workable alternative technologies. We offer complete design services to our clients. These services are geared to environmental planning, both on the site and in the building design. We work with people's concerns and help them generate livable solutions.

679. NEILL & GUNTER INC.
P.O. Box 1959
Portland, Maine 04104

Curt Bartram
(207) 781-3001

Neill & Gunter Inc. is a 50-man consulting engineering firm. We are interested in solar energy and actively search for solar energy design work. We are currently working on a solar energy feasibility study for the U.S. Navy.

680. NEW DAY BUILDERS
Route 2, Box 140-D
Doniphan, Missouri 63935

Norman L. Guittar
(314) 593-4608

We design and build space and energy-efficient homes using all means of conservation. The homes are designed to save 80% of the total home energy budget and are priced below conventional.

681. NEW JERSEY INSTITUTE OF TECHNOLOGY
CENTER FOR TECHNOLOGY ASSESSMENT
323 High Street
Newark, New Jersey 07102

Dr. Sanford Bordman
(201) 645-5610/645-5611

One aspect of our energy conservation program is to study the feasibility of solar equipment on various institutions such as schools, hospitals, nursing homes, municipal buildings, etc., as an energy conservation measure. Major emphasis is placed on the retrofit of solar equipment on existing buildings.

682. DANIEL A. NOBBE & ASSOCIATES, INC.
1408 Arborview
Ann Arbor, Michigan 48103

Dan Nobbe
(313) 663-0730

DESIGN—ARCHITECTURE & ENGINEERING 239

Design consultant for solar energy management in buildings—residential, commercial and industrial. Evaluation and research.

683. A. J. NYDAM CO. INC.
2634 S. Division Avenue
Grand Rapids, Michigan 49507

James A. Nydam
(616) 452-8745

Residential, commercial and industrial design—active and passive systems for new and existing structures.

684. OLMON & HUTCHINSON ARCHITECTS
9-1/2 South Market Street, P.O. Box 339
Troy, Ohio 45373

Gary Olmon, AIA
(513) 339-2543

Engaged in general practice of energy-oriented architecture, both in new construction and in historic preservation.

685. OLSON AND ASSOCIATES ENGINEERING, INC.
P.O. Box 1006
Yreka, California 96097

Robert Bly
(916) 842-1661

Primary activity is in the design and consultation phase of solar engineering—residential, commercial and industrial applications of active and passive solar, new and retrofit.

686. OLSON AND MACDONALD, INCORPORATED
1025 North Dutton Avenue
Santa Rosa, California 95401

Bill Mattinson
(707) 546-4600

Olson and MacDonald is a group of architects and planners designing passive solar and energy conserving buildings throughout Northern California. A number of our passive solar homes have been built in Sonoma County, and we are currently planning and designing two passive solar subdivisions of 25 and 280 units in Santa Rosa, and a 120-unit passive condominium project in Ukiah. Additionally, we have designed two passively heated, naturally cooled, day-lighted office buildings in Santa Rosa and are working on a mixed-use auto repair/dwelling project in Mountain View.

687. AL PAAS AND ASSOCIATES
2378 E. Stadium Boulevard
Ann Arbor, Michigan 48104

Richard Chadwick
(313) 971-8111

Architectural design of buildings utilizing solar energy, and engineering of solar heating and domestic hot water systems, primarily passive and hybrid systems.

688. PACIFIC SUN INCORPORATED
439 Tasso Street
Palo Alto, California 94301

Harry T. Whitehouse, Ph.D., P.E.
(415) 328-4588

Pacific Sun specializes in energy system analysis and design for commercial and industrial clients. Pacific Sun offers a broad range of conventional engineering and scientific services. Emphasis is on the reduction of energy use in commercial and industrial environments—particularly in large buildings and in industrial processes.

Energy Analysis—System analyses are based on careful consideration of all energy-related interactions. Manual and/or computer-based calculations pinpoint crucial energy flows and their controlling parameters. Projects utilizing solar energy are modeled with a variety of software including: the University of Wisconsin's TRNSYS or F-Chart computer programs, SOLCOST, and custom software developed by Pacific Sun.

Economic Studies—feasibility studies emphasize formal engineering economy techniques. Computer-generated summaries present both pre-tax and after-tax cash flows, payback period, and return on investment.

System Design—Pacific Sun provides a full range of mechanical design services including: system sizing, layout, equipment specification, control strategy, and contract documentation. Areas of special expertise include: process instrumentation, monitoring and appraisal, advanced simulation studies of complex energy systems, and high-level product R&D assistance.

689. PAINTRIDGE DESIGN & DEVELOPMENT, INC.
803 Russell Boulevard, No. 6
Davis, California 95616

Judy Painter
(916) 756-2148

Architectural design, production building of passive solar homes. Design and land use planning consultants.

690. DAVID PANICH, ARCHITECT
SOLAR DESIGNS
Route 5, Box 88C
Athens, Ohio 45701

David Panich
(614) 593-8686

Projects that I have produced as Solar Designs or as project architect for Koe/Krompecher Architects in Athens include passive greenhouse for Pike County Community Action Agency, Piketon, OH (1976); passive house and studio, Athens, OH, direct gain with attached sunspace, active backup (1977); freestanding passive greenhouse for Camp Woodland Altars, Peebles, OH — earth-sheltered three sides; five identical sun tempered 3-bedroom residences (1978); underground/ passive residence, 2-bedroom, with 1-1/2 feet of earth on roof; Sugar Grove Lodge (Peebles, OH), 24-person lodge with meeting area, passive system, 5 trombe walls and direct gain (1979); 100,000 board-foot solar lumber kiln for Sherwood Forest Products, Inc., Waverly, OH. Active —this project has been submitted for funding by DOE small-tech program (1980).

691. PASSIVE SOLAR ALTERNATIVES
302 Denver, No. 203
Rapid City, South Dakota 57701

Craig Johnson/Gail Johnson
(605) 341-3816

We are a construction, information and consulting firm, with our primary focus being all practical uses of passive technology in the built environment. We are currently building a project with a DOE Small-Scale AT Grant.

692. PAULSON ENGINEERING
110 W. Bennett
Glendora, California 91740

O. B. Paulson
(213) 963-5513

Civil and structural engineering. Designed and built low-energy home in the high desert. Provide cost-effective energy designs and calculations to clients. Created compound arch design for higher, wider, more open design for commercial and residential buildings. Consult on retrofit energy design. Mechanical engineering associate.

693. ARTHUR HALL PEDERSEN
34 North Gore Avenue
Webster Groves, Missouri 63119

Arthur Hall Pedersen, P.E.
(314) 962-4176

Design and consulting engineers—specializing in single- and multi-family residential applications, new and retrofit.

694. EDWARD PEDERSEN ARCHITECT/PLANNER
109 Haffenden Road
Syracuse, New York 13210

Ed Pedersen
(315) 472-5016

Consultant/architect for 120-ft 300-hp wind generation system, 80-ft dia, three-bladed configuration driving opposed piston, swash plate, adiabatic, air compressor providing for secondary oil recovery. Client—American Ultramar, Ltd., Mt. Kisco, NY (1974—76, design development).

Consultant/architect for 170-ft high 300-hp vertical axis, Darrius, two-bladed configuration. Client—American Ultramar, Ltd. (1974—76, schematic design).

Consultant/architect for design of three single-family residences, Scriba, NY. Solar air system, rock storage, wood furnace supplemental heat with electric backup.

Architect—retrofit of 1900 American barn, Cuddebackville, NY. For solar air system with silo rock storage with wood furnace supplemental with oil furnace backup.

Consultant/architect, Carrier Corporation, Energy Systems Division, Syracuse, NY. Working drawings for liquid solar heating/cooling prototype system, Naples, FL, for a single-family residence.

Architect for the John Ben Snow Foundation, Fayetteville, NY. Carried out feasibility study for nine buildings for Pulaski, NY, for implementing alternate energy systems (solar, wind, hydro, methane, natural gas wells) for the purpose of reducing tax base for utility expenses.

Architect for the Pulaski school board for installing Phase I of the feasibility study for the above-mentioned project. Consisted of a weather recording station located at a local school. Information gathered will be utilized in selecting solar/wind system installation.

695. PEOPLE'S SUN SOLAR CONSTRUCTION CO.
P.O. Box 5187
Santa Rosa, California 95402

Stephen Pursell
(707) 542-5439

Primary focus is the construction of custom, solar single-family residences and the development of solar subdivisions. Active and passive systems; new and retrofit design.

696. MORRIS RICHARD PERKINS
ARCHITECT/P.A.
245 N. Hillside
Wichita, Kansas 67214

Morris R. Perkins
(316) 685-9781

Feasibility and cost studies—commercial and industrial solar applications.

697. PETERS—WILLIAMS—KUBOTA, P.A.
2500 W. 6th
Lawrence, Kansas 66044

Richard Peters
(913) 843-5554

Provides complete solar architectural services for single- and multi-family residences, commercial and industrial facilities (new and retrofit). Recent projects include Citizens Mutual Savings and Loan Association building in Leavenworth, KS, and the GSA federal office building, courthouse and parking facility in Topeka, KS.

698. DAVID L. PETITE & ASSOCIATES
Route 1, Box 884
Shingle Springs, California 95682

(916) 622-2696

Consulting, mechanical and civil engineers. Design services—active and passive systems for residential and commercial/industrial applications (new and retrofit).

699. PIONEER SOLAR CONSTRUCTION
P.O. Box 213
Cloverdale, California 95425

Arlie Coplin
(707) 894-4118

Currently we are doing design/construction (including consultation) of existing as well as new structures. Domestic hot water systems are included. Our main emphasis is passive structures. We continue to expand our operation into wind energy, photovoltaics, 12-volt systems. Also, we design and construct underground houses and passive/hybrid structures.

700. **PITTMAN EARTHWORKS**
P.O. Box 5031
Santa Fe, New Mexico 87502

Scott Pittman
(505) 988-1635

Pittman Earthworks is primarily involved in design and construction of earth-sheltered housing, using a passive gain system for heat. We have provided for 90% heating efficiency in three earth-sheltered homes built in Santa Fe.

701. **LINCOLN A. POLEY, ARCHITECT**
118 West Washington Street
Marquette, Michigan 49855

Lincoln Poley, AIA/Lloyd Bloom
(906) 226-2424

We are presently working on a project in Idaho which will employ active and passive solar systems. The building is a meeting and service center for a religious retreat complex. The active system employs roof-mounted collectors and subfloor storage of gained heat. The passive system employs a solar corridor on the south elevation. We are also working on solar greenhouses, air envelope concepts and subterranean structures.

702. **PRADO**
Box 1128
Madison, Wisconsin 53701

Don Schramm
(608) 256-4908

The Prairie Research and Design Office (PRADO) is a group of architects engaged in natural solar energy conscious design and research. PRADO has been in business since 1977. In December 1978, the submission to the First National Passive Solar Design Competition received a design award and was later selected as one of ten homes in the U.S. to receive an instrumentation package to measure system performance. PRADO received a second design award in the Residential Solar Demonstration Program administered by HUD in 1979.

PRADO is active in both residential and commercial solar design. Projects include work in new construction and remodeling for energy efficiency. In addition to single building projects, PRADO has designed an energy conserving Planned Unit Development for 300 townhouses. The combination residential/commercial project includes several innovative approaches to larger scale energy management, such as wind energy conversion systems, energy storage techniques, earth-sheltering and active/passive solar systems.

QUALITY ENERGY SYSTEMS, *See (843)*

703. REAL GAS & ELECTRIC COMPANY, INC.
P.O. Box F
Santa Rosa, California 95402

Wayne Griffith/Solomon Kagin
(707) 526-3400

The Real Gas & Electric Company, Inc. has been selling, installing and servicing alternative energy systems since 1969. In our work with wind, solar and other energy conserving systems, we have accumulated vast experience with their care and handling in a variety of environmental conditions, including the most extreme. We have completed installations from Alaska to Virginia in rural, suburban and urban locales and have shipped our products all over the U.S. and Canada.

In addition to our wind energy systems, Real Gas & Electric Co. provides complete solar energy implementation services. We can provide feasibility studies, computerized engineering, appropriate equipment necessary, and installation services, for such purposes as domestic hot water, pool heating, and space heating and cooling.

Product lines include established solar panel manufacturers. Provide complete turn-key installations for any solar application, or supervisory services for owner-completed installations.

DESIGN—ARCHITECTURE & ENGINEERING

704. REFINE BUILDING & CONSTRUCTION CO.
24607 W. Warren
Dearborn Heights, Michigan 48127

Dennis Bishop, P.E.
(313) 565-1800

Design and build single-family residences and commercial structures; new and retrofit solar system design.

705. REISZ ENGINEERING COMPANY
2607 Leeman Ferry Road
Huntsville, Alabama 35805

Emily D. Craven
(205) 533-4613

Design of active and passive systems for residential, commercial and industrial application. Research and development of new energy systems with particular emphasis on agricultural and industrial applications.

706. REMMERS ENGINEERING
175 Pawnee Drive
Boulder, Colorado 80303

Harry E. Remmers
(303) 499-5217

Mr. Remmers is a registered Professional Engineer with 18 years of solar-related experience, primarily in heat transfer. Since 1975, Mr. Remmers has been a consulting engineer specializing in technical services including grant writing, feasibility studies and the design and development of solar/thermal conversion systems, ranging from residential and commercial solar heating and cooling (HVAC) projects including heat pumps, to solar thermal powerplants for generating electricity. Solar instrumentation and data acquisition services are also provided. Clients have included private industry (e.g., Ball Brothers Research Corp., Beech Aircraft, CMK Industries, and Rockmont Envelope) and city, state and federal agencies (e.g., DOE, HUD, Water & Power Resources Service, Environmental Protection Agency, University of

Colorado, and Housing Authorities in Boulder, Colorado Springs, Denver and Littleton).

707. **RESTORATION PRESERVATION ARCHITECTURE, INC.**
51 South Ritter
Indianapolis, Indiana 46219

Larry B. Justice
(317) 353-9808

Although most of our experience is with restoration and preservation projects, RPA also has designed solar greenhouses and passive/active systems for residential and commercial application. We have been active in lobbying for solar legislation and other incentives.

708. **RHEMCO**
RAY HUFFMAN ENERGY MANAGEMENT COMPANY
4688 Oregon Street
San Diego, California 92116

Wally Hartwell
(714) 280-1234

Active and passive system design for residential and commercial application.

709. **ROCKRISE/ODERMATT/MOUNTJOY ASSOCIATES (ROMA)**
405 Sansome Street
San Francisco, California 94111

Wm. Stevens Taber, Jr., AIA
(415) 392-3730

Architecture, urban design, land planning. Projects include residential, commercial and institutional buildings, as well as large planning projects, utilizing natural energy systems for heating, cooling, lighting, wastewater, storm drainage, etc.

DESIGN—ARCHITECTURE & ENGINEERING

710. ROGGENSACK INSULATION & SOLAR INC.
55 Prairiewood Drive
Fargo, North Dakota 58103

Jim Baum
(701) 280-2904

Insulation/solar heating contractor involved in design for new and existing housing.

711. RoKi ASSOCIATES, INC.
P.O. Box 301, Route 113
Standish, Maine 04084

Roger D. Beaulieu
(207) 675-3341

RoKi Associates is principally in the passive solar business. The bulk of the business involves designing and contracting passive solar single-family homes. But retrofits, especially of the greenhouse/sunspace type are also an important facet of the business. Because energy conservation and efficiency are of paramount importance to the effectiveness of any energy system, we also do energy audits, consulting, site and blueprint analysis to enhance the efficiency of a particular energy system. Every energy-expending aspect of the home or building is considered before final plans are drawn up in order to optimize the conservation of energy. We sell some energy conserving devices, fixed glass, and arrange building materials dealings for our clients.

712. PAUL ROSENBERG ASSOCIATES
330 Fifth Avenue
Pelham, New York 10803

Dr. Paul Rosenberg
(914) 738-2266

Consulting physicists and engineers involved in solar research and development—air and liquid collector technology, photovoltaics, and design of wind generation systems.

713. ROSS & BARUZZINI, INC.
7912 Bonhomme
St. Louis, Missouri 63105

D. K. Ross, President
(314) 725-2242

Ross & Baruzzini is a full-service electrical, mechanical and industrial engineering and architectural firm. For over a decade it has specialized in energy conservation related to engineering, and scientific work. R&B were the consultants to the State of Missouri creating the first Energy Conservation Plan. Consultants to the State of California, City of Seattle, GSA, and the Internal Revenue Service for lighting standards in new buildings. They are working with ASHRAE and Brookhaven National Laboratories on heat loss in buildings.

Engineering work covers both systems retrofit and new construction design with emphasis on commercial buildings, institutions, warehouses and manufacturing facilities.

R&B has designed and installed both active and passive solar systems. Hold patents on an economy model solar collector system. Are active in solar research with concentration on solar energy storage systems.

714. ROTH ECKERT & JACOBSON, INC.
227 West Crawford Street
Findlay, Ohio 45840

Paul Jacobson
(418) 424-1681

Design of residential and commercial buildings using active and passive solar.

715. R. P. WOODCRAFTERS
1614 Westminster Drive
Columbia, South Carolina 29204

Mike Craig
(803) 256-0403

Currently designing and building greenhouse for natural foods co-operative store and restaurant.

DESIGN—ARCHITECTURE & ENGINEERING

716. EDWARD SALTZBERG & ASSOCIATES
14733 Oxnard Street
Van Nuys, California 91411

Edward Saltzberg
(213) 873-4752

Consulting mechanical engineer involved in all aspects of solar utilization for plumbing, heating and air conditioning applications for all types of installations.

717. THE SCHEMMER ASSOCIATES INC.
Architects, Engineers, Planners
10830 Old Mill Road
Omaha, Nebraska 68154

David G. Adams, P.E., Manager—Mechanical Engineering
(402) 333-4800

Architectural engineering firm with commercial, institutional and industrial emphasis.

718. SCHIPKE ENGINEERS
3101 West 69th Street
Minneapolis, Minnesota 55435

(612) 925-3131

Engineering services specializing in commercial/industrial application—new and retrofit.

719. E. GARY SCHLOH, AIA
213 Bean Avenue
Los Gatos, California 95030

Gary Schloh
(408) 354-4551

Office is involved in the design and construction of pleasant passive solar residences. Though most of our work is residential, we are currently working on a solar hot water system for a restaurant we are restoring. This past year, we received an award from the California Energy Commission for a passive solar house under construction in Modesto, CA.

252 SOLAR CENSUS

720. SCHMIDT/CLAFFEY ARCHITECTS, INC.
333 North Pennsylvania Street
Indianapolis, Indiana 46204

Don R. Claffey
(317) 634-8100

Design of residential, commercial and industrial structures. Member: Hoosier Solar Energy Association, Indiana Solar Resources Advisory Panel, Construction Specifications Institute, Indiana Society of Architects, American Institute of Architects, and Indiana Energy Exposition Planning Committee.

721. RICHARD A. SCHRAMM, ARCHITECTURE, INC.
2001 South 4th Street
Kalamazoo, Michigan 49009

Richard A. Schramm, AIA
(616) 375-2472

Architectural practice specializing in energy-efficient designs, active or passive solar, earth-sheltered design, superinsulated structures, high mass structures, and other systems. Practice ranges from single-family residences to large-scale commercial or industrial projects integrating energy efficiency with good design aesthetics.

722. LEE SCHRIEVER — ARCHITECT
R.F.D.
Bennet, Nebraska 68317

Lee Schriever
(402) 782-2497

Residential and light commercial work with emphasis on passive systems. Range from freestanding greenhouses, greenhouse additions, major passive solar additions to existing homes, new homes and small office buildings.

723. SEAgroup
SOLAR ENVIRONMENTAL ARCHITECTURE
418 Broad Street
Nevada City, California 95959

David Wright, AIA/Dennis A. Andrejko, AIA
(916) 265-9458

SEAgroup is a small, creative architectural design firm specializing in environmental design and passive solar architecture. Since 1973, design approach has evolved from a small number of direct-gain passive adobe projects in the Southwest to a national practice involving a multitude of passive solutions for heating, cooling, ventilation and humidity control. The locations of our projects are diverse, spanning 22 states to date. The scale of our work has grown from single-family residential to include commercial and multi-family designs as well.

SEAgroup's design process emphasizes natural solar architecture, stemming from the principles of microclimate design. Our goal is to create energy-conscious architecture that metabolizes with the environment. We are dedicated to the ideals of environmental planning, design and information dissemination of energy-conscious, people-responsive human environments.

Our service includes the five phases of an architect's basic service: schematic design, design development, construction documents, bidding, and construction observation. Additionally, we perform a thermal analysis on each building design which anticipates the natural heating and cooling potential. We follow each project from the initial programming stages through completion, with continual interaction with the client.

724. SENNERGETICS
18621 Parthenia Street
Northridge, California 91324

James C. Senn
(213) 885-0323

Design active solar heating systems, including solar-assisted heat pump systems. Conduct seminars for designers and installers. Publish solar heating training manual. Wholesale distributor, solar equipment and systems.

725. **WILLIAM R. SEPE**
Architect & Planner
One Maple Street
Camden, Maine 04843

Bill Sepe
(207) 246-8231

Design of low-energy passive and active solar buildings—residential and commercial.

726. **ROBERT FOOTE SHANNON, ARCHITECT**
49 Garden Street
Boston, Massachusetts 02114

Robert Foote Shannon
(617) 367-6691

Design of passive solar residences, single- and multi-family, new and retrofit. Also incorporation of active DHW systems.

727. **SHERIDAN SOLAR**
8429 North Monticello
Skokie, Illinois 60076

C. Sheridan
(312) 677-4694

Sheridan Solar is a small carpentry contracting firm. The bulk of our work is building repair and remodeling. We are presently building a prototype passive solar "tract home" in Wisconsin.

728. **E. F. SIEGEL AND ASSOCIATES LTD.**
Consulting Engineers
7104 Milford Industrial Road
Baltimore, Maryland 21208

E. F. Siegel
(301) 484-1100

Consulting engineers specializing in the field of energy conservation. We review plans, design and specify systems (active and passive), inspect existing structures, recommend energy conservation measures, including solar systems, and prepare plans and specifications for retrofitting structures—residential, commercial and industrial.

729. SIERRA SOLAR SYSTEMS, INC.
12050 Charles Drive
Grass Valley, California 95945

Karl R. Stewart
(916) 272-3444

Active and passive solar system application to residential and commercial structures, new and retrofit. Consulting and energy audit services; also, component and system distribution.

730. SIMMONS & SUN, INC.
SOLAR EARTH CONSULTANTS, INC.
P.O. Box 1497
High Ridge, Missouri 63049

Lon B. Simmons, President
(314) 677-3969

Designer/builder of passive solar earth-sheltered homes. We are currently building in six states and will begin construction in several more states this year.

731. JOHN SKUJINS, AIA
2300 East 22nd Street
Minneapolis, Minnesota 55406

John Skujins
(612) 340-0590

Planner and architect of single- and multi-family solar homes. Also, commercial and industrial solar system application, new and retrofit.

732. SKYTHERM PROCESSES & ENGINEERING
2424 Wilshire Boulevard
Los Angeles, California 90057

Harold R. Hay
(213) 389-2141 ext. 114/704

Passive design, new and retrofit—residential, commercial and industrial. Utilize roof ponds and movable insulation.

733. SLACK ASSOCIATES INC.
540 South Longwood Street
Baltimore, Maryland 21223

Slack/Milton
(301) 566-2520

Design single-family residential and industrial structures—passive, active, hybrid—new or retrofit. Research and development. Technology generation. Prototyping.

734. SLEMMONS ASSOCIATES ARCHITECTS, P.A.
1 Townsite Plaza, Suite 1515
Topeka, Kansas 66603

Robert S. Slemmons
(913) 235-9244

Successfully functioning solar energy system designs include active and passive systems, solar energy space heating, cooling and domestic hot water systems, solar-assisted heat pump systems, installations built under grants from DOE and HUD and privately financed projects. Special emphasis on building design to integrate collection equipment as an aesthetic component of structure. Applications include banks, office buildings and multi-family housing. Largest system designed to date has 6000 square feet of flat plate collectors for a hydronic heat pump installation in an administrative central building with 75 apartments in a retirement/health care complex.

DESIGN—ARCHITECTURE & ENGINEERING

735. DAN SMITH — ARCHITECT
164 East Main
Box 536
Valley City, North Dakota 58072

Dan Smith
(701) 845-4099

Residential design services, for six years, incorporating air, liquid, vacuum tube and trombe wall collector systems. Certified Energy Auditor by the North Dakota Office of Energy Management & Conservation, Bismarck, ND. Coordinated efforts for a solar heating and hot water heating feasibility study for the North Dakota Soldiers' Home. Submitted grant proposal for an energy efficient house for HUD's Residential Solar Energy Program—Cycle 4. Submitted grant proposal for solar heat collection and utilization system for a commercial greenhouse for DOE Appropriate Technology Small Grants Program 1980. Working on technical assistance program for an elementary school and a junior/senior high school, which is to comply with federal standards.

736. SOLARAIRE
2459 Nicholasville Road
Lexington, Kentucky 40503

Jim Watkins
(606) 277-0112

Specializes in retrofit of single-family residences.

737. SOLAR BUILDING CORPORATION
1004 Allen
St. Louis, Missouri 63104

John Newman
(314) 772-6369

Solar Building Corporation is a full-service architectural/engineering firm and is also engaged in wind/solar system research.

738. SOL-ARC
2040 Addison Street
Berkeley, California 94704

David Baker
(415) 548-7327

Sol-Arc is an architectural energy consulting group. We specialize in the evaluation and design of all energy-use systems of a building or proposal. The firm combines energy system designers, mechanical engineers, lighting experts and computer programmers to offer complete energy-related services to architects and builders.

The basic philosophy of Sol-Arc is to reduce energy consumption through careful and informed architectural decisions. Workable, cost-effective and long-lasting solutions begin with simple opportunities found in the building design itself. Fundamentals such as orientation, mass and natural lighting have significant and interdependent effects upon energy-use and human comfort. We examine how basic building conditions can be turned to the service of energy conservation and then explore the range of specific energy systems appropriate to the project design.

Analysis services: local climate and microclimatic analysis; siting, landscaping and building geometry energy evaluation; feasibility reports for retrofitting existing structures or designs with new energy systems; alternate energy systems comparisons and evaluations; HVAC systems evaluations and computer analysis; overall energy consumption profiles for buildings and systems; building thermal dynamics computer analysis; insulation and materials evaluation; lighting needs analysis; lighting systems efficiency evaluation and computer graphics analysis; cost/benefit analysis for component systems; and life-cycle costing for overall buildings.

Design services: energy-efficient HVAC systems design; passive and active solar energy systems design; thermal storage systems design; specialized alternate energy systems design (wind power, waste disposal); building envelope design (insulation, glazing and cladding combinations); integrated natural and artificial lighting design; suncontrol and sun-shading design; complete lighting systems design; energy conscious siting and landscaping design.

739. SOLAR CLIME DESIGNS
Box 9955
Stanford, California 94305

Jim Berk
(415) 321-9953

Designer of convective loop housing. Specialize in consultative and tutorial service in the use of energy in the human habitat (workplace and living quarters). Practical designer of passive and active systems using water or air for any solar application.

740. SOLARCON, INC.
607 Church Street
Ann Arbor, Michigan 48104

Dr.-Ing. Roderich W. Graeff
(313) 769-6588

Development and manufacture of software for solar energy calculations.

741. SOLARCRETE OF SOUTHERN CALIFORNIA
P.O. Box 476
Wrightwood, California 92397

Rick Thorngate
(714) 249-3406

Design and build passive and hybrid buildings; commercial, residential, industrial. Also, retrofit existing structures. Total solar engineering and construction.

742. SOLAR DAKOTA
Box 1394
Minot, North Dakota 50701

Dave Bachmeier
(701) 852-2431

Design and install active systems for new and existing residential and commercial structures.

743. SOLAR DESIGN ASSOCIATES INC.
205 West John
Champaign, Illinois 61820

Rob Gorham
(217) 359-5748

SDA is an association of a plumbing/solar installation/solar service firm, a licensed architectural firm (residential, commercial, industrial projects). Also a solar equipment sales firm, and provides the SDA 500 Solar Service Kit. The components of the SDA 500 Solar Service Kit allow for accurate performance testing of RTD and thermistor sensors, and for checking the sensor line continuity of your solar system.

744. SOLAR DESIGN CONSULTANTS INC.
2540 Walnut Hill Lane, Suite 166
Dallas, Texas 75229

(214) 350-2306

Solar Design Consultants, Inc. is a team of professional architects and engineers highly experienced in the area of energy conservation and solar energy utilization. Active and passive systems for residential, commercial and industrial application.

745. SOLAR DESIGN GROUP, LTD.
821 5th Street
Lyons, Colorado 80540

Mark S. Weber
(303) 447-8854

Projects include three completed residential (passive) buildings. Passive retrofit—greenhouses, trombe and water walls. Passive solar medical clinic. Residential, thermal envelope buildings—three in construction.

746. SOLAR DESIGNS
2864 Ray Lawyer Drive, Suite 209
Placerville, California 95667

Jim Finley
(916) 626-3345

Architectural and engineering solar design services. Solar construction—new structures and retrofit. Greenhouses.

747. SOLAR ENERGY ASSOCIATES, LTD.
5614 Western Avenue
Omaha, Nebraska 68132

Bill Holmes, AIA
(402) 553-4002

East Coast Office:
2316 39th Street
Washington, D.C. 20007

(202) 337-0754

West Coast Office:
1226 Villanova Drive
Davis, California 95616

(916) 758-1332

Solar Energy Associates (SEA) is a private, full-service solar energy design and consulting group. We specialize in active, passive and hybrid solar energy systems design. We also do energy consulting in education, economics, computer applications, instrumentation, research and development.

SEA consults in all areas of solar design and engineering. Emphasis is placed at present in the areas of passive solar, earth-sheltered construction, natural cooling, complete residential design, low-cost storage systems, systems feasibility studies, systems verification, computer modeling and design, electronic instrumentation and controls, energy economics, solar economics through life-cycle costing and pay-back studies.

748. SOLAR ENERGY DESIGN CORPORATION OF AMERICA
400 Remington Street
P.O. Box 1943
Fort Collins, Colorado 80522

Producers of design (sizing) aids for active and passive solar energy systems, residential and commercial. Our products include G-chart™ and P-Chart™, which give optimal (cost-effective) passive system size given various parameters selected and location.

749. SOLAR ENERGY DESIGN, INC.
3822 Prince William Drive
Fairfax, Virginia 22031

R. M. Cohen, P.E.
(703) 978-0873

Engineering services for solar and alternative energy systems. Complete systems engineering from conceptual design through turn-key operation: feasibility studies, economic studies, contract designs, detail designs, inspection and test programs, installation instructions, operating and maintenance manuals, maintenance planning and programs.

Solar systems: domestic or service hot water, space heating, process heating, space cooling. Alternative energy systems: wood or waste heat, specialty hydronic systems.

Architectural engineering for energy conservation in new or existing buildings.

750. SOLAR ENERGY ENGINEERING/SOLAR ENERGY SYSTEMS
13450 Northland Drive
Big Rapids, Michigan 49307

Howard W. Beatty, Jr., P.E./Paul E. Beatty
(616) 796-5501

Design and engineer DHW, space heating, backup systems for single- and multi-family residential, commercial, industrial applications. Industrial process heating systems. Provide computerized analyses (technical and economic). Conduct energy audits with emphasis on using solar energy.

Also wholesaler of components and systems. Provide dealer with technical and marketing support.

751. SOLAR ENERGY ENGINEERING
1838 Alverne Drive
Poland, Ohio 44514

Gene Ameduri
(216) 757-8687

We have designed and built fourteen solar homes utilizing over 6000 square feet of flat plate air collectors for space and domestic water heating systems. We have also designed a concentrating collector and built a 3000 square foot system for industrial process water heating. In addition, design of a flat plate air collector has recently been completed by our firm.

752. SOLAR ENERGY EQUIPMENT CORPORATION
18368 Bandilier Circle
Fountain Valley, California 92708

Marsh Hauge
(714) 964-2239

Design and install solar pool heating and DHW systems (residential and commercial).

753. SOLAR HOME SYSTEMS INC.
8732 Camelot Drive
Chesterland, Ohio 44026

Joseph Barbish
(216) 729-9350

Design active and passive systems for residential application (single- and multi-family).

754. **SOLAR INTERNATIONAL**
A Division of Taliman USA INC.
366 Bel Marin Keys
Novato, California 94947

Fred Varastell
(415) 883-2321

7246 Bellaire Avenue
North Hollywood, California 91605

John Magill
(213) 765-2764

We have been active in solar design, sales and installation for the last three years. Primarily, complete solar packages for swimming pools and domestic hot water systems; also, solar-assisted heat pumps.

755. **SOLARPAK, INC.**
6516 West Higgins Avenue
Chicago, Illinois 60656

T. McNamara
(312) 775-9717

We designed one of the earliest and largest active solar energy systems in the country, at the Museum of Science and Industry, Chicago, IL. The Museum of Science and Industry project consists of a retrofit solar energy space heating and space cooling system integrated into existing mechanical heating, cooling and ventilating systems. The installation demonstrates practical use of solar energy to provide space heating and cooling, utilizing existing, typical commercial-type ventilating systems.

756. **SOLAR PATHWAYS ASSOCIATES**
3710 Highway 82
Glenwood Springs, Colorado 81601

Robert Clarke, President/John Ehlers
(303) 945-6503

Solar Pathways was formed in the fall of 1977 to provide support services to architects, engineers, builders and developers

DESIGN—ARCHITECTURE & ENGINEERING 265

on the Western Slope of Colorado. An in-house library, open for public use, provides reference material for the consulting services. An in-house computer (micro with modem interface to large machines) performs simulations of buildings for architects and code review officials in the area.

Projects under the auspices of Solar Pathways Associates vary widely from low-temperature agricultural and commercial solar applications, to thermal building code revision for a town at 8600 feet (altitude), to passive design recommendations and simulations. A principal service that is anticipated to be available soon is a four-level architect support package, to provide any or all of the following: site analysis; initial design "rules-of-thumb"; simulations of varying sophistication; product and system specification.

757. SOLAR PRODUCTS INFORMATION & ENGINEERING (SOLAR P.I.E.)
P.O. Box 506
Columbus, North Carolina 28722

Richard Pratt
(704) 859-9287

We are an appropriate technology shop. We help individuals size, design and install solar space and water heating systems (pool and domestic for water). We have developed a kit of solar instruments used to evaluate potential solar sites. The solar site kit is low-cost and designed as a give-away to prospective customers. We also think we are the only people to be operating a solar-powered chain saw.

758. SOLAR SAUNA
P.O. Box 466
Hollis, New Hampshire 03049

(603) 465-7811

Provides plans for constructing an attached greenhouse of double-glazed, Kalwall or Filon® glazing, and pressure-treated wooden frame.

759. SOLAR SERVICES, INC.
1029 North Wichita, Suite 5
Wichita, Kansas 67203

Edward Martin, President
(316) 265-0701

Registered architect in solar business since 1975, with solar system design and consulting services available. Manufacturers representative and/or wholesale distributor offering a broad range of components and full systems to meet most solar application needs. We do not do installation/service work, rather work through dealers and contractors.

760. SOLAR STRUCTURES INC.
Wenning Associates, Architects—Engineers
221 Mill Street
Poughkeepise, New York 12601

Harry Wenning, AIA
(914) 471-7280

We design, distribute and install solar systems. We also design and construct solar buildings—residential, commercial and industrial—new and retrofit.

761. SOLAR TECHNOLOGIES INC.
P.O. Box 8175
Shreveport, Louisiana 71108

J. A. Walters

Design of total solar systems for residential, commercial and industrial applications in the southern U.S. Also design and fabrication of special system controls.

762. SOLAR TECHNOLOGY ASSOCIATES
7348 Bonita Way
Citrus Heights, California 95610

Edwin J. Merrick
(916) 723-5719

Design of active and passive systems for residential, commercial and industrial application to new and existing structures. Also specialize in community shared systems.

763. SOLARTEK
P.O. Box 298
Guilderland, New York 12084

SUNSIM-1, solar home simulation program. Calculates sun's energy in hourly intervals and demonstrates its use for domestic space heating and cooling and water heating.

764. SOLAR-THERM-EL
Energy Savers, Inc.
1829 Northwood Boulevard
Royal Oak, Michigan 48073

Tom Grady
(313) 399-7062

Conservation Energy Consultant. Energy Auditor. Design—residential and commercial—new and retrofit. General contractor.

765. SOLAR/WIND ENERGY SYSTEMS INC.
10833 Farmington Road
Livonia, Michigan 48150

Joe St. Cyr, AIA
(313) 421-6000

Pioneers in energy conservation. First installation of "regenerative wheel" in round school 1962. Have designed over 15 solar (active and passive) projects since then. Currently designing a Catholic church utilizing passive and active systems. Have under construction "Solar House II" on Sanibel Island, FL (active, passive, wind).

268 SOLAR CENSUS

766. SOLAR WORKS INC.
3296 McConnell
Charlotte, Michigan 48813

Dale Goodrich
(517) 645-2713

Residential design—passive and active systems.

767. SOL DEV CO INC.
P.O. Box 43125
Tucson, Arizona 85733

Dr. T. E. Shrode
(602) 743-0721

Involved in nearly all facets of solar legislation, design and development. Primary emphasis on development of residential and commercial properties with passive design integrated into structure in an aesthetically pleasing manner. Consultant to builders, developers, and financial institutions involved or contemplating solar design.

768. PAOLO SOLERI ASSOCIATES, INC.
COSANTI FOUNDATION
6433 Doubletree Road
Scottsdale, Arizona 85253

Jeff Stein
(602) 948-6145

The not-for-profit Cosanti Foundation, under the direction of architect Paolo Soleri, is pursuing the research and development of an alternative urban environment. Given that the ecological, logistic, economic, cultural and energy problems of present cities are closely interwoven, it is the foundation's understanding that long-term solutions must be sought within a comprehensive perspective. For the past 15 years, the Cosanti Foundation has been experimenting with an urban reorganization of highly integrated three-dimensional complexes called ARCOLOGIES, urban concepts reinforcing the interdependence between population, resources and diverse urban functions.

Present related research investigates the use of solar energy within arcologies. The focus is on the use of peripheral, extensive greenhouses employed for both food production and as solar collectors from which heat is directed to the town complex to meet heating and cooling needs on a community scale.

The construction of Arcosanti began in 1970 in the mesa country of central Arizona, 70 miles north of Phoenix. When completed the town of 4500–5000 people will rise 25 stories, cover 13 acres of an 860-acre land preserve, and will serve as a study center for the social, economic and ecological implications of its architectural framework.

769. SOLSTICE DESIGNS, INC.
Box 2043
Evergreen, Colorado 80439

Malcolm Lillywhite/Lynda Lillywhite
(303) 674-1597

Solstice Designs, Inc. has grown out of a decade of grass roots solar energy experience and has matured as a renewable energy technology small business involved with low-cost and passive solar systems. Out of this movement to develop technologies which give individuals, families, and communities more control over their lives have come some promising, practical and proven technologies. It is these renewable energy tools that Solstice Designs, Inc. packages and distributes through its mail-order division, Solstice Publications.

Solstice prepares and distributes renewable energy technology and energy conservation materials designed in cooperation with the Domestic Technology Institute. These designs have been constructed, tested and evaluated by the institute and have been optimized for both thermal and cost-effective performance. Plans are provided to enable individuals or small businesses to construct and install these systems. Solstice also offers consulting and design services to insure proper design construction, installation and operation of these and other systems.

270 SOLAR CENSUS

770. SPACE/TIME DESIGNS, INC.
P.O. Box 4229
Bellevue, Washington 98009

George Reynoldson
(206) 746-0870

Space/Time Designs, Inc. is the publisher of *Let's Reach for the Sun*, a book containing basic information on siting, choosing a design, and building your own solar home. The 144-page volume contains 30 unique and unusual active/passive solar designs in styles ranging from neo-Victorian to ultra-contemporary. The company will also do original custom design work for both residential and light commercial construction.

**771. THE SPITZNAGEL PARTNERS, INC.
ARCHITECTS, ENGINEERS, PLANNERS,
ENERGY CONSULTANTS**
1112 West Avenue, North
Sioux Falls, South Dakota 57104

Stephen H. Randall
(605) 336-1160

TSP is a multidiscipline firm of architects, engineers and planning consultants dedicated to principles of energy-efficient design and construction. Houses, churches, schools, banks ... all are designed with tested and proven methods of reducing energy needs.

TSP is meeting the challenge of reducing our need for non-renewable energy resources through: passive solar design, active solar design, earth-sheltered design, wind power generation, energy management programming, and energy conservation.

Services include: energy audits, energy grant applications, structural, civil, mechanical and electrical design of all active system types, including wind power generation, feasibility studies, computerized building analysis, system design and life cycle costs, site planning and landscaping for energy conservation.

DESIGN—ARCHITECTURE & ENGINEERING

772. SPRING MOUNTAIN ENERGY SYSTEMS, INC.
2617 San Pablo Avenue
Berkeley, California 94702

A. Krebs
(415) 841-3000

Design, sales and installation of solar equipment (active and passive systems) for residential and light commercial purposes.

773. S S SOLAR INC.
16 Keystone Avenue
River Forest, Illinois 60305

D. T. Sedgwick
(312) 771-1912

S S Solar Inc. provides full consulting and design services using computer analysis for both process and residential heating applications. In addition, S S Solar will provide a total solar system using both inhouse manufactured components and those available from the market. Our staff consists of individuals trained and degreed in business, engineering, construction and marketing. S S Solar provides lecturers and consultants in the field. The company is concentrating on the north central U.S.

774. STANLEY CONSULTANTS, INC.
Stanley Building
Muscatine, Iowa 52761

A. G. Bardwell, Sr. Vice-President
(319) 264-6600

Stanley Consultants, Inc. designs active and passive systems for commercial, office, and industrial buildings utilizing available state-of-the-art components.

775. E. J. STANLEY ENTERPRISES
2730 E. Broadway
Long Beach, California 90803

E. J. Stanley
(213) 434-5151

System engineering of solar installations for hot water, space heating, and swimming pools.

776. STEED/HAMMOND/PAUL
10 Court Street
Hamilton, Ohio 45011

G. Hammond
(513) 863-5441

Design incorporating active and passive solar systems—residential and commercial.

777. GERALD B. STEEL ENGINEERING CONSULTANT
P.O. Box 2369
Rancho Santa Fe, California 92067

Gerald B. Steel
(714) 789-0014

Engineering consultant currently writing two design manuals on passive solar design. Also involved in land use planning, water development, and wastewater reclamation.

778. THE STEIN PARTNERSHIP
588 Fifth Avenue
New York, New York 10036

Carl Stein
(212) 757-0284

Passive design for residential, commercial, and industrial application—new and retrofit.

779. **EDWARD V. STELLA**
 208 Chapel Drive
 Camden, Delaware 19934

 Edward V. Stella, P.E.
 (302) 697-1421

 Professional engineer whose activities involve assistance to clients in selection of systems; structural, mechanical and electrical design for integration with backup system; contractor coordination during construction; and follow-up service for startup and system performance evaluation and troubleshooting.

780. **STRAUB, VAN DINE & DZIURMAN ASSOCIATES, INC.**
 1441 East Maple
 Troy, Michigan 48084

 (313) 689-2777

 Development and design of earth-sheltered structures.

781. **R. D. STRAYER, INC.**
 ARCHITECTS, ENERGY CONSULTANTS, PLANNERS
 5490 Ashford Road
 Dublin, Ohio 43017

 Richard D. Strayer
 (614) 889-1288

 Solar architectural design—residential, commercial and industrial. Engineering design of air, liquid and trombe wall collection systems, natural cooling systems, storage systems. Twenty-three years experience in architectural, structural and product engineering. Specializing in energy conservation and renewable energy systems. Received HUD National Passive Solar Design Award. Selected as one of 20 firms nationally to design an energy-efficient townhouse for the American Institute of Architects Research Corporation.

782. JAMES R. STUTZMAN ARCHITECTS
P.O. Box 55111
Indianapolis, Indiana 46205

James R. Stutzman, AIA
(317) 251-0533

Architectural services, solar design and consulting. Involved in passive and hybrid solar space heating, earth-sheltered architecture, and summer space cooling and venting via solar space design. Research and development for low-cost manufactured housing utilizing highly energy-independent designs, i.e., conservation, glazing, site planning standards, night insulation, etc.

783. R. S. SUBJINSKE & ASSOCIATES, INC.
1826 South Main Street
Akron, Ohio 44301

Russell S. Subjinski
(216) 773-0326

Design engineering—solar assisted heat pump systems (water source) and domestic hot water systems, commercial and residential. Capacity/knowledge for larger, more complicated systems.

784. SUNDANCE BUILDING CONTRACTORS
724 Main Street
Norway, Michigan 49870

Ric Sanderson
(906) 563-5771

We design and build energy-efficient structures specializing in active and passive solar systems. We have been involved actively in the solar field since 1974, and have been involved in the installation of numerous active solar systems. We have published articles concerning our work in *Solar Age* (January 1979), *House and Garden* (Spring 1977), *MSEA Bulletin* (July 1978), and *Energy & Alternatives* (Fall 1978).

785. SUNDESIGN SERVICES INC.
Route 2
Cobden, Illinois 62920

Don Vogenthaler/Phil Randall
(618) 893-4088

Solar design and construction. Involved in 12 building/design projects present and past. Consultation with other non-solar oriented firms for solar applications. Public speaking and education of passive design techniques. Residential construction includes: convective loop, indirect gain greenhouses, underground, earth-sheltered, and superinsulated direct gain.

786. SUN HARVESTERS OF OREGON
P.O. Box 5481
Eugene, Oregon 97405

Tom Scott
(503) 345-3111

Passive and active system design for residential, commercial and industrial application—new and existing structures. Working with the University of Oregon Solar Center, coordinated the design of the award-winning passive solar heated homes in HUD's Passive Design Competition.

787. SUNLIGHT ENERGY SYSTEMS INC.
17961 Cowan Street
Irvine, California 92714

Douglas Warnock
(714) 979-0722

Engineering, selling and installing solar space, hot water, and pool heating systems.

788. SUN MIZER INC.
275 East Pleasent Valley Road, P.O. Box 146
Camarillo, California 93010

Kenneth Mize
(805) 659-2677

276 SOLAR CENSUS

Design/construction company—both residential and commercial projects. Design and install most energy saving devices. Have installed over 900 systems since 1975.

789. SUNRISE SOLAR PRODUCTS
A Division of ENERSPAN, INC.
14168 Poway Road
Poway, California 92064

J. Coxsey
(714) 748-1810

Active solar system design for single-family residences—new and retrofit applications.

790. SUNSEED ENERGY
G-4 Koshland
Santa Cruz, California 95064

Dale Hathaway
(408) 425-5071

General design and consulting services, especially passive residential heating, new or retrofit. Also design and sell passive hot water systems. Experienced in teaching solar design principles.

791. SUNSHELTER DESIGN
1209 Hillsborough Street
Raleigh, North Carolina 27603

Mike Funderburk/John Meachem
(919) 834-2170

Specialize in regionally appropriate designs for single-family residences as well as some multi-family units. We are concerned with providing livable spaces with heating and cooling systems characterized by simplicity.

792. SUNSPACE UNLIMITED
34 Silver Hill Road
Weston, Massachusetts 02193

Keith B. Gross
(617) 894-0573

Designer/builder—all aspects of passive and active solar heating. Have available catalog of plans of sunspaces, houses, retrofit solar heating devices. Firm is also involved in educational workshops.

793. **SUNSTRUCTURES**
 SUNSTRUCTURES CONSTRUCTION
 201 East Liberty
 Ann Arbor, Michigan 48104

 Bruce McCullen
 (313) 994-5650

 Sunstructures is involved in the design of energy-conserving and solar-heated buildings; energy audits, solar feasibility studies on existing buildings; educational activities, such as lectures, workshops and seminars on alternative energy. Sunstructures Construction installs solar DHW systems and constructs solar greenhouses.

794. **SUNSYSTEMS**
 Weld Road
 Dryden, Maine 04225

 Conrad Heeschen

 Architectural design, passive solar design, consulting, review of plans, lectures and slide shows (passive solar, solar greenhouses), workshops, site analysis, evaluation of solar retrofit potential, calculator energy analysis and simulation.

 Honorable Mention 1980 Maine Solar Building Competition; Design award in HUD/DOE Residential Passive Solar Design Competition (1978); Design Assistance and Construction Grants in HUD Cycle 5 Residential Solar Demonstration Program (1979, with Starbird's); National Center for Appropriate Technology Passive Solar Heating Demonstration Grant (1979, with Rural Community Action Ministry).

795. **SUNTAPPERS**
 17501 Irvine Boulevard, No. 3
 Tustin, California 92680

Currently active in five areas: retail; engineering services; commercial; builder/contractor; and retrofit.

796. **SUNTROLLER PRODUCTS**
6384 Rockhurst Drive
San Diego, California 92120

Richard E. Watt
(714) 582-1215

Engineering of solid-state control systems for active systems; also collector system design.

797. **SUNWAY HEATING SYSTEMS, INC.**
409 Lawrence Street
Petoskey, Michigan 49770

John B. Whitmore, President
(616) 347-6213

Emphasis is on simplicity, cost-effectiveness, and large solar contribution. Our primary activity involves the design of air thermosiphoning solar space heating systems, fabrication of components, and installation of complete systems within our service area (approximately 100 miles from Petoskey). We offer a mail-order solar survey and design service to those outside our service area, allowing do-it-yourselfers or hired labor to fabricate and install remote systems. The systems normally provide 100% of the heat load for 8–12 hours on sunny days in mid-winter without added thermal storage. Systems installed on well-insulated buildings having high thermal mass provide the full heat load for several days.

Most installations have been retrofits and range from mobile homes to luxury dwellings, offices, and industrial buildings. Many systems have also been installed on new buildings where the collector replaces conventional siding materials and therefore results in considerably lower installed cost. There are currently over 60 of our systems in operation and in addition to the Petoskey service area, six similar operations have recently been established in other areas of the state.

798. SUN-WEST SOLAR ENERGY SYSTEMS INC.
20445 Walnut Drive
Walnut, California 91789

Glen Schaffer
(714) 594-2749

Sun-West Solar is a designer/distributor of solar systems, both liquid and air type. We have a dealer organization and installation company for the larger projects we contract. We do all of our own design and engineering and have the capabilities to handle almost any situation that has to do with energy. We also have a branch office in San Diego, and so can offer our services throughout Southern California.

799. SURVIVAL CONSULTANTS
P.O. Box 21
Rapidan, Virginia 22733

Angus MacDonald
(703) 672-5440

We provide standard low-cost passive solar in-ground house plans to the mail-order market. We also offer consulting services where standard plans are inadequate for this type of construction. An expanded line of plans will soon include low-cost windmills, watermills, and solar greenhouses.

Angus MacDonald, Architect. Projects include: solar house and geothermal application for 60 apartments in Orange, VA; six passive solar apartments in Burlington, VT.

800. JACK W. SWANEY
In the Forest at 1511 Bellevue Boulevard, North
Bellevue, Nebraska 68005

Jack W. Swaney, AIA
(402) 733-6346

Design compatible energy-efficient structures. Self-supportive, self-contained units (singular or group beneficial): earth-sheltered, passive/active (biomass backup), orientation, design, development, implementation.

801. SYNAGO CORPORATION
P.O. Box 444
Rapid City, South Dakota 57709

Richard Pearson
(605) 342-9444

Architectural design of single-family dwellings and commercial buildings, new and retrofit.

802. SYNERGY, INC.
515 W. Maplewood Drive
Ossian, Indiana 46777

Mr. Gene E. Bell
(219) 622-7122

Synergy has designed and engineered 45 air-type solar systems, including space heating systems, domestic hot water systems, swimming pool heating, commercial and industrial process heating. Geographically, systems are installed in Indiana, Ohio, Michigan, Illinois and Kentucky. System sizes range from 64 square feet to array sizes up to 1500 square feet.

803. PHILIP S. TAMBLING, ARCHITECT
42 Middle Street
Portsmouth, New Hampshire 03801

P. S. Tambling, AIA
(603) 436-4274

General practice of architecture with employment of passive solar techniques whenever possible. Active solar occasionally for domestic water heating.

804. EATON W. TARBELL & ASSOCIATES, INC.
84 Harlow Street
Bangor, Maine 04401

Stephen B. Rich
(207) 942-8227

The firm of Eaton W. Tarbell & Associates, Inc. was founded in 1944 and has provided full services within the disciplines

DESIGN—ARCHITECTURE & ENGINEERING 281

of architecture and engineering since that time. Specifically, we concentrate on complete architectural services; mechanical, electrical and structural engineering; energy analysis and value engineering; life-cycle costing; estimating; inspection; construction management; and historic preservation consultation.

Since 1944 our firm has consistently offered progressive and imaginative solutions to design problems, each as individual cases. As a basic example of historical foresight and concern common to our firm's policy, we have since 1944 incorporated the principles of passive solar heat in our building designs. Examples include: Vine Street School (1951), Fruit Street School (1954), Bangor High School (1962), and All Souls Church Addition (1953) — all of Bangor, ME; Dryden Terrace Apartments (1944), Orono, ME; and Waterville Junior High School in Waterville, ME (1977).

Recent work of the firm in energy conservation and renovations includes: U.S. Navy renovations, Cutler, ME (1978) — insulation, weatherization, mechanical renovations and modernization. Waterville City Hall, Waterville, ME (1977) — EDA, insulation, mechanical adjustments and renovations, ANSI accessibility. Piscataquis County Courthouse Addition, Dover-Foxcroft, ME (1977) — EDA, expansion, insulation, mechanical adjustments, ANSI accessibility. Somerset County Extension Service Building, Skowhegan, ME (1977) — EDA, new structure, basic passive principles, wood and electric heat. Pooler Pavilion Renovations, Bangor Mental Health Institute (1977) — ward modernization, mechanical renovations and adjustments.

805. TECHNICAL VENTURES
P.O. Box 06175
Portland, Oregon 97206

Robert C. Brown, Sr.
(503) 224-2056/775-8572

Design of residential (single- and multi-family), commercial and industrial structures. Solar applications—new and retrofit. Also, siting and feasibility studies.

806. TELLURIDE DESIGNWORKS
222 W. Colorado Avenue, P.O. Box 1248
Telluride, Colorado 81435

Eric Dood
(303) 728-3303

Telluride Designworks is an architectural firm that has been involved since 1974 with the development of concepts for low-cost solar equipment, designing, installing and testing hardware through various projects.

The heart of our solar system is an integral roof solar hot air collector of our own design. The collector cost has been minimized in order to make the system economically viable, yet the collector is efficient, durable, easily fabricated and installed, and generally designed for long useful life in the severe climate of mountainous and high desert regions. Because it is an integral roof, the cost of an equivalent conventional roof partially offsets the cost of the collector roof.

Current thinking and costs have led to our increased emphasis on passive solar heat collection and storage. Passive collection has been shown effective on our large projects (schools, warehouses) as well as smaller (residential) applications.

807. TENNESSEE VALLEY AUTHORITY ARCHITECTURAL DESIGN BRANCH
400 Commerce
Knoxville, Tennessee 37902

Withers Adkins, Chief, ADB
(615) 632-2973

Design of energy-efficient buildings for seven-state federal agency. Technical assistance to private building community in energy-efficient and solar designs—new and retrofit systems for residential, commercial and industrial application.

808. TERRA-DOME CORPORATION
14 Oak Hill Cluster
Independence, Missouri 64057

Nancy E. Day, Public Relations Director
(816) 229-6000

We specialize in our own patent pending construction system. It is a modular system using 24 x 24 foot modules, each with its own domed roof. Each module is poured in one monolithic pour of concrete (reinforced with steel) so it has no cold joint at the roof line. Our structures, because of our mechanized system, can be built quickly and efficiently, cutting labor and bringing costs down to that of conventional housing. It is an extremely variable system; modules can be placed together in any number or arrangement and any architectural or interior design style can be used.

Our dealership program began in late 1979 and has expanded rapidly. We maintain a road crew that can travel anywhere in the United States to build our homes.

809. **TERRA-SOL**
S. R. 2242 A
Woodland Park, Colorado 80863

Lee B. Cerioni
(303) 687-3104

Design and build passive solar homes and greenhouses. An envelope home recently completed that utilizes some of the newest concepts in passive design. Also, home energy audits for existing structures; retrofit with passive solar greenhouses and passive solar additions.

810. **TERRIEN ARCHITECTS**
5 Moulton Street
Portland, Maine 04101

George B. Terrien
(207) 774-6016

Architectural practice of diversified projects—residential and commercial, active and passive systems.

811. **THEOREM SOLAR ENERGY**
697 Brokaw Road
San Jose, California 95112

Active and passive system design for domestic hot water, space and pool heating.

284 SOLAR CENSUS

812. JOSEPH B. THOMAS IV, ARCHITECT
Box 428
N.E. Harbor, Maine 04662

Joseph B. Thomas IV, AIA
(207) 244-3394

Residential design specializing in passive solar with wood backup.

813. PAUL TIMBERMAN, ARCHITECT
Box 130
Occidental, California 95465

Paul Timberman
(707) 874-3325

Have completed 23 buildings with passive/hybrid systems. Residential, commercial and industrial applications.

814. TIOS CORPORATION
740 South 300 West
Salt Lake City, Utah 84101

Mr. M. A. Salim
(801) 363-3661

Tios Corporation is an architectural/engineering firm which is actively engaged in efficient energy utilization, solar system design and implementation, energy management (including energy audits), and the handling and sales of related products.

815. TOTAL ENVIRONMENTAL ACTION, INC.
Church Hill
Harrisville, New Hampshire 03450

Dolores Wolfe
(603) 827-3375

Design and Consulting Services. The design group is trained in architecture and engineering, and is experienced in energy efficient techniques. We design residential and commercial buildings, evaluate plans for retrofits, and offer consulting services for thermal engineering and renewable energy systems design. Additionally, we advise on site planning, provide economic assessments of energy options, and assist with the communication of technical subjects.

Research Services. The research group engages in a wide range of environmental research. Past projects include the design of a low-cost, site-fabricated solar collector, a policy study of the public utility/solar interface, curriculum development, a detailed statistical analysis of the costs and performance of passive solar buildings, materials research, and the development of computer programs for solar performance simulations.

Information Services. The information services group produces reprints of technical papers by TEA's staff, educational solar slide kits, and calculator programs for passive solar design.

816. UNISOL ENERGY CORPORATION
36 N. Day Street
Orange, New Jersey 07050

(201) 674-1948

Complete solar system design and installation—water heaters, space heating and air conditioning—for commercial/industrial complexes.

817. UNIVERSAL DESIGN AND CONSTRUCTION
Rural Route 9, Box 160
Columbus, Indiana 47201

Daniel Gobin
(812) 342-6178

Universal Design and Construction has been active in solar design and construction for over six years. Our main activity centers on the design of truly energy-efficient solar architecture and its solar systems. Our experience and expertise lies in residential and commercial application of solar systems, underground and earth-tempered buildings. Other activities include design of solar collectors for manufacturers, solar/wood system research and development (DOE grant pending), and solar systems installation.

818. VACHON, NIX & ASSOCIATES
6855 Jimmy Carter Boulevard
Norcross (Atlanta), Georgia 30071

Dr. R. I. Vachon
(404) 448-5235

Solar engineering and architectural design of active and passive systems in the industrial, commercial, agricultural and residential sectors. Applications include: heating water, heating and cooling buildings, agricultural conditioning, industrial process heat, agricultural drying, desalination, power, and pumping. Experienced in photovoltaics.

A team of 100 solar specialists provides total project management of engineering and architectural projects from feasibility to the completed project. Projects include: Alabama Power Company, 1580 M^2 office building with solar-assisted heating and cooling system, one of ERDA's first demonstration projects. Also, U.S. EPA office and laboratory (4831 M^2) with solar-assisted rankine cycle heating and cooling system.

819. VIKING ENERGY CORPORATION
4709 Baum Boulevard
Pittsburgh, Pennsylvania 15213

Dr. Jack Saluja
(412) 683-6440

Solar Thermal services include: solar hot water systems for small rural hospitals; solar crop drying systems for tropical countries; feasibility of using solar energy in the food processing industry; solar thermal power applications for irrigation systems; solar-assisted organic digesters.

Solar Photovoltaics: farm irrigation systems; feasibility of small-scale PV electrical power technology in LDCs.

Geothermal: feasibility of geothermal energy applications for small-scale electrical power production in LDCs; application of geothermal energy resources in rural and agricultural communities.

Biomass—methane digestion systems.

DESIGN—ARCHITECTURE & ENGINEERING

820. JOSEPH VON ARX PASSIVE SOLAR CUSTOM HOMES
5104 Saddlewood Street
Sacramento, California 95841

Joseph Von Arx
(916) 334-4967

Design, engineer and build passive solar homes; new single-family homes and retrofit.

821. SEISS WAGNER, ARCHITECT
1309 25th Street
Sacramento, California 95816

Seiss Wagner
(916) 446-0434

At this time, most of our solar applications involve solar heating of swimming pools, trombe walls, daylighting and proper orientation of buildings on site. In the future, we hope to provide more sophisticated applications of solar energy.

822. WALGAMUTH & ASSOCIATES
608-1/2 Columbia Street
Lafayette, Indiana 47901

Tom Walgamuth
(317) 742-0688

All buildings are designed with passive in mind. Active systems on request of client. We are presently initiating design work on earth-sheltered buildings.

823. JAMES A. WARD, P.E.
10731 Treena Street, Suite 100
San Diego, California 92131

Jim Ward
(714) 578-8240

Design and performance analysis. Consulting services, feasibility and performance studies.

824. C. EDWARD WARE ASSOCIATES, INC.
415 Y Boulevard
Rockford, Illinois 61107

C. Edward Ware
(815) 963-8407

Provide design services for projects involving active and passive solar energy—residential, commercial and industrial applications.

825. ANDRÉ R. WARREN CONSULTANT & DESIGN INC.
P.O. Box 718
Brunswick, Maine 04011

A. R. Warren/Loren Arford
(207) 725-6223

Firm consists of a designer, architect and engineer (Colorado State Solar School, Ft. Collins) involved in the engineering and architectural design of active and passive systems for residential applications—new and retrofit.

826. DONALD WATSON
P.O. Box 401
Guilford, Connecticut 06437

Donald Watson, AIA
(203) 453-6388

Donald Watson is an architect and Visiting Professor of Architectural Research at the Yale School of Architecture. He is also chairman of the school's Master of Environmental Design Program.

In his professional practice, Donald Watson has been designer or consultant for over 100 solar housing projects. His designs have received awards from the Connecticut Society of Architects, the Michigan Product of the Year, the U.S. Department of Housing and Urban Development and the New England Council (AIA).

He has served as consultant for the United Nations, the World Bank, the U.S. Department of Energy, the AIA Research Corporation, the National Association of Home Builders Research Foundation, and numerous private organizations. He is an advisory board member of Consumer Action Now of Washington, DC, the Energy Task Force of New York City, and the Connecticut Governor's Solar Energy Alliance. He currently serves as Chairman of the Solar Energy Task Group of the American Institute of Architects.

827. **WEST CAP (WEST CENTRAL WISCONSIN COMMUNITY ACTION AGENCY)**
525 2nd Street
Glenwood City, Wisconsin 54013

Gary Titus
(715) 265-4271

West CAP has been actively involved in alternative energy and energy conservation since 1974. Over 70 solar systems have been installed on low-income homes. Of those 70, only 20 have been commercially bought systems (all solar hot water systems). The rest have been active and passive solar systems designed and built by West CAP, including 20 solar greenhouses and 20 air-to-water solar hot water heaters. Additionally, West CAP has an extensive consumer education component with expertise on wood, wind, water, alcohol, methane and energy conservation. Over 1200 low-income homes have been weatherized in the seven counties served by West CAP.

828. **WESTINGHOUSE ELECTRIC CORPORATION**
Solar Heating & Cooling Systems
5205 Leesburg Pike, Suite 201
Falls Church, Virginia 22041

Pamela Charles
(202) 833-5950

Westinghouse Solar Heating & Cooling Systems performs solar system design and engineering of space heating and hot water systems for: residences, buildings, and industrial processes up to 220F. In addition, we identify and specify optimum system components and configurations with assistance of computer

290 SOLAR CENSUS

programs for each application based on maximum return on equity, ROI and undiscounted life-cycle savings.

829. WHATWORKS
P.O. Box 588
Calistoga, California 94515

Douglas Hayes
(707) 942-5107

Specialize in passive solar buildings that do not look solar. Residential and commercial application with emphasis on totally self-sufficient systems.

830. JOHN WHIPPLE, ARCHITECT
386 Fore Street
Portland, Maine 04101

Ed Detmer
(207) 775-2696

Active and passive design, residential and commercial; new and retrofit solar applications.

831. S. WIEL & ASSOCIATES
940 Matley Lane, Suite 8
Reno, Nevada 89502

Steve Wiel
(702) 323-0238

Our firm is involved in emergency energy planning and energy conservation planning. We design retrofit solar systems and new construction solar systems. We do infiltration tests on buildings and energy audits for residences and commercial establishments.

832. RAY A. WILEY, ARCHITECT/ENGINEER
541 Willamette, Suite 308
Eugene, Oregon 97401

Ray Wiley
(503) 686-1892

DESIGN—ARCHITECTURE & ENGINEERING 291

Specialize in energy-efficient (passive solar) buildings and systems. Oregon Energy Conservation Board Chairman. Developed the Eugene Water & Electric Board's Energy Efficient Building Standards (so far the most energy efficient buildings in the Northwest). Extensive research in the areas of superinsulation, infiltration, condensation control, vapor barriers, ventilation, heat recovery, indoor air quality, passive heating and cooling. Developed the "Cloudy Day Solar Home." Developed the "Super Ark," a home that uses only one-fourth the total energy of a conventional home.

833. **WILSON & COMPANY**
ENGINEERS & ARCHITECTS
P.O. Box 1648
Salina, Kansas 67401

J. M. Hall III
(913) 827-0433

Wilson & Company provides complete engineering and architectural services through use of inhouse staff members who are on call to plan and design a wide variety of function/cost-effective facilities to serve the needs of our clients. Our computer simulation of system operation can help evaluate alternative plans.

In addition to architectural services, Wilson & Company offers the following fields of engineering: civil, mechanical, electrical, chemical, structural, sanitary, and industrial. A complete chemical laboratory is available to provide testing and analytical service.

834. **WINZLER AND KELLY**
P.O. Box 1345
Eureka, California 95501

T. Westlund
(707) 443-8326

Design of active and passive systems—residential, commercial and industrial—new and retrofit. Our major solar project was the installation of solar collectors on two residence halls at Humboldt State University. The design included approximately

10,000 square feet of solar collectors with the heat used for space heating and hot water preheating.

835. WOODFORD/SLOAN, AIA ARCHITECTS
150 Green Street, Penthouse B
San Francisco, California 94111

Bruce G. Sloan, AIA
(415) 788-6418

General architectural practice utilizing new technology for residential and commercial applications.

836. WOODS AND ASSOCIATES, INC.
112 Water Street
Naperville, Illinois 60540

Kenneth R. Woods
(312) 355-2360

Architectural projects completed include commercial, institutional, industrial, and apartment buildings and single-family residences. Active and passive solar system applications, new and retrofit.

837. WORMSER SCIENTIFIC CORPORATION
88 Foxwood Road
Stamford, Connecticut 06903

Eric M. Wormser
(203) 329-2001

Wormser Scientific Corporation is a company dedicated to designing and implementing the highest quality solar systems and providing research and development on new concepts. We offer in-depth experience in both commercial and residential applications.

Each project is thoroughly researched. We bring to bear all available energy conservation and passive solar strategies prior to any consideration of active solar systems. These include: use of maximum cost-effective insulation; window placement for maximum thermal benefit; judicious use of building mass to reduce air temperature swings arising from local heat gain; space arrangements to provide thermal buffering; recovery of waste heat; efficient lighting design; building siting and landscaping for minimum exposure to winter winds; optimum control of space temperature and ventilation. Such strategies reduce thermal requirements. The remaining load can be handled by a cost-effective active solar system design.

We combine new concepts in optics with established solar techniques. Wormser makes a maximum effort to integrate the system architecturally to achieve pleasing aesthetics.

Services: complete active solar system design; building design for passive solar heating; instrumentation system design; supervision of installation; system startup, check out and debugging; heat recovery and energy conservation studies; development and testing of new concepts; optics and infrared consulting; feasibility and economic payback studies; grant applications; state and federal program proposals.

838. **WILLIAM J. YANG & ASSOCIATES**
1258 North Highland Avenue, Suite 302
Los Angeles, California 90038

William J. Yang
(213) 465-3191

Consulting engineering firm, established in 1966, specializing in design and preparation of construction documents for industrial exhaust, air and water pollution control, industrial waste treatment, heating and air conditioning, boiler central plant, plumbing process piping, fire protection, cost estimates, economic feasibility studies, energy analysis and solar energy systems for all types of structures. Provides engineering and design services nationwide.

839. **JOHN YELLOTT ENGINEERING ASSOCIATES INC.**
901 West El Caminito Drive
Phoenix, Arizona 85021

(602) 943-5805

Develop and design solar-powered space heating and cooling, water and pool heating systems.

840. HERBERT C. YOUNGER, P.E.
5782 Turquoise Avenue
Alta Loma, California 91701

Herbert C. Younger, P.E.
(714) 980-0191

Designer/installer of custom data acquisition systems and data loggers for monitoring temperatures and other performance data. Also control systems for solar systems integrated with other equipment for energy conservation, environmental control, intrusion and fire alarms, etc.

841. ZAUGG & ZAUGG ARCHITECTS, AIA
60 South Main Street
Mansfield, Ohio 44902

Thomas Gene Zaugg
(419) 524-8054

This firm has been involved with solar design since 1947 (first passive system). We have designed and built three solar houses (1975,76,77). Have taught solar courses at local university, and own a small development firm, Design Development Corporation, which does hands-on solar research. We have been active in product design, but are largely interested in design application.

842. BRATTLEBORO DESIGN GROUP
113 Main Street, P. O. Box 235
Brattleboro, Vermont 05301

James I. Williams
(802) 257-7644

Building design and construction services. Product research and development. Construction projects integrating component research in movable insulation, site-built air collectors, trombe walls in various stages of completion.

DESIGN—ARCHITECTURE & ENGINEERING 295

843. QUALITY ENERGY SYSTEMS
7411 Broadway
Lemon Grove, California 92045

Richard M. Cartmell
(714) 464-3733

Consultant design and engineering, and installation and service of conventional heating and air conditioning, sheet metal, energy management, solar space, pool, DHW and spa heating systems for commercial and residential use. Also, energy audits and systems analysis.

843. ENVIRONMENTAL CONSULTING & TESTING SERVICES
(A)   ECTS — Energy Division
383 North Kings Highway
P.O. Box 3521
Cherry Hill, New Jersey 08034

Nathan R. Frenkel
(609) 779-1195

* * *

Also see the following entries—

1.   A.A.I. CORPORATION
4.   ACUREX SOLAR CORPORATION
5.   ADVANCED ENERGY SYSTEMS INC.
7.   ADVANCE ENERGY TECHNOLOGIES INC.
14. ALTERNATIVE ENERGY RESOURCES, INC.
22. AMERICAN SUN CORPORATION
23. AMERICAN TIMBER HOMES/SOLARTRAN DIVISION
61. CONTEMPORARY SYSTEMS, INC.
67. CREIGHTON SOLAR CONCEPTS, INCORPORATED
73. del SOL CONTROL CORPORATION
80. DIXIE ROYAL HOMES, INC.

296 SOLAR CENSUS

103. THE ENERGY FACTORY
109. ENERTECH CORPORATION
122. FIRST SOLAR INDUSTRIES, INC.
142. HALEAKALA SOLAR RESOURCES, INC.
150. HELIO THERMICS INC.
161. HYPERION, INC.
163. ILI, INC.
165. IMPAC CORP.
170. INTERTECHNOLOGY/SOLAR CORPORATION
192. LEWIS & ASSOCIATES/ SOLAR SITE SELECTOR
211. MONOSOLAR, INC.
227. NOVAN ENERGY, INC.
228. NRG COMPANY
239. PIPER HYDRO INC.
254. REACTION RESEARCH LABORATORIES
268. SANTA CRUZ SOLARWORKS INC.
278. SKILLESTAD ENGINEERING, INC.
283. SOLARA ASSOCIATES INC.
288. SOLAR COMFORT SYSTEMS MANUFACTURING CO.
296. SOLAR DEVELOPMENT INC.
308. SOLAR FAB, INC.
312. SOLAR HEATER MANUFACTURING CO.
322. SOLARNETICS CORPORATION
323. SOLARON CORPORATION
327. SOLAR PRODUCTS, INC.
334. SOLAR SEARCH CORPORATION
345. SOLAR UNLIMITED INC.
349. SOLAR WORLD, INC.
362. SOUTHWEST ENER-TECH, INC.
369. SUNDUIT, INC.
375. SUNMASTER CORPORATION
377. SUN-PAC, INC.
379. SUN RAY OF CALIFORNIA
381. SUN-RAY SOLAR HEATERS
385. SUN STONE COMPANY, INC.
389. SUNTRON

- 403. THOMASON SOLAR HOMES, INC.
- 404. SVEN TJERNAGEL SOLAR SYSTEMS
- 411. UNITED ENERGY CORPORATION
- 421. VULCAN SOLAR INDUSTRIES, INC.
- 423. WEATHER ENERGY SYSTEMS, INC.
- 427. YING MANUFACTURING CORP.
- 845. UNIVERSITY OF ALABAMA
- 858. ATEK DESIGN
- 862. BIO-GAS OF COLORADO, INC.
- 868. CASCADE SOLAR TECHNICS
- 898. ENERGY SYSTEMS GROUP/DIVISION OF ROCKWELL INTERNATIONAL
- 900. ESCHER:FOSTER TECHNOLOGY ASSOCIATES INC.
- 928. NATIONAL CENTER FOR APPROPRIATE TECHNOLOGY
- 935. OAK RIDGE NATIONAL LABORATORY
- 945. RADIANT EQUIPMENT COMPANY
- 963. SOLAR ENVIRONMENTAL ENGINEERING COMPANY
- 966. SOLAR SURVIVAL
- 986. VIRGINIA POLYTECHNIC INSTITUTE AND STATE UNIVERSITY
- 996. BIG FIVE COMMUNITY SERVICES, INC.
- 1001. CORNERSTONES ENERGY GROUP, INC.
- 1002. DOMESTIC TECHNOLOGY INSTITUTE
- 1012. NEW MEXICO SOLAR ENERGY ASSOCIATION
- 1020. SHELTER INSTITUTE
- 1023. SOLAR GREENHOUSE ASSOCIATION

RESEARCH and DEVELOPMENT

844. AESOP INSTITUTE
SUNWIND LTD.
P.O. Box 880
Sebastopol, California 95472

Mark Goldes
(707) 525-1111

Areas of research include bioconversion, concentrating collectors, alcohol distillation, engines, ocean thermal gradient systems, photovoltaic cells, fuel cell storage, wind generation systems, thermoelectrics, and solar-powered vehicles (windmobiles).

845. UNIVERSITY OF ALABAMA
JOHNSON ENVIRONMENTAL & ENERGY CENTER
P.O. Box 1247
Huntsville, Alabama 35807

David L. Christensen
(205) 895-6257

The Kenneth E. Johnson Environmental and Energy Center (JEEC) at The University of Alabama in Huntsville is a multifaceted research organization, heavily involved in addressing the energy, transportation and environmental problems that face society.

In the nine years since its creation, the Johnson Center has acquired a national and international reputation for excellence in research, testing, development, and demonstration of technologies designed to cope with or alleviate these concerns.

The Energy Research and Technology Office has demonstrated expertise in solar research and development, certifying solar equipment, compiling meteorological data, performing computer analyses, writing publications and coordinating meetings.

Approximately $2 million in government and industrial solar energy related contracts have been awarded to JEEC, some $1.2 million from the Solar Heating and Cooling Office at the NASA-Marshall Space Flight Center.

JEEC operates an American Refrigeration Institute-certified solar energy test facility. It is equipped to test liquid- and air-type solar systems, tracking collectors, and domestic hot water systems. A computerized data acquisition system, with a 96-channel processing and recording system, is utilized in testing programs. A meteorological station is operated in the facility and its daily data-taking is overseen by the State Climatologist, a JEEC staff member.

A comprehensive data base and special collection of documents on solar energy programs and technical activities is maintained at JEEC.

Among the tasks performed for the Solar Heating and Cooling Office at Marshall Space Flight Center was the development of a massive, nationwide data base to effectively analyze the viability of solar energy. The base includes climatological data. architectural details, heating and cooling systems and components, their costs and performance parameters, energy availability and related costs. Other tasks include development of visual aids, a literature survey of solar radiation measuring instruments, identification of sources of solar radiation data, designing and developing a modular solar space heating system, and preparing special reports, reviews and surveys. A project, under the direction of the Solar Energy Research Institute, in which JEEC will gather solar radiation data from around the world and compile and analyze it, is now under way. This contract is a direct outgrowth of a similar contract for gathering such data throughout the U.S.

The Johnson Center coordinated and hosted the First and Second Solar Heating and Cooling Commercial Demonstration Program Contractors' Reviews and produced their preliminary and final proceedings. Many other national, state and local workshops, seminars and short courses have been coordinated by the JEEC.

The center has been collecting and analyzing natural gas and electric rates from all over the U.S. and has issued a report, part of the SOLCOST program to compare the cost effectiveness of solar energy systems. Other activities include efforts to accelerate the commercialization of solar energy products and speed the development of the export market for U.S. solar equipment and services.

846. ALTAS CORPORATION
500 Chestnut Street
Santa Cruz, California 95060

F. de Winter
(408) 425-1211

Altas Corporation was founded in 1974 as an outgrowth of the consulting practice of Francis de Winter. Most of Altas' contractural work to date has been for organizations such as the U.S. Energy Research and Development Administration, the U.S. Department of Energy, the Solar Energy Research Institute, the Gas Research Institute, the Copper Development Association, the Electric Power Research Institute, and the California Energy Commission.

A highly qualified staff specializes in: heat transfer, thermodynamics, fluid mechanics, instrumentation, hardware design, computer simulation techniques, data acquisition and reduction, economic analysis, controls, and prototype fabrication and testing.

Clients in government and industry are provided expertise in design and analysis of thermal systems: solar heating and cooling of buildings, industrial process heat, energy conservation, power, refrigeration and heat pump cycles.

Inhouse research efforts include flat plate collectors and heat exchangers, solar-assisted heat pumps, phase-change energy storage systems, and the testing and evaluation of both active and passive solar systems and equipment. Presently, Altas is developing a gas-fired backup heat system for DOE and GRI.

847. ALTERNATE ENERGY RESEARCH
AND DEVELOPMENT
P.O. Box 77
Atlanta, Michigan 49709

William Huey
(517) 742-3347/785-4718

Research areas include domestic hot water systems, flat plate and concentrating collectors, and backup wood furnace systems outside the building, insulated, thermostat-controlled with the average fill lasting one to two weeks.

848. AMAF INDUSTRIES
Box 1100
Columbia, Maryland 21044

Research areas: solar environmental systems; heat pump systems; energy conversion.

849. AMITY FOUNDATION
P.O. Box 7066
Eugene, Oregon 97401

Virginia Bulger
(503) 484-7171

The Amity Foundation is a small nonprofit research and educational foundation established in 1976. We presently offer workshops, tours and other education programs in the areas of small-scale aquaculture, solar greenhouse design and urban agriculture. We also do applied research in these areas. In the past, Amity has been involved in the design and construction of alternative domestic waste treatment systems and in the concept of integrated systems housing. Our emphasis is on developing and communicating the technologies needed for energy-efficient and self-reliant lifestyles.

850. **ANACHEM INC.**
Princeton Park of Commerce
3300 Princeton N.E.
Albuquerque, New Mexico 87107

Researches materials for solar heating and cooling. Investigating biomass conversion.

851. **ANA-LAB CORP.**
2600 Dudley Road
Kilgore, Texas 75662

Solar pond research.

852. **ARGONNE NATIONAL LABORATORY**
Solar Energy — Building 362
9700 South Cass Avenue
Argonne, Illinois 60439

William Schertz
(312) 972-6230

The National Solar Energy Development program includes solar hot water heating, solar residential and commercial building heating and cooling, solar industrial process heat, solar thermal energy conversion, fuels from biomass (bioconversion), and investigation of the satellite power systems concept.

The current ANL solar energy program involves some aspects of all of the above solar energy research and development activities. Through a matrix management system, resources, facilities and personnel from eight ANL divisions are being utilized in the program. Expertise established in the past in other energy technologies is readily brought to bear on the solution of solar energy utilization problems.

ANL became involved in solar energy research and development in FY 1974 in a joint effort with the University of Chicago to develop a new solar concentrator concept called the Compound Parabolic Concentrator (CPC). Since that time, the level of effort applied to solar energy R&D at ANL has risen from the initial six manyears in FY 1974 to 40 manyears in FY 1978 (and an expected 80 manyear level).

Programmatic involvement has been expanded during this period from the initial concentrator development work to include efforts in thermal energy storage, ocean thermal energy energy conversion, solar heating and cooling, solar power satellite, bioconversion, solar thermal power, and photovoltaics. The principal thrusts of these programs are:

— Solar concentrator development for heating and cooling, industrial process heat, solar electric and photovoltaic applications.

— Solar heating and cooling development including thermal energy storage, demonstration program support, and reliability analyses.

— Ocean thermal energy conversion technology development.

— Materials development for photovoltaic, heating and cooling, solar thermal power, and OTEC programs.

— Environmental and technology assessment for bioconversion, OTEC and satellite power system programs.

The solar energy R&D activities at ANL are described in the body of report number ANL-79-16, available through NTIS, U.S. Department of Commerce, 5285 Port Royal Road, Springfield, Virginia 22161. Specific information regarding DOE sponsors, ANL principal investigators and contacts for further information are contained at the end of each paper of that report.

853. UNIVERSITY OF ARIZONA
Solar Energy Research Facility
Civil Engineering Building, Room 206
Tucson, Arizona 85721

Mr. Charles Glickman
(602) 626-4965

Researching liquid and concentrating collectors, insulation, heat exchangers, and photovoltaics.

854. ARIZONA STATE UNIVERSITY
Center for Solid State Science
Tempe, Arizona 85281

Prof. J. B. Wagner, Jr.
(602) 965-4544

Research activities center on solid electrolytes for storage batteries; corrosion—especially high-temperature oxidation.

855. ARIZONA STATE UNIVERSITY
Laboratory of Climatology
Tempe, Arizona 85281

Dr. A. J. Brazel
(602) 965-6265

Solar Energy Assessment and Climatology: houses the Office of the State Climatologist, analyzes state solar radiation data, calibrates solar sensors in the state, installed and maintains the state solar monitoring network, have done DOE solar projects, solar consultations to governmental agencies and industries, cooperative solar monitoring with the U.S. Water Conservation Laboratory, Solar Energy Commission Spectral Project—urban effects in the desert.

856. ASSOCIATED ENTERPRISES
2832 Montclair Drive
Ellicott City, Maryland 21043

Researching solar concentrator-assisted production of fuel by converting solar energy to liquid hydrogen.

857. ASTRO SOLAR CORP.
1602 Clare Avenue
West Palm Beach, Florida 33401

Researching flat plate collectors.

858. ATEK DESIGN
3 Crawford Street
No. 3
Cambridge, Massachusetts 02139

David Walden
(617) 354-5427

Author of *Resonant Solar Collector* submitted to office of energy-related inventions. The paper deals with applying laser technology to solar, i.e., a solar-powered laser.

Atek operated in Denver for two years before moving to Cambridge.

859. BALDWIN SOLAR TECHNOLOGIES
5088 Laurel Drive
P.O. Box 5104
Concord, California 94524

Investigating air and liquid collection and storage systems, as well as solar installation construction.

860. BARBER-NICHOLS ENGINEERING CO.
6325 West 55th Avenue
Arvada, Colorado 80002

Daryl Prigmore
(303) 421-8111

Barber-Nichols is a development engineering company specializing in energy conversion systems. Its technical staff consisting of specialists in the fields of thermodynamics, turbomachinery, and power transmission engineers has been associated with the solar energy field since the 1950s when they worked on NASA programs to utilize solar energy for space power systems. More recent Barber-Nichols solar activity involves terrestrial application. Specifically, Barber-Nichols is applying turbomachinery and Rankine cycle experience it has acquired over the years, to convert solar heat into mechanical power to drive air conditioners, water pumps, and electric generators.

861. BATTELLE MEMORIAL INSTITUTE
COLUMBUS LABORATORIES
505 King Avenue
Columbus, Ohio 43201

D. Karl Landstrom
(614) 424-4803

Battelle Columbus Laboratories has been active in solar energy research since 1952 when a study was made of the use of solar parabolic cylinders to generate steam for multi-stage distillation of seawater plants. Since then, the staff has worked on a variety of solar applications, including solar heating and cooling of buildings, solar industrial process heat, solar powered engines, wind generators, solar cells, and the bioconversion of crops to fuel. Following are examples of Battelle solar research:

Solar Heating and Cooling: study for ERDA of a solar-operated heat pump which uses a low-friction rotary-vane expander coupled to a refrigeration compressor; feasibility studies of solar collectors for heating large building complexes; mathematical models for designing and analyzing solar solar heated and cooled buildings; investigation of a "black" fluid concept for a low-temperature plastic solar collector for DOE; development of heat pipes for passive solar heating; investigation of corrosion in solar collectors from aqueous solutions, DOE; photochemical materials for energy storage; evaluation for NSF of buried water tanks for heat storage for solar-assisted heat pumps.

Solar Thermal Power: development of a 50-hp solar-powered irrigation pump; a solar Stirling cycle engine for production of power up to 2 hp; and a novel solar steam-operated residential power unit.

Solar Process Heat: solar distillation of sea water to fresh water; designs and cost estimates of solar parabolic cylinders for use in generating steam at temperatures up to 600F; a survey of applications of solar thermal energy systems to industrial process heat.

Photovoltaics: design of a 150-kW concentrator photovoltaic system on an industrial building; develop materials for the dies and containers used to make silicon crystals for solar cells; encapsulation of low-cost silicon solar arrays; design of a flat plate photovoltaic system to supply power to Battelle's machine shop; development of chemical processes for production of low-cost silicon for solar cells.

862. BIO-GAS OF COLORADO, INC.
5620 Kendell Court
Arvada, Colorado 80002

John Downs/Susan Schellenbach
(303) 422-4354

Since its inception, Bio-Gas of Colorado, Inc. has directed its operations toward agricultural waste treatment and resource recovery. The primary thrust of the research activities has been the conversion of animal waste material into methane gas, animal protein feed and fertilizer.

Bio-Gas of Colorado has received a patent for a process utilizing solar energy to heat the liquid in the digester. The principal objective of the invention covered by the patent was to use solar energy, rather than burning the biogas to heat the process thereby making the process more efficient. Another patent application is being filed for a waste treatment and gas-producing system to be utilized in animal confinement buildings.

Based on its work to date, Bio-Gas of Colorado has received considerable praise and support from various governmental agencies and personnel. The company was cited with an "Excellence in Government" award for "Achievement in Energy Resource Development," and was the recipient of the American Consulting Engineers' Grand Conceptor Award for its role in the design of the Lamar bioconversion utility plant.

863. BOSTON COLLEGE
Chestnut Hill, Massachusetts 02167

P. H. Fang
(617) 969-0100 Ext. 3578

Research and development of (1) thin film polycrystal and amorphous silicon solar cells, and (2) silicon ribbons produced at high rate.

864. BROOKHAVEN NATIONAL LABORATORY
Upton
Long Island, New York 11973

Researching design of low-cost solar-assisted heat pump systems. Investigating ground-coupled heat pump performance and economical thin-film flat plate collectors.

865. BROWN UNIVERSITY
Division of Engineering
Providence, Rhode Island 02912

Prof. J. J. Loferski
(401) 863-2674

Areas of research are air collectors and photovoltaic cells and panels.

866. UNIVERSITY OF CALIFORNIA
LAWRENCE LIVERMORE NATIONAL LABORATORY
P.O. Box 5500
Livermore, California 94550

Robert L. Lormand
(415) 422-5833

The laboratory has a number of solar projects under study at this time. The main emphasis is on photovoltaic systems (i.e., photocell design and materials testing) and on solar ponds. In addition, there is interest in alternative fuel production and utilization and electric vehicle design.

867. CARY ARBORETUM OF THE NEW YORK BOTANICAL GARDEN
Box AB
Millbrook, New York 12545

Mr. Winfried U. Schubert
(914) 677-5343

The Cary Arboretum was founded in 1971. Its major research and education themes are ecology and the conservation of natural resources. The site of the Plant Science Building was chosen with this environmental commitment in mind. The structure has been placed into the earth—about two-thirds of it is underground, insulated against wintry winds and summer heat. Energy from the sun provides virtually all of the heat in the building by means of rooftop solar collectors. The solar energy is either fed directly into the building's heating system or indirectly by means of a water-to-water heat pump. Water from deep wells is used both as the backup source of heat for the heat pump and to cool the interior of the building in the summertime. Throughout the structure, a variety of special systems work both to conserve heat and natural resources and to minimize energy consumption.

868. CASCADE SOLAR TECHNICS
216 Paxton Avenue
Salt Lake City, Utah 84101

David Weston
(801) 467-3714

Solar research areas include: agricultural applications, backup systems, air and liquid collectors, controls and instrumentation, DHW systems, greenhouses, heat exchangers, insulation, photovoltaics, liquid and solid storage, as well as residential and commercial design applications.

869. THE CENTER FOR MAXIMUM POTENTIAL BUILDING SYSTEMS
8604 FM 969
Austin, Texas 78724

Pliny Fisk, III
(512) 928-4786

The Center for Maximum Potential Building Systems is devoted to a search for low-cost, labor-intensive, environmentally sound energy and resource alternatives. It functions as a regionally based appropriate technology laboratory concerned with environmental and resource issues, but not apart from the socioeconomics of a region in its own right.

870. **CENTRAL MONTANA HUMAN RESOURCES DEVELOPMENT COUNCIL**
Dist. VI
604 W. Main
Lewistown, Montana 59457

Jim Mayec
(406) 538-7488

Areas of research include air and liquid collectors, DHW systems, greenhouses.

871. **CHAMBERLAIN MANUFACTURING CORPORATION**
845 Larch Avenue
Elmhurst, Illinois 60126

Solar energy research and development, and solar panel testing.

872. **UNIVERSITY OF CHICAGO**
ENRICO FERMI INSTITUTE
5630 S. Ellis Avenue
Chicago, Illinois 60422

J. O'Gallagher/R. Winston
(312) 753-8637

The solar energy group at the University of Chicago specializes in the design, prototype fabrication, and testing of nonimaging concentrating collectors. These collectors provide wide acceptance angles for solar concentration and reduce or eliminate the need for tracking. The most well-known versions of these collectors are the Compound Parabolic Concentrators (CPCs). An informal collaboration with the solar group at Argonne National Laboratory is maintained.

873. **WILLARD D. CHILDS**
582 Rancho Santa Fe Road
Encinitas, California 92024

Researching alternate energy conversion—solar, wind, ocean thermal.

874. **UNIVERSITY OF COLORADO**
Civil, Environmental & Architectural Engineering
Boulder, Colorado 80309

Hugh Davis
(303) 492-7315

The Department of Civil, Environmental and Architectural Engineering is involved in system evaluation and monitoring and has undergraduate and graduate programs in energy engineering which stress energy conservation and solar energy utilization. As part of this program, projects of design, analysis and research are performed as sponsored research.

875. **COLORADO STATE UNIVERSITY**
SOLAR ENERGY APPLICATIONS LABORATORY
Fort Collins, Colorado 80523

Dr. S. Karaki
(303) 491-8617

Solar Space Heating and Cooling: Development of practical, commercially available systems since 1973. Several types have been shown to be reliable and cost-effective in many areas of the U.S. Research and development continues for new heating and cooling systems. Passively heated building with many types of solar gain devices is being constructed for experimental testing during the winter of 1980-81 and in following years.

Agricultural Applications: Production of liquid and gaseous fuels for on-farm use is an active research program in the Agricultural and Chemical Engineering Department at CSU. Uses of active solar systems for crop drying, shelters and farm buildings also being developed.

Photovoltaics: Development of low-cost solar cells for production of electricity, onsite and control station. Primary emphasis is development of efficient solar cells.

High-Temperature Solar Conversion: Endothermic/exothermic reactions to separate chemical compounds with solar heat, recombine to release heat called "Solar Chemical Research" is also underway.

Wind Systems for Rural Areas: Development of wind energy conversion systems for rural uses is a part of CSU's R&D effort.

876. **UNIVERSITY OF CONNECTICUT ENERGY CENTER**
Box U-139
Storrs, Connecticut 06268

David R. Jackson
(203) 486-3478

The Energy Center began as the Solar Energy Evaluation Center which was established in response to a governmental request to evaluate the performance of three solar collector manufacturers who were bidders on a solar-heated housing project for the elderly in Hamden, CT. Since that time, the center has tested 35 solar panels for various industries. The center is now one of 11 nationally recognized solar energy test laboratories (National Bureau of Standards Report NBSIR 78-1535).

The Energy Center has been continuously growing and is engaged in a number of long-term research projects. In 1978 the center installed a solar hot water system and data acquisition facility at the Visitor Center located in Willington, CT. This research is to act as a model for all rest area solar augmentation in the state.

The Center has engaged in air infiltration research resulting in two published papers and established an air infiltration facility on the campus for further studies into this important area.

The Center has installed a domestic hot water system and instrumentation facility at a dwelling in Ellington, CT, which is currently undergoing long-term degradation research (3 years at present). The Center has also started a long-term research project on the study of urea formaldehyde regarding the long-term thermal degradation of this material in situ.

In January 1980 the Center contracted with the Connecticut Department of Transportation to provide system design and detailed monitoring for a passive solar retrofit of a State Highway Maintenance Facility.

The Solar Energy Evaluation Laboratory is currently providing an independent laboratory test service to manufacturers of solar collectors and other energy devices and to anyone else desiring test results for equipment submitted for evaluation. Currently available tests include: Thermal Performance Evaluation (ASHRAE 93-77), Development Testing, and 30-Day Exposure Testing (HUD 4930.2).

877. **COPPER DEVELOPMENT ASSOCIATION, INC.**
1011 High Ridge Road
Stamford, Connecticut 06905

Raymond A. Weisner
(203) 322-7639

Publications, communications, educational activities, and research and development of applications of copper tube and sheet for solar systems—primarily active.

878. **CORPORATE ENERGY DEVELOPMENTS INC.**
P.O. Box 332
Hermitage, Tennessee 37076

Gene Russell, Research Director
(615) 883-0472

Researching use of ground tubes for cooling and heating. Design solar collectors and provide plans to mail-order houses.

RESEARCH & DEVELOPMENT 315

879. CRYSTALITE EMBEDMENTS INC.
6 Industrial Drive
Smithfield, Rhode Island 02917

Research and development concerns medium-temperature flat plate collector systems.

880. CRYSTAL SYSTEMS, INC.
Shetland Industrial Park
35 Congress Street
Salem, Massachusetts 01970

Dr. Chandra P. Khattak, Director R&D Division
(617) 745-0088

We are a materials supplier and are actively working on technologies to reduce costs of photovoltaic materials. Our unique Heat Exchanger Method (HEM) converts polycrystalline silicon into square cross-section, single-crystal ingots. It has been demonstrated that high-efficiency devices can be fabricated using low-cost polycrystalline silicon feedstock and HEM processing. For converting silicon ingots into wafers, we use Fixed Abrasive Slicing Technique (FAST) which combines best material utilization with highest quality wafer at low-cost.

881. DECKER MANUFACTURING
Impac Corp.
312 Blondeau
Keokuk, Iowa 52632

Areas of research: photovoltaics and passive solar design.

882. UNIVERSITY OF DELAWARE
INSTITUTE OF ENERGY CONVERSION
One Pike Creek Center
Wilmington, Delaware 19808

Joseph L. Rykiel
(302) 995-7155

The Institute of Energy Conversion of the University of Delaware is an applied research institute committed to making a measurable contribution to the solution of specific energy problems by developing new technologies which can be taken from the laboratory to the marketplace. The work of the Institute is organized to achieve excellence by concentration of effort in four specific areas:

Solar Cell Development — the direct conversion of sunlight into electricity. The Institute has earned a reputation as a world leader in the development of low-cost, thin-film photovoltaic solar cells.

Unit Operations Laboratory — the building and study of a prototype process to acquire the information necessary to take solar cells from bench-scale development to commercial-scale mass production.

Thermal Energy Storage — the retention of energy in a patented mixture of salt hydrates for use when needed. The Institute has developed, and is evaluating through demonstration projects, energy storage systems for both solar and off-peak electrical applications.

Energy Information Service — the dissemination of reliable, technically accurate energy information to consumers. The Institute's information staff responds to more than 15,000 requests annually for energy information.

883. DENVER RESEARCH INSTITUTE
University Park
Denver, Colorado 80208

Dr. R. H. Cornish, Associate Director
(303) 753-2271

The Denver Research Institute (DRI), a department of the University of Denver, is the leading multidisciplinary research center in the Rocky Mountain region and one of the significant contract research institutions in the U.S. There are seven operating divisions at the Institute: Chemical, Electronics, Industrial Economics, Laboratories for Applied Mechanics, Metallurgy and Materials Science, the Office of International Programs and Social Systems Research and Evaluation. With its broad range of professional disciplines, DRI is able to work on newly emerging problems of national and regional importance such as energy research, socioeconomic studies, economic and management programs, and international management training and assistance programs.

The Chemical Division has pioneered in the investigation of new fuel and energy sources including oil shale research, as well as research on fuel gas production from sanitary landfills, tertiary oil recovery by micellar flooding, geothermal brine, gasification of coal and oil shale, solvent extraction of tar sands, and direct conversion of solar to electrical energy.

The energy and environmental activities of the Laboratory for Applied Mechanics cover a broad area which includes research in resource recovery, fossil fuels, geothermal, water quality, air pollution, transportation and safety.

The Metallurgy Division has been active in corrosion research since its inception. Environmental embrittlement has been studied in liquid metal, liquid salts, organic and inorganic solutions, and various gaseous atmospheres. Stress corrosion of fabricated components is studied as a function of the mode of fabrication. Recently, an instrument called a corrodoscope was developed by DRI to study the protection afforded by various coatings. Corrosion in heat exchangers and desalination plants is another area that has been studied.

**884. DETROIT LAKES TECH
AREA VOCATIONAL-TECHNICAL INSTITUTE
Highway 34 East
Detroit Lakes, Minnesota 56501**

Jim Wylie, Horticulturist
(218) 847-3129

We are working with the Minnesota Solar Greenhouse Project in design and construction of vegetable and floral crop production units throughout a four-county area. We have 15 units to date. We present lectures or workshops to groups and have a vocational home study course available to interested persons. We are engaged in research and development of solar greenhouses and greenhouse vegetable production.

DEVON D INC., *See (991).*

885. DSET LABORATORIES, INC.
Box 1850, Black Canyon Stage
Phoenix, Arizona 85029

M. W. Rupp
(602) 465-7356

DSET is the leading solar collector test laboratory in the nation. Our facilities include six individual hydronic test loops, two separate air collector test loops, and two individual high-temperature nonaqueous fluid test loops. The heart of each test station is a specially designed sun-tracking altazimuth mount that greatly reduces turnaround time. This means that results are highly accurate and are delivered on a timely basis. DSET has completed over 1000 individual solar collector thermal performance tests for a wide variety of collector types including typical flat plates and highly sophisticated tracking concentrators.

Multiple options are available for developing specific test programs aimed at specific needs. DSET is approved or accredited by a number of governmental agencies and industry associations to test in support of their programs, such as the HUD testing programs, DOE testing programs, and the California TIPSE program. Our test results are accepted by the Florida Solar Energy Center for the Florida Certification Program.

We are prepared to test in support of other programs as they are developed and implemented. Operational testing of components and systems plays an important role in DSET's overall solar operations. This type of testing service is available for both short- and long-term programs. Reliability of performance and long-term durability are identified and assessed periodically using various performance oriented diagnostic tools.

DSET is also active in an associated solar technology—namely, photovoltaics. We have developed a line of services for testing photovoltaic performance characteristics and for testing long-term durability/reliability characteristics.

Thermal Performance Testing of Solar Collectors (flat plate through high-temperature concentrators); Durability/Reliability Testing of Solar Collectors; Systems Testing; Photovoltaics Performance Characteristics (cell, module, array); Conventional and Accelerated Outdoor Weathering (durability/reliability of materials); Property Assessment Services (emittance, absorptance, reflection, transmittance).

886. **DUNHAM ASSOC.**
SYMCOM INC.
528 Kansas City Street
Rapid City, South Dakota 57701

Control, instrumentation and simulation research for geothermal, wind and solar energy.

887. **EARLE ENGINEERING**
P.O. Box 850
Alpine, California 92001

John Earle

While still producing windmill regulators for battery charging with car alternators, most of our present effort is in developing systems and concepts and patents. Products being developed but not presently available include: (1) fossil or combustible fuel, external combustion engine; (2) low-pressure steam engine; (3) energy storage system; (4) high-speed windmill.

888. **EARTHMIND**
4844 Hirsch Road
Mariposa, California 95338

Michael Hackleman

Earthmind is a public nonprofit corporation for research and education, working with natural energy sources, organic gardening, and other aspects of more self-reliant living. Present research interests include wind generation systems and electric vehicles.

889. ECOENERGETICS INC.
180 Viewmont Avenue
Vallejo, California 94590

D. Benemann
(707) 644-6938

Consulting, design/engineering, research and development, commercialization.

Biofuels: (a) methane from agricultural wastes, specializing in farm systems; (b) hydrogen from water and/or wastes and sunlight; (c) aquatic plant energy systems.

Aquaculture: (a) chemicals from algae; (b) wastewater treatment with algae and aquatic plants; biotechnology.

890. ECOTRONIC LABORATORIES, INC.
TALLEY INDUSTRIES
7745 E. Redfield Road
Scottsdale, Arizona 85260

Lane Garrett, President
(602) 948-8003

Research, development and manufacturing of battery chargers and regulators used with photovoltaic panels. Photovoltaic systems, controls for thermal systems. Microprocessor based systems. Instrumentation and data recording systems. Digital thermometers. Energy management systems.

891. EIC CORPORATION
55 Chapel Street
Newton, Massachusetts 02158

Research centers on energy conversion (geothermal and solar) and energy conservation.

892. ELAND ELECTRIC CORP.
1841 Morrow Street
Green Bay, Wisconsin 54304

J. M. Eland
(414) 432-8286

Research activities include photovoltaics, backup systems, and the design of multi-family residential and commercial structures.

893. ENERAD INC.
P.O. Box 3982
Bellevue, Washington 98009

Dr. R. R. Brown
(206) 746-4672

Development of solar thermophotovoltaic generators for terrestrial use. Solar energy conversion to electrical power.

894. ENERGY DEVELOPMENT COMPANY
179 East Road 2
Hamburg, Pennsylvania 19526

Wind energy system design; also, solar, methane and water power generation.

895. ENERGY DYNAMICS CORPORATION
6062 E. 49th Avenue
Commerce City, Colorado 80022

Investigating residential uses of solar power.

896. ENERGY RESEARCH INSTITUTE OF DELAWARE
500 Homestead Road G-1
Wilmington, Delaware 19805

Edmund Banas
(302) 654-5426

We are a small research institute. Several fields are presently under exploration, as described below. Findings will be scaled up for industrial and for domestic use.

Solar Collector: a hermetically sealed, very low-cost, high-efficiency, sheet metal construction collector. The units are filled with special heat exchange fluid. Operation is similar to a heat plate, except our product is flat. Expected production: 1981 by Energy Research Institute of Delaware.

Circulating Pump: This product is a circulating pump which has no seal; therefore no leak can develop. The operating characteristics of the pump are similar to the characteristics of ordinary centrifugal pumps. This product is specifically designed for solar application; to operate with any heat transfer fluid for years without maintenance. At present, only a laboratory model is available. Expected production: 1981.

Heat Pump: this device is operated by heat; it requires no electricity. It can be used with gas, oil furnaces, or with wood or coal stoves. Two versions are under design. Both versions use the heat generated by a burner. The device will add a certain amount of heat (25–50%) to the theoretical heat energy, generated by the burner. Version A — low cost, for domestic use. Version B — higher efficiency and output for domestic and industrial use. Expected production: 1981–1982.

Solar Desalination: a special cascade-vacuum distillation process. Very high in efficiency, moderate to low in investment cost. The device has practically unlimited life. Expected production: 1982.

Gas Analyzer: a low-cost instrument aimed to use for Leak Test Gas Analysis for manufacturers of heating and cooling devices.

897. ENERGY SCIENCE CORPORATION
6211 Covington Way
Goleta, California 93017

Research area: solar-electric power plants.

898. ENERGY SYSTEMS GROUP
DIVISION OF ROCKWELL INTERNATIONAL
8900 DeSoto Avenue
Canoga Park, California 91304

Tom H. Springer
(213) 341-1325

The Energy Systems Group (ESG) is presently engaged in two conceptual design studies to determine feasibility of utilizing energy from a solar receiver system, using liquid sodium as the heat transport fluid, to repower existing fossil fueled power plants in Texas.

The first study, sponsored by the Department of Energy, involves Texas Electric Service Company's (TESCO) Permian Basin Steam Electric Station Unit No. 5, near Monahans. Using an advanced solar central receiver system, the unit would help save natural gas. The unit has a net power output of 115 megawatts (MWe) and would be solar repowered at about 50 MWe.

The second study is of West Texas Utilities (WTU) Company's Paint Creek Power Station Unit No. 4, near Lake Stamford. Work and funding is being performed by the utility company and ESG, with the support of the University of Houston, the Boeing Company, and Sargent & Lundy Engineers. The plant operates on fossil fuel at 110 MWe and would be solar repowered at 60 MWe.

The conceptual design of the repowering system consists of a single tower and surrounding field occupying around 380 acres adjacent to the plant unit. Sodium would be transported to the top of the tower by an 11,000-gallons-per-minute pump at the base of the tower.

Successful completion of the conceptual design studies could lead to construction of the solar plants commencing around mid-1981, with operations targeted for around mid-1985.

ESG has also completed conceptual studies for the DOE, using the central receiver concept, toward the storage-coupled or stand-alone application in plant sizes of 100 to 500 MWe and at capacity factors of 80%. Also, recently completed for the DOE was a conceptual study of the so-called hybrid solar plant. Two basic types were investigated—one with thermal storage and one without storage, each employing sodium as the heat transport fluid. In the hybrid plant—the solar loop is combined with a fossil-fueled power plant greatly reducing or eliminating need for thermal storage; the fossil fuel serves essentially as the "stored energy." Thus, the plant has a high availability as an energy source and can have a constant 24 hour per day and 365 day per year (neglecting planned maintenance and forced outages). Hybrid plant sizes of 100 to 615 MWe were investigated. It is particularly noteworthy that the 615-MWe solar hybrid was found to be competitive with a new coal plant for a 1990 date for the start of operation for an assumed cost of coal of $1.40/MBtu.

ESG is also funding research in the areas of solar photovoltaics, wind energy systems, and thermal storage systems.

899. ENVIRONMENTAL RESEARCH INSTITUTE OF MICHIGAN (ERIM)
P.O. Box 8618
Ann Arbor, Michigan 48107

Reed Maes
(313) 994-1200

Research and development of low-energy commercial-sized greenhouses.

900. ESCHER:FOSTER TECHNOLOGY ASSOCIATES, INC.
P.O. Box 189
St. Johns, Michigan 48879

William J. D. Escher
(517) 224-3268

Escher:Foster Technology Associates, Inc. has been active in the field of energy technologies and energy systems engineering for a number of years. E:F has demonstrated capabilities in a number of specific disciplines: alternative fuels, community energy systems, solar energy applications, and transportation energy systems.

We conduct assessments and design studies of advanced technology solar energy systems, e.g., solar/hydrogen systems. We are also conducting R&D on an advanced solar collector unit capable of matching up with an advanced "all services" solar energy system. Our work remains proprietary at this writing.

901. FERMI NATIONAL ACCELERATOR LABORATORY (FERMILAB)
P.O. Box 500, MS—115
Batavia, Illinois 60510

Penelope Horak, Assistant Head for Technical Services
(312) 840-3393

Solar research and development (new and retrofit applications): heating—process, space, water; flat panels and mirror augmented, liquid; passive walls, space heating; greenhouse design and development, residential/industrial applications.

Energy conservation—retrofit: insulation, building skins, glazings, vegetation.

902. UNIVERSITY OF FLORIDA
SOLAR ENERGY AND ENERGY CONVERSION LABORATORY
Room 325 MEB
Gainesville, Florida 32611

E. A. Farber/H. A. Ingley
(904) 392-0812

The research and development activities are key elements in the educational aspects of alternative energy at the University of Florida. Graduate and undergraduate students assist in the laboratory and with field work. The students design, construct and test many of the projects which lead to the development of equipment and systems capable of converting solar energy into energy forms that will meet a variety of needs. The research of the Solar Lab is conducted at the University's Energy Research and Education Park. This park is a 23-acre facility set aside for researching questions of concern and importance in the following areas: liquid collectors, absorbers, glazing, heat exchangers, insulation, DHW systems, chillers, low-temperature cooling systems, liquid and phase-change storage, distillation, methane digestion systems, bioconversion, solar ponds, and photovoltaics.

One new element has recently been added to the Lab's activities—the education and training of persons from developing countries in the use of alternative energy. This program is totally funded by USAID and is entitled "Training in Alternative Energy Technologies." Two 15-week sessions are held each year with a class of 30 participating in lectures, laboratory and field activities.

903. FLORIDA SOLAR ENERGY CENTER
300 State Road 401
Cape Canaveral, Florida 32920

Joan Morin
(305) 783-0300

The Florida Solar Energy Center (FSEC) was created by the state's legislature in 1974 and became operational in 1975, serving as the nucleus of Florida's solar activity. FSEC's activities and goals, broadly stated, are: (1) commercialization of solar energy; (2) programs for solar energy equipment testing, certification and standards; (3) educational services and dissemination of information about solar energy; (4) solar demonstration projects; (5) advancement of research and development.

The Testing and Laboratories Division's testing facility, one of few such operations in the nation, is equipped with six sun-tracking stands capable of simultaneously testing six flat plate liquid-type collectors—the kind most commonly used for solar water heaters. Since its inception in October 1976, the testing program has been voluntary; however, it has recently become mandatory that solar devices made or sold in Florida be certified by FSEC.

Much of the Research and Development Division's activity centers around the new experimental solar laboratory where FSEC scientists and engineers conduct in-depth research on solar domestic hot water systems aimed at reducing costs and increasing reliability. They also are working with applications of photovoltaic cells.

904. GEOSCIENCE LTD.
410 South Cedros Avenue
Solana Beach, California 92075

C. M. Sabin
(714) 755-9396

Geoscience Ltd. conducts research and development in heat and power engineering, including a variety of solar-heated systems. The testing division performs certification tests on insulation of all kinds, on solar and infrared transmittance,

RESEARCH & DEVELOPMENT 327

emissivity, shading coefficient, specific heat, etc. Testing procedures have been certified by the State of California, ICBO, the U.S. Navy and others. Geoscience participates in the National Voluntary Laboratory Accreditation Program.

Consultation and design services are available on control systems, specialized instrumentation, heat transfer processes. These services are confidential, protecting the customer's proprietary ideas.

905. HAWAII NATURAL ENERGY INSTITUTE
2540 Dole Street
Honolulu, Hawaii 96822

Mary Troy
(808) 948-8890

The Hawaii Natural Energy Institute (HNEI) was established as an institute of the University of Hawaii at Manoa to provide leadership, focus and support for natural energy research, development and demonstration.

HNEI has participated in the development of geothermal energy, from the drilling of the only fluid-producing well in the state to the present wellhead electric generator being constructed. Additionally, HNEI is involved with studies designed to locate other geothermal resources in Hawaii and to develop nonelectric uses of geothermal energy, including agricultural applications.

HNEI's Seacoast Test Facility is a major ocean energy test center. The development of synthetic liquid fuels is vital to Hawaii's plan for energy self-sufficiency, and emphasizes the importance of a proposal for a feasibility study on the production of gasoline from biomass being prepared by HNEI. In other biomass energy projects, HNEI is working on giant koa haole energy tree farms, the production of ethanol from molasses and many others.

HNEI has strongly supported the wind energy program in Hawaii. By the end of 1980, approximately one dozen wind energy conversion systems will be operating. The largest solar energy project in Hawaii, the photovoltaic system for Kauai Wilcox Hospital, is another activity in which HNEI is participating, as well as continued work in photovoltaics.

328 SOLAR CENSUS

The breadth of HNEI's involvement with natural energy resource development is evidenced by its work on the electric vehicle demonstration, the studies of an undersea electric cable, and the national conference on renewable energy sources.

906. **HITTMAN ASSOCIATES, INC.**
9190 Red Branch Road
Columbia, Maryland 21045

William Scherkenbach
(301) 730-7800

In the area of solar energy, Hittman Associates provides professional services to government agencies, architects, consulting engineers, and planning consultants. These professional services include technical and economic studies, concept analysis, system modeling, and control optimization, and cover the full range of solar energy applications including heating and cooling of buildings, water heating, power generation, and water desalting.

907. **HONEYWELL INC.**
Rosedale Towers
1700 W. Highway 36
St. Paul, Minnesota 55113

Glen Merrill
(612) 378-4477

R&D and systems engineering of total systems. Development of control systems for H/C systems and manufacture of trackers for single-axis concentrating systems. Also, controls for ethanol plants.

908. **UNIVERSITY OF IDAHO**
College of Engineering
Moscow, Idaho 83843

Richard Williams
(208) 885-6479

Research, development, demonstration and technology transfer in the areas of solar heating and cooling, agricultural and industrial applications of solar energy, biomass-derived fuels, and solar power. Also, education and training in these areas.

909. INDEPENDENT POWER DEVELOPERS INC.
P.O. Box 1467
Noxon, Montana 59853

Researching solar and wind power generation.

910. JBF SCIENTIFIC CORPORATION
2 Jewell Drive
Wilmington, Massachusetts 01887

Investigating alternate energies, encompassing solar, wind, biomass, and ocean thermal power.

911. KAMAN AEROSPACE CORPORATION
Old Windsor Road
Bloomfield, Connecticut 06002

B. A. Goodale
(203) 242-4461

Kaman developed preliminary designs of 500-kW and 1500-kW wind turbine generators (WTG) for electric utility operations under NASA contract as part of the DOE wind energy program. These designs provided the basis for detail design, fabrication and experimental demonstration of large WTG units at selected utility sites. It was shown that WTGs can be designed in the size ranges studied, using presently existing technology, to meet specified operating environments and life requirements. Investments and operating costs appear competitive with other alternate energy sources when in high-volume production.

Kaman participated in a DOE-sponsored design study of multi-unit wind energy conversion systems (WECS) installations in the U.S. offshore and Great Lakes areas as a subcontractor to Westinghouse Electric Corporation. In this program, Kaman developed a series of conceptual WECS designs, and provided parametric design tools down to the interface with the platform and the distribution and transmission systems, for technical and economic assessments. Both horizontal axis and Darrieus-type wind turbines of various sizes were studied.

Kaman plans to develop and commercialize a product line of intermediate size (20 to 100 kW) wind turbine systems for U.S. and international applications. In addition, Kaman plans to develop and provide turbine blades for large wind systems.

Kaman, under contract to Enertech Corporation, has participated in their development of a small high-reliability WTG rated at 2 kW in winds of 20 mph, performing reliability analyses and design consulting for the Rockwell/Rocky Flats program sponsored by DOE. Kaman is also providing consulting services to Virginia Polytechnic Institute on their USDA/DOE program for wind energy applications to crop storage. Kaman offers its consulting expertise to others on small and large wind energy programs.

912. **UNIVERSITY OF KANSAS CENTER FOR RESEARCH, INC.**
Applied Energy Research Program
2291 Irving Hill Road
Lawrence, Kansas 66045

Mr. Robert F. Riordan
(913) 864-4078

913. **LIPE-ROLLWAY CORPORATION**
806 Emerson Avenue
P.O. Box 1397
Syracuse, New York 13201

Researching energy conversion, with an emphasis on wind energy systems.

914. **LONG ISLAND LIGHTING CO. (LILCO)**
250 Old Country Road
Mineola, New York 11501

Fred P. Avril
(516) 228-2042

LILCO is conducting a solar demonstration program which consists of marketing and monitoring 600 domestic hot water systems in residences within our service area. We expect to complete installation of the 600 systems by the end of 1980.

Each installation includes meters which measure hot water and backup electric consumption. These data are collected and reduced monthly in order to calculate average solar contribution and cost savings compared to fossil fuels. Since oil is used in 78% of the homes on Long Island, these savings are substantial.

At this time, we expect to continue with some form of solar program beyond this demonstration.

915. LOS ALAMOS SCIENTIFIC LABORATORY
Solar Energy Group, Q-11
P.O. Box 1663, MS 571
Los Alamos, New Mexico 87545

LASL is engaged in the testing of air and liquid collectors and high-temperature collectors. We are also engaged in a small effort in coatings on glass tubes. Our main effort is in passive systems, where we study 14 reconfigurable test rooms for two main purposes. We test various materials and constructions and compare their performance with known materials and configurations; we also use these test rooms to provide real data against which to validate computer simulation models for the design of passive buildings. We have written Volume Two, *Passive Solar Design Analysis,* of a two-volume set, *Passive Solar Design Handbook*, for the Department of Energy available through the National Technical Information Service as DOE/CS-0127/2.

LASL is also working on the optimum mix of conservation measures and passive solar features in any climate of the U.S.; economics of solar; interzone transport of thermal energy in passive buildings; and off-peak use of electricity in passive homes.

916. MacDONALD ENGINEERING
1235 Ashland Avenue
Wilmette, Florida 60091

Investigating wind energy systems development.

917. UNIVERSITY OF MAINE
Department of Industrial Cooperation
109 Boardman Hall
Orono, Maine 04469

Richard C. Hill, Director

With funding from ERDA and DOE an experimental program was conducted at the University of Maine at Orono to develop a combustion system for residential furnaces that solves the traditional problems of wood burning: inefficiency, air pollution and fire hazard. The program led to the designs that are now covered by a patent application and are being manufactured by three companies. Stages were: design and test various configurations for stick-wood combustion schemes; integrate designs into a heat exchanger and storage system; refine design and construct sample units; construct production prototype and testing.

918. MECHANICAL TECHNOLOGY INCORPORATED
968 Albany-Shaker Road
Latham, New York 12110

Thomas Marusak/Herman Leibowitz
(518) 456-4142/785-2390

Solar Chiller: Under subcontract to the energy systems division of the Carrier Corporation, Mechanical Technology is performing the design and development of the heat engine subsystem for a high-temperature, solar-powered, air-cooled water chiller for a 25-ton commercial or multi-family dwelling applications. The program is being funded by the U.S. Department of Energy. The system consists of a high-temperature solar-driven Rankine cycle heat engine powering a vapor compression chill water loop. Solar simulated testing of the system is scheduled.

Solar Turbine Generator: Sandia Laboratories of Albuquerque, NM, has contracted Mechanical Technology to design, fabricate and test a high-speed, two-stage axial turbine-generator unit. The steam turbine-generator set will become part of the solar total energy system experiment at Shenandoah, GA. The total energy system will provide 400 kWe of electric power,

1380 lb/hr of process steam and plant air conditioning during daily operation of a knitting mill.

Dispersed Solar Electric Power System: Mechanical Technology has completed a design study of a 15-kW free-piston Stirling engine (FPSE) linear alternator under a contract from NASA-Lewis Research Center. The solar-powered linear alternator under study would be a part of the dispersed solar electric power system concept.

Also under current development at Mechanical Technology Incorporated is a free-piston Stirling engine generator concept, under private sponsorship, which could find application in a solar thermal system.

919. **UNIVERSITY OF MIAMI**
CLEAN ENERGY RESEARCH INSTITUTE
School of Architecture and Engineering
P.O. Box 248294
Coral Gables, Florida 33124

Dr. T. Nejat Veziroglu
(305) 284-4666

The Clean Energy Research Institute was established in 1974 by the Department of Mechanical Engineering of the School of Engineering and Environmental Design in order to serve as the focal point of energy-related activities at the University of Miami. Its goals are: to conduct research and to generate research proposals to investigate clean energy problems; to organize seminars, workshops and conferences using researchers within and without the university; to assemble, compile, publish and disseminate information related to every aspect of clean energy problems; and to cooperate with other organs of the university, other academic institutions, government and private organizations in connection with the above-mentioned activities.

920. **MISSISSIPPI STATE UNIVERSITY**
Drawer ME
Mississippi State, Mississippi 39762

Dr. Richard Forbes
(601) 325-4915

Research activities dealing with agricultural applications. Development of low-cost, site-built air heaters. Trombe and water wall research applications. Undergraduate and graduate courses offered in solar energy. Advanced degrees awarded with solar specialty.

921. **MOBIL CORPORATION**
Montgomery Ward Testing Laboratories
RD 5-5 Montgomery Ward Plaza
Chicago, Illinois 60671

Test solar energy equipment.

922. **MODUPAC INC.**
8309 Monroe, P.O. Box 12897
Houston, Texas 77017

R. C. Macaulay
(713) 941-8570

Research activity concerns the automatic thermal blending of liquids—large volumes, 5000 gpm up.

923. **MONTANA STATE UNIVERSITY**
Mechanical Engineering Department
220 Roberts Hall
Bozeman, Montana 59717

Dr. William Martindale
(406) 994-2203

We are currently involved in developing analytical models for direct gain passive solar systems and the design of low-cost solar collectors. We are also constructing a nationally certified solar collector testing laboratory.

924. **MOSFET MICROLABS INC.**
Penn Centre Plaza
Quakertown, Pennsylvania 18951

Research area: photovoltaics.

RESEARCH & DEVELOPMENT 335

925. MOTOTRON INC.
292 Montgomery Avenue
Bala Cynwyd, Pennsylvania 19004

Solar and wind energy research, both applied and product.

926. M&W ENTERPRISES
720 W. Jefferson
Ann Arbor, Michigan 48103
Mary E. Mandeville
(313) 994-1511

Horticulture programming for solar greenhouses. Experience in managing a northern solar garden and in developing horticultural curricula for handicapped populations, i.e., geriatric physically handicapped, psychiatric, substance abuse, mentally retarded, and correctional clients. Registered as a Horticultural Therapist Master with the National Council for Therapy and Rehabilitation through Horticulture. Provide extension horticulture program from a botanical garden. Biological control of insect pests for solar greenhouses.

927. NATIONAL AERONAUTICS AND SPACE ADMINISTRATION (NASA)
600 Maryland Avenue, SW
Washington, D.C. 20546
Public Affairs Office
(202) 755-2497

Research and development in support of U.S. Department of Energy in the following areas: absorbers, collectors, chillers, heating and cooling systems, DHW systems, coatings, controls and instrumentation, fuel cell storage, photovoltaics, wind generation systems.

928. NATIONAL CENTER FOR APPROPRIATE TECHNOLOGY
P.O. Box 3838
Butte, Montana 59701

Larry Palmiter
(406) 494-4572

One of the most active passive solar research teams in North America, we have an active solar test facility for both water heaters and air panels.

The National Center's technical staff conducts extensive research utilizing passive and active solar test facilities, a solar greenhouse and other resources at its Montana headquarters, and at other sites around the country. The technical program is focused on applied research—developing and testing model applications of appropriate technologies with potential for immediate impact on the problems of low-income consumers and their communities.

Our R&D staff consists of 35 persons of various technical and scientific disciplines, divided into four sections: passive solar; renewable resources, including passive and active solar systems, wind generation and small-scale hydro power; building technology—home weatherization, energy conservation, housing rehabilitation, and transportation planning; agriculture, biofuels and recycling—with concentrations in greenhouse and organic farming methods, community gardening, methane production and wood heat.

929. NATIONAL INSTITUTE OF CREATIVITY INC.
P.O. Box 44067
Tacoma, Washington 98409

Wind energy research.

930. NATIONAL PATENT DEVELOPMENT CORPORATION
Energy Systems Inc.
1455 Research Boulevard
Rockville, Maryland 20850

Solar collector research—absorber coatings, inhibitor fluids, photovoltaic cells.

931. THE NEW ALCHEMY INSTITUTE
237 Hatchville Road
East Falmouth, Massachusetts 02536

John Wolfe
(617) 563-2655

Our work includes agriculture (gardening, tree crops, sandy soil research, pest control, and windbreaks), aquaculture (dome fishery, solar algae ponds, cage culture, fish feeds and waste disposal), wind systems (sailwing water pumper, windmill design modeling, windmill refrigeration) and solar designs.

932. NEW LIFE FARM, INC.
Drury, Missouri 65638

(417) 261-2553

New Life Farm is an independent, tax-exempt corporation located in the south central Missouri Ozarks. The nonprofit research and education project is dedicated to improving the health of the natural environment through the use of appropriate technology and energy alternatives. Concepts such as conservation, self-reliance, bio-regionalism, and continuing education are first practiced at home, then taught to others. New Life strives to provide high quality, low-cost assistance to individuals, families, groups (especially low-income) and communities.

The center of the farm is a restored farmhouse that demonstrates several energy-saving features, including insulation and storm windows, wood-fired furnace and water heater, solar water heater, composting toilet, and gas appliances for using methane generated at the farm. There are plans to add an attached solar greenhouse for raising vegetables and for providing extra space heat during winter. Other resources on the farm include a large barn, a methane digester, garden, hydraulic ram-powered water supply.

New Life Farm's pioneer appropriate technology project was a 16,000-gallon methane digester constructed over a period of four years. The data and expertise gained in designing, building and operating this research digester attracted a federal grant to develop experimental small farm units on a cooperative basis with low-income farmers throughout the country. These second-generation digesters are helping to determine the feasibility of locally produced farm fuel. Information gained from the project will also guide future investigations into other promising bio-fuels, such as alcohol.

933. **NIELSEN ENGINEERING & RESEARCH, INC.**
510 Clyde Avenue
Mountain View, California 94022

Richard D. Schwind
(415) 968-9457

Performing mixed free and forced convection research as it pertains to large solar central receivers.

934. **UNIVERSITY OF NORTH DAKOTA**
Engineering Experiment Station
P.O. Box 8103
University Station
Grand Forks, North Dakota 58202

Prof. Don V. Mathsen
(701) 777-3132

The Engineering Experiment Station is active in the design, development and testing of alternate energy systems in the areas of solar, wind and biomass utilization. The station is currently involved with three solar-assisted residential sites, each with a different application of ice-making heat pump systems (the Annual Cycle Energy System). Development work is ongoing in the area of multiple-source heat pump systems using combinations of direct solar, air and earth energy inputs. A ground source heat pump test facility is operational. Monitoring programs are active in regard to ground water heat pump system performance.

Wind monitoring data in the Grand Forks region has been collected and small wind conversion devices have been tested. Alcohol processes using waste potato resources are being studied in operational facilities.

935. OAK RIDGE NATIONAL LABORATORY
P.O. Box X
Oak Ridge, Tennessee 37830

Stephen I. Kaplan
(615) 574-5819

Oak Ridge National Laboratory (ORNL) is a multidisciplinary energy laboratory operated by Union Carbide Corporation for the U.S. Department of Energy. ORNL conducts several types of solar energy projects including fundamental research, applied R&D, external project monitoring and evaluation, and systems design and analysis. Basic studies include investigations of electron transport in photovoltaic semiconductor materials and characterization of the basic biophysical processes occurring in biophotolysis of water. Applied studies planned or underway in photovoltaics include evaluation of laser annealing techniques for improved photovoltaic cell production, characterization of structural ceramic materials for containment of solar-driven chemical fuel production processes and the study of atmospheric contaminant effects on cells operating in concentrated daylight.

Applied research is also conducted in the design and economics of passive solar structure and in mechanisms for improved passive solar heating and cooling

ORNL also conducts and manages research on environmental effects of woody biomass and crop biomass harvesting for fuels and provides monitoring and evaluation for DOE on two college photovoltaic system demonstration projects. The lab assists the U.S. Department of Agriculture in monitoring and evaluating some 50 solar-related agricultural research and demonstration projects throughout the U.S., involving livestock shelters, greenhouse applications and crop drying. Assistance in national program planning for these activities is also provided. Another area involving external subcontract management and evaluation is the development of open-cycle concepts for ocean thermal energy systems.

The Laboratory is studying the economics of solar-powered district heating systems for urban communities. It also engages in community solar conversion planning, in cooperation with community management and civic groups, and prepares social impact evaluations related to these planning activities.

936. UNIVERSITY OF OKLAHOMA
Division of Economics
Norman, Oklahoma 73019

Dr. David A. Huettner

Activities include economic assessment of: active and passive solar systems for residential and commercial use; central station solar plants; and the rate design and other problems encountered when integrating solar units into utility systems or the systems of utility customers. This work has been done for the U.S. Department of Energy, the Solar Energy Research Institute, and various private corporations.

937. OKLAHOMA STATE UNIVERSITY
ARCHITECTURAL EXTENSION
118 Architecture Building
Stillwater, Oklahoma 74078

Jody Proppe
(405) 624-6266

The Office of Architectural Extension at Oklahoma State University has been a leader in the field of earth shelter and has a team of professionals who have been conducting impressive seminars throughout the nation for over three years. A national technical conference was held in April 1980 at Oklahoma City on "Earth-Sheltered Building Design Innovations." Co-sponsors, along with Architectural Extension, were the U.S. Department of Energy and *Earth Shelter Digest*. The conference drew an audience of 255 participants representing 37 states, France and Canada. Twenty-two papers were presented at the conference.

938. OKLAHOMA STATE UNIVERSITY
School of Mechanical Engineering
Stillwater, Oklahoma 74078

Jerald D. Parker
(405) 624-5900

Work has been primarily with solar heating of asphalt storage tanks and in solar-augmented ground source heat pumps.

RESEARCH & DEVELOPMENT

**939. UNIVERSITY OF OREGON
SOLAR ENERGY CENTER**
Department of Architecture & Allied Arts
Eugene, Oregon 97403

John Reynolds
(503) 686-3696

We act as a source of information about solar and alternative energy production. Research carried out associated with the University of Oregon. We have access to local experts and try to answer specific questions or direct questions to qualified personnel. Some general solar booklists available; also some workshops held, lectures scheduled.

940. ORLANDO LABORATORIES INC.
Box 8008
Orlando, Florida 32856

Investigating solar-powered hot water systems.

941. PARKWAY WINDOWS
127-17 20th Avenue
College Point, New York 11356

Research interest: thermal glass closures, with an emphasis on heat transfer and energy saving.

942. PASSIVE SOLAR INSTITUTE
P.O. Box 722
Davis, California 95616

David Bainbridge
(415) 526-1549

Research areas include: passive design—residential, commercial and industrial, new construction and retrofit. Also, trombe wall and water wall design, backup systems, super-insulation details, glazing. Also, manufacture instruments.

342 SOLAR CENSUS

943. THE PENNSYLVANIA STATE UNIVERSITY
Department of Engineering Science and Mechanics
227 Hammond Building
University Park, Pennsylvania 16802

Stephen J. Finasl
(814) 865-4931

Research and development in basic solar cell device physics and materials.

944. POLYTECHNIC INSTITUTE OF NEW YORK
SOLAR ENERGY APPLICATIONS CENTER
333 Jay Street
Brooklyn, New York 11201

Dr. Richard S. Thorsen
(212) 643-5044

The Solar Energy Applications Center at Polytechnic Institute of New York (SEAC/POLYTECHNIC) was established in 1977 to provide leadership in the solar commercialization process. Funding is provided by the New York State Energy Research and Development Authority and supported by the U.S. Department of Energy via the Northeast Solar Energy Center.

The two major institutional functions of SEAC, planning and analysis and certification and testing, are intended to accelerate solar commercialization in New York and the Northeast by identifying optimum applications of solar energy in the context of local and regional resource-demand mixes, and by building user confidence in solar hardware through assistance in product improvement and identification of products in compliance with applicable standards.

In addition to its institutional roles, SEAC has undertaken a number of projects of a research, development or demonstration nature, including: the Westchester Solar Energy Demonstration and Evaluation Project; the HUD Solar Hot Water Initiative, to provide technical evaluations and central coordination for the installation of 10,000 solar hot water systems; development of an improved solar powered cooling system;

testing and evaluation of performance potential of high-performance solar collectors; and data acquisition and analysis for a solar-assisted Annual Cycle Energy System.

945. RADIANT EQUIPMENT COMPANY
1798 Panda Way
Hayward, California 94541

Don J. Amo
(415) 537-6581

Radiant Equipment Company is a pioneer in the energy conservation field (founded 1963) with the emphasis on hard to heat industrial buildings.

The RDT Division (Research, Development, Testing) of Radiant is actively researching and will soon begin adapting sensible, mainly passive with some active solar equipment to industrial and commercial buildings.

946. RADIATION RESEARCH ASSOCIATES INC.
3550 Hulen Street
Ft. Worth, Texas 76107

Solar power systems analysis.

947. F. C. RADICE
RESEARCH & DEVELOPMENT
13515 N.E. 70th Street
Redmond, Washington 98052

Investigating alternate sources of energy, with an emphasis on solar and wind power.

948. RCA CORPORATION
RCA Laboratories
Princeton, New Jersey 08540

Brown Williams
(609) 734-2535

Research centers on photovoltaic cells and panels.

949. REA ASSOCIATES INC.
Faulkner Street
North Billerica, Massachusetts 01821

Research and development of thermal conductors for solar energy systems.

950. RECTECH INC.
421 East Beaver Avenue
P.O. Box 177
State College, Pennsylvania 16801

Researching solar energy and biomass conversion.

951. RESEARCH & DESIGN INSTITUTE
P.O. Box 307
Providence, Rhode Island 02901

Investigating energy conservation and conversion, especially solar and wind.

952. RESEARCH INSTITUTE FOR ADVANCED TECHNOLOGY
U.S. Highway 190 West
Killeen, Texas 76541

John J. Kincel
(817) 526-1171

Research Institute for Advanced Technology (RIAT), formerly American Technological University, was founded in 1973 as a private nonprofit educational institution providing upper division and graduate technical programs leading to bachelor's and master's degrees. As a matter of policy the University is committed to research services accommodating worldwide instructional systems. It provides worldwide technological support services for over 100 operations in Europe, Latin America, the Middle East and the Far East.

RIAT has been actively involved in the national energy program since 1974 when it began the Fort Hood Solar Total Energy Military Large-Scale Experiment (LSE-1). Part of the experiment involves the Solar Engineering Test Module (SETM) which provides insolation data. The University's involvement and commitment in the solar energy field is exemplified by the decision of AS/ISES to relocate its national headquarters to RIAT. In keeping with its aggressive stance in the energy field, RIAT has developed a curriculum in Energy Management Sciences within the existing Management Sciences Program. Courses in solar energy have been included in this program.

RIAT has access to support functions that include systems analyses, consulting services, staff development, program development, management and evaluation. A public service television station, a data processing operation, and library facilities are also available.

Areas of solar research include: lightweight plastic absorbers, selective black absorber coatings; air and liquid collectors, trombe walls, recuperators, agricultural applications (greenhouses, stills), photovoltaics, phase-change storage, and structure design.

953. **ROCHESTER INSTITUTE OF TECHNOLOGY**
INSTITUTE FOR APPLIED ENERGY STUDIES
1 Lomb Memorial Drive
Rochester, New York 14623

Robert Desmond
(716) 475-2151/475-2162

We have a solar-assisted heat pump system in a single-family residence on campus. Also, faculty are involved in methane production from animal wastes and in various wind generation systems.

954. **SAF ENERGY CONSULTANTS, INC.**
P.O. Box 3052
Peoria, Illinois 61614

Dr. Y. B. Safdari
(309) 691-7505

Designed, installed and tested five solar homes in Illinois. Presently working on the design of swimming pool and DHW system for a high school under a DOE grant. Dr. Safdari has a patented air collector (SAF™).

955. SCHWINGHAMERS
3310 North 27th Avenue
Phoenix, Arizona 85017

Research interests: solar-energized structures; heating systems.

956. SCIENTIFIC BUILDING AND ENERGY CONSULTANTS, INC.
Route 1, Box 314X
Jefferson, Wisconsin 53549

Peter J. Lorenz, President/General Scientist
(414) 674-4748

We are a research and development corporation involved in practical science applications to energy problems. We do some manufacturing and licensing of our ideas. We work nationwide and have been established since 1977. We also work in other related fields and provide emergency service for severe building problems and natural disasters.

957. SHELTON ENERGY RESEARCH
P.O. Box 5235
Santa Fe, New Mexico 87502

Jay Shelton
(505) 983-9457

Shelton Energy Research is an independent organization specializing in providing high quality technical information on residential solid-fuel heating systems.

Our core activities are research, product development and testing. We work mostly in areas where there are not yet any standards, areas requiring formulating new test procedures. Examples include testing the effectiveness of chemical chimney cleaners, research on catalytic combustion, comparative testing of factory-built chimneys for their creosote-accumulating tendencies, developing and proving design modifications in heating equipment, testing the effects of wood fuel moisture content and species on creosote accumulation. Much of the laboratory's work is related to generating the information needed before a standard can be written.

RESEARCH & DEVELOPMENT 347

958. SHOCK HYDRODYNAMICS DIVISION
WHITTAKER CORPORATION
4716 Vineland Avenue
North Hollywood, California 91602

Dr. Emil Lawton/Larry N. Bohanan
(213) 985-6940

Shock Hydrodynamics is a R&D company with a California TIPSE and SEIA Solar Collector Testing Facility, and a vast background in solar simulation and photovoltaics.

959. SOLAR AGE DESIGNERS
39108 Charbeneau Street
Mt. Clemens, Michigan 48043

(313) 468-3225

Firm does field audits with a team of engineering students. Literature and applied research in solar construction, photovoltaics, alcohol distillation (agricultural), greenhouses, winter gardening and wind energy. Also see, Detroit Environmental Control Engineers (506).

960. SOLAR DEVELOPMENT CO.
Solar Energy Research & Development Project
P.O. Box 208
Herald, California 95638

5860 Callister Avenue
Sacramento, California 95819

Alvin L. Gregory
(916) 455-3100

Alvin L. Gregory has been active continuously in the solar energy field for over 25 years. He has had practical experience in almost all types of solar energy means and usage (both active and passive). He is a Solar Energy Consultant (Solar Development Co.) and Director of Solar Energy Research & Development Project (Herald, CA). While he has practical experience in almost all solar technologies, his main activities are now in the development of water and power through solar means. Through his R&D program, a new type of thermal engine series was patented.

961. SOLAR ENERGY RESEARCH INSTITUTE (SERI)
1617 Cole Boulevard
Golden, Colorado 80401

Jerome Williams
(303) 231-1236

SERI is operated for the U.S. Department of Energy by the Midwest Research Institute. It is the major center for research and development of solar energy technologies in the U.S. In addition to pursuing technological breakthroughs and improvements, SERI's staff is working to help reduce legal, economic, and institutional barriers to widespread solar use. In this capacity, SERI works closely with the four Regional Solar Energy Centers commercializing solar technologies. SERI also coordinates information dissemination with the Regional Centers, the National Solar Heating and Cooling Information Center, state and local governments, universities, corporations, and other groups involved in solar energy development.

SERI has been structured around 15 program areas: photovoltaics; biomass energy systems; wind energy systems; active solar heating and cooling; passive technology; industrial process heat; solar thermal technology; ocean systems; solar energy storage; advanced solar energy research; information systems; academic and university research programs; commercialization activities; international programs; planning, analysis and social science research.

SERI's primary mission is the research and development of solar energy technologies. Much of the research is conducted in SERI's own laboratories. A major activity at SERI, however, involves the subcontracting of research, development and demonstration of solar-related activities. Approximately two-thirds of SERI's total budget is passed through to subcontractors throughout the U.S.

SERI has already made a number of significant contributions to the advancement of solar technology:

SERI's research scientists have conducted the first known measurements of the chemistry of grain boundaries in polycrystalline photovoltaic materials (including cast silicon, thin-film gallium arsenide, and indium phosphide). They have also introduced a method for determinations of local structure in

RESEARCH & DEVELOPMENT

hydrogenated amorphous silicon. This work will help lower the cost and increase the reliability of solar cells.

SERI has developed a Solar Education Database covering 3200 institutions in 50 states.

SERI has developed a family of new instruments for measuring solar radiation. These include a high spectral resolution spectro-radiometer, an automatic solar tracker for direct beam measurements, and an economical multiple surface radiometer for simultaneously measuring insolation on various tilts and azimuths.

SERI has prepared a detailed study of State Solar Energy Incentives, providing the first direct feedback to state governments on the successes of various approaches to implementing state incentives. The results have aided in the design and amendment of legislation in several states.

SERI's U.S./Saudi Arabian SOLERAS program—a $100 million jointly funded research and development solar energy program—has awarded contracts for development of a photovoltaic-powered Saudi Arabian village and for a solar cooling engineering field test. In addition, SERI has cooperative programs with South Korea, Mali, Spain, France, Italy, Israel, Australia, Japan and Brazil.

962. **SOLAR ENTERPRISES HAWAII INC.**
P.O. Box 27031
Honolulu, Hawaii 96827

Researching alternate energy generation and management with an emphasis on solar.

963. **SOLAR ENVIRONMENTAL ENGINEERING COMPANY, INC.**
2524 E. Vine Drive
Fort Collins, Colorado 80524

William G. Huston
(303) 221-5166

Research and development activity oriented toward engineering, design and installation of solar hot water systems, and

350 SOLAR CENSUS

active and passive solar heating and cooling systems. Market hand-held calculator programs for the TI-59, HP-67/97/41C and for microprocessors. Technical manager for DOE for the SOLCOST Mini Center. Developer and manager for DOE for the Solar Index.

964. SOLAR PATENTS
P.O. Box 21
Turtle Creek, Pennsylvania 15145

We are conducting research and development in the solar energy field. Our first published report is U.S. Patent No. 4,184,479, a combination revolving greenhouse and parabolic trough concentrator. It breaks the cost barrier for concentrators by sharing costs of land, structure and maintenance with the greenhouse.

965. SOLAR SUNSTILL
15 Blueberry Ridge Road
Setauket, New York 11733

Researching particulate, condensation and light control by coatings on glass and plastic surfaces.

966. SOLAR SURVIVAL
Cherry Hill Road
Box 275
Harrisville, New Hampshire 03450

Gretchen Poisson
(603) 827-3797

We are a small decentralist group working on a cottage industry scale. All our activities are focused on total solar living concepts. Our solar priorities are: food production, food preservation, food processing, energy conservation, solar heating, communication, and appropriate technology.

967. SOLAR SUSTENANCE TEAM
P.O. Box 733
El Rito, New Mexico 87530

Leslie Davis
(505) 581-4454

The Solar Sustenance Team is a nonprofit educational organization specializing in research and training in appropriate technology. Much of our work has been done with heat and food-producing solar greenhouses. The team, a group of eight professionals, has conducted construction workshops nationwide and has trained over 30 teams around the country in the design, construction and operation of solar greenhouses and in the organization of construction workshops. (Blueprints for portable greenhouses are available.) Bill Yanda, Director of the Solar Sustenance Team, has lectured widely, published and consulted and has done work for UNESCO in Europe and presentations to the Architectural Society of London.

968. SOLAR THERMAL TEST FACILITIES USERS ASSOCIATION
First National Bank Building East
Suite 1204
Albuquerque, New Mexico 87108

Marylee Adams, Project Administrator
(505) 268-3994

We are a DOE-funded organization with a contract through the Solar Energy Research Institute (SERI) and employed by the University of Houston. Our major role is to seek out, review and recommend for funding experiments to be done on solar thermal test facilities such as the Sandia 5-MW Central Receiver Test Facility, the Georgia Tech 400-kW Advanced Component Test Facility, the Jet Propulsion Lab Parabolic Dish Test Site, the U.S. Army White Sands Solar Facility, and several different solar furnaces in Odeillo, France. We also are responsible for dissemination of information on these facilities and sponsor workshops for people interested in using these high-temperature facilities for research experiments.

969. SOL ENERGY CORPORATION
761 Rohde
Hillside, Illinois 60162

Areas of research: photovoltaics, agricultural applications of solar energy, methane production and utilization.

970. SRI INTERNATIONAL
333 Ravenswood Avenue
Menlo Park, California 94025

Dr. Jeffrey G. Witwer
(415) 326-6200 Ext. 4647

Economic and technical studies, new product development, market studies, basic and applied research, experimental testing, environmental and policy assessments of all forms of solar energy including thermal, photovoltaics, wind, biomass, and ocean thermal conversion.

971. STANFORD RESOURCES, INC.
1095 Branham Lane, Suite 201 (95136)
P.O. Box 20324
San Jose, California 95160

Dr. Joseph A. Castellano
(408) 448-4440/268-5452

We perform research and development in the area of photovoltaic materials and devices. Our work is concerned primarily with new types of organic and inorganic solar cells.

Another aspect of our activities is the development of solar-powered electronic displays for use in outdoor application (i.e., scoreboards, traffic signals, signs).

We also perform research into new materials for hydropiezo-electric energy conversion from sea waves.

972. STANFORD UNIVERSITY
Department of Materials Science and Engineering
Stanford, California 94305

Professor Richard H. Bube
(415) 497-2534

We are engaged in a basic materials research investigation related to the properties of heterojunction thin-film photovoltaic cells. The absorber materials are p-type cadmium telluride, indium phosphide and zinc phosphide. The large bandgap window materials are n-type cadmium sulfide, zinc oxide, and indium tin oxide. Film deposition methods are vacuum evaporation, rf sputtering, magnetron sputtering, spray pyrolysis, and chemical vapor deposition. Purpose of the research is to investigate the materials interactions at heterojunction interfaces and their role in determining the properties of photovoltaic cells. Optical, electrical and electron microscopic techniques are being applied.

973. STERN RESEARCH CORPORATION
Route 3, Box 274, Orcutt Road
San Luis Obispo, California 93401

Robert L. Stern
(805) 543-4444

R&D—water pumping photovoltaic systems.

974. SUN CYCLE CO./SOLAR R&D
P.O. Box 2111
El Centro, California 92243

Richard Dessert
(213) 399-8751

Solar domestic hot water and space heating. R&D on passive liquid collectors and solar-assisted vehicles.

975. SUNPOWER INC.
6 Byard Street
Athens, Ohio 45701

Craig Kinzelman
(614) 594-2221

R&D in the area of heat engines with special emphasis on Stirling engines and the design of small decentralized power sources.

Sale of the following research tools: 70-watt free piston Stirling engine/alternator with complete instrumentation; Duplex Stirling/Stirling heat pump (100 watts heat-lifted).

Also, 70-watt, propane-fired free piston Stirling engine/ alternator battery charging system; and free cylinder water pump capable of operating on concentrated solar energy, biofuels or conventional fuels.

976. **SUNPOWER RESEARCH & DEVELOPMENT CORP.**
200 Madison Avenue
New York, New York 10016

Mr. Masieri
(212) 889-4889

Research and development—industrial applications of solar energy. Also, agricultural applications; bioconversion; distillation.

977. **SUNSEARCH, INC.**
669 Boston Post Road, P.O. Box 275
Guilford, Connecticut 06437

Peter D. Clark, Vice-President
(203) 453-6591

Sunsearch is a solar energy research firm. Most of our work is related to the direct conversion of sunlight to heat. Most of the technology developed by Sunsearch is licensed to other companies which manufacture and market products developed by Sunsearch.

Products currently under development include: air-cooled flat plate collector; evacuated collectors; suntracking carriage for flat plates, single-axis concentrators and photovoltaics; improved domestic hot water systems; thermal storage by phase-change and heat of solution; improvements in the efficiency and reliability of passive/hybrid space heating systems.

In addition to developing the above products, Sunsearch manufactures support racks for solar collectors and uses its computer simulation programs to assist architects and engineers in evaluation of design options for both new and existing structures.

978. **SUNTEK RESEARCH ASSOCIATES**
506 Tamal Vista Plaza
Corte Madera, California 94925

Research interest: passive solar systems.

979. UNIVERSITY OF TENNESSEE SOLAR PROGRAM
Energy, Environment and Resources Center
329 South Stadium Hall
Knoxville, Tennessee 37916

Dr. Robert L. Reid
(615) 974-4251

Experimental projects currently in progress: solar-augmented heat pump system; modular passive houses; ground-coupled heat pump; stratified thermal storage system; solar grain drying; utility management of wind energy systems; passive cooling of buildings; dessicant cooling.

980. TETRA CORPORATION
P.O. Box 4369
Albuquerque, New Mexico 87196

Stephen Gendron
(505) 255-4313

We are active in the research and development of photovoltaic water pumping. We are also active in passive solar building research in the area of retrofit.

981. THERMO ELECTRON CORPORATION
101 First Avenue
Waltham, Massachusetts 02154

Ronald S. Schallack
(617) 890-8700

Research areas include: air, concentrating and liquid collectors, controls and instrumentation, photovoltaics, and Fresnel lenses.

982. UCE INC.
24 Fitch Street
Norwalk, Connecticut 06855

R. Borstelmann
(203) 838-7500

Advanced photovoltaic components in heterojunction and organic/inorganic silicon solar cells for low-cost, high-value utilization and conversion of energy.

983. **UNDERGROUND SPACE CENTER**
221 Church Street
University of Minnesota
Minneapolis, Minnesota 55455

Glen Strand
(612) 376-1200

Researching the area of earth-sheltered single-family residenrial, commercial and industrial structures.

984. **USDA—SEA—AR**
RURAL HOUSING RESEARCH UNIT
Box 792
Clemson, South Carolina 29631

Jerry Newman
(803) 654-3646

Solar research interests include: plywood and aluminum absorbers; electric backup systems; coatings; high-temperature flat plate air collectors; thermostats; single-family residential design (solar attic, greenhouse, solar/earth); panelized DHW systems; tempered glass glazings; air to water heat exchangers; and insulation.

985. **UNIVERSITY OF UTAH**
Mechanical and Industrial Engineering Department
Salt Lake City, Utah 84112

R. F. Boehm
(801) 581-6441

We have several research projects related to active and passive solar energy topics, as well as power generation from salt-stratified ponds and focusing collectors. We also have two formal courses in solar energy.

986. VIRGINIA POLYTECHNIC INSTITUTE AND STATE UNIVERSITY
Agricultural Engineering Department
311 Seitz Hall
Blacksburg, Virginia 24061

Our solar research, all for agricultural applications, includes the following: (1) solar-assisted heat pump, which uses flat plate, water-cooled solar collectors and thermal storage in inexpensive, below-grade, covered and insulated ponds, for heating a pig nursery building; (2) an air-type solar collector which uses a metal lathing absorber over plywood painted flat black and fiberglass covering, for drying peanuts, soybeans and grain; (3) several other applications with the solar collector integrated into the roof of an adjacent farm shed for drying grain; (4) a solar heating system for heating hot water for use in the seafood processing industry; and (5) an application of wind power, using a 10-kW wind generator with battery and thermal storage (in ice), for providing the power for a refrigerated apple storage building.

987. UNIVERSITY OF WASHINGTON
Aerospace and Energetics Research Program
FL–10
Seattle, Washington 98195

Professor Abraham Hertzberg
(206) 543-6321

Solar-powered satellite, solar-powered aircraft, high-efficiency solar engines, advanced forms of chemical processing, and innovative concepts related to solar central systems.

988. WESTERN ILLINOIS UNIVERSITY
Department of Agriculture
Macomb, Illinois 61455

Project AlEnAg
(309) 298-1231

Project AlEnAg (Alternative Energy in Agriculture) is designed to actively demonstrate renewable alternative energy sources utilized in agriculture. The demonstration site is located at Western Illinois University in Macomb, Illinois.

The objective of AlEnAg is to develop an integrated "energy farm" for the demonstration of all alternative energy sources on the farm. Specific activities will be solar livestock confinement heating, solar grain drying, alcohol production, solar pond, methane generation, greenhouse heat, biomass burning, photovoltaic generation of electricity and various applications of wind generated energy as well as a solar (passive) heated house and "underground" visitor center. The integrated energy farm will use existing technology available from a variety of sources and suppliers. It will also provide a site for future agricultural energy research and educational opportunities and activities at WIU.

989. WINDEPENDENCE ELECTRIC POWER CO.
9080 Beeman Road
Waterloo, Michigan 48118

Craig Toepfer
(313) 475-9669

Founded in 1975, Windependence Electric Power Company conducts wind electric product and concept testing and development. It offers the following wind electric products and services:

— Used wind electric generators produced during the 1930s and 40s for use on farms prior to rural electrification.

— Wind electric power plants based on the Jacobs Wind Electric Plant manufactured in Minneapolis between 1931 and 1957. Completely remanufactured and modernized, each power plant is custom-designed for the intended application and includes complete installation services.

— Dealer for the new Jacobs Wind Energy System.

990. UNIVERSITY OF WISCONSIN—MADISON SOLAR ENERGY LABORATORY
1500 Johnson Drive
Madison, Wisconsin 53706

J. A. Duffie
(608) 263-1586

Research on simulation and design of solar thermal processes.

991. DEVON D INC.
P.O. Box 227
Fillmore, Utah 94631

Gordon D. Griffin
(518) 392-3923

Development of lower cost collecting units and solid-state storage of collected heat for homes and offices as retrofit to existing buildings.

More recent concentration has been in very low-cost (portable) heat collecting units especially suitable for supplying hot water to swimming pools.

* * *

Also see the following entries—

1. A.A.I. CORPORATION
7. ADVANCE ENERGY TECHNOLOGIES INC.
18. AMERICAN HELIOTHERMAL CORPORATION
22. AMERICAN SUN CORPORATION
27. APPROTECH SOLAR PRODUCTS
30. ARCO SOLAR INC.
31. ARKLA INDUSTRIES INC.
52. CHICAGO SOLAR CORPORATION
64. COPPERSMITH'S
67. CREIGHTON SOLAR CONCEPTS, INCORPORATED
79. DEVICES & SERVICES CO.
95. ELECTRA SOL LABS INC.
96. ELECTROLAB INC.
103. THE ENERGY FACTORY
106. ENERGY SYSTEMS, INC.
119. FAFCO, INC.
123. FISCHER SUN COOKER

127. FREE ENERGY SYSTEMS INC.
161. HYPERION, INC.
164. ILSE ENGINEERING, INC.
170. INTERTECHNOLOGY/SOLAR CORPORATION
186. WM. LAMB & CO.
190. LENNOX INDUSTRIES INC.
195. LODI SOLAR CO. INC.
211. MONOSOLAR, INC.
215. MOTOROLA SEMICONDUCTOR GROUP
218. NATIONAL SEMICONDUCTORS LTD.
226. NORTH WIND POWER CO., INC.
227. NOVAN ENERGY, INC.
243. PROGRESS INDUSTRIES INC.
249. RAMADA ENERGY SYSTEMS, INC.
258. REVERE SOLAR AND ARCHITECTURAL PRODUCTS
260. RICHDEL, INC.
274. SERA SOLAR CORPORATION
288. SOLAR COMFORT SYSTEMS MANUFACTURING CO.
296. SOLAR DEVELOPMENT INC.
297. SOLAR DYNAMICS OF ARIZONA
304. SOLAR ENGINEERING AND MANUFACTURING
305. SOLAR EQUIPMENT CORPORATION
306. SOLAREX CORPORATION
319. SOLAR KING INTERNATIONAL, INC.
320. SOLAR MAGNETIC LABS
326. SOLAR POWER CORPORATION
327. SOLAR PRODUCTS, INC.
333. SOLAR RESOURCES, INC.
334. SOLAR SEARCH CORPORATION
336. SOLAR SPECTRUM INC.
349. SOLAR WORLD, INC.
352. SOLECTRO-THERMO, INC.
363. SOUTHWEST SOLAR CORPORATION
364. SPECTROLAB

369. SUNDUIT, INC.
375. SUNMASTER CORPORATION
385. SUN STONE COMPANY, INC.
389. SUNTRON
411. UNITED ENERGY CORPORATION
430. ZIA ASSOCIATES, INC.
446. ANCO ENGINEERS INC.
464. MARK BECK ASSOCIATES
470. BERNHEIM, KAHN & LOZANO
478. BURT HILL KOSAR RITTELMANN ASSOCIATES
485. CENTRAL STATES ENERGY RESEARCH CORP.
495. CONKLIN & ROSSANT
499. CROFT & COMPANY
502. DAS/SOLAR SYSTEMS
503. DAVIS ENGINEERING, INC.
509. DIMETRODON
516. DUBIN-BLOOME ASSOCIATES, P.C.
520. ECOTOPE GROUP
523. EKOSE'A, INC.
531. ENERGY APPLICATIONS
536. ENERGY ENGINEERING GROUP, INC.
539. ENERGY PLANNING & INVESTMENT CORPORATION
550. ESR — ENVIRONMENTAL SYSTEMS RESEARCH
573. GO SOLAR INC.
577. CARLETON GRANBERY, ARCHITECT, FAIA
583. GLENN F. GROTH ARCHITECT AIA
587. HAMMER, SILER, GEORGE ASSOCIATES
601. IBE ENERGY
602. i e associates, inc.
603. UNIVERSITY OF ILLINOIS
609. INTEGRAL DESIGN
612. IONIC SOLAR INC.
628. KESSEL INSOLAR DESIGNS
647. LIVING SYSTEMS

653. PAUL H. LUTTON
658. MASSDESIGN ARCHITECTS & PLANNERS, INC.
665. MILTON INTERNATIONAL, INC.
673. MORELAND ASSOCIATES
674. MORMEC ENGINEERING, INC.
688. PACIFIC SUN INCORPORATED
694. EDWARD PEDERSEN, ARCHITECT/PLANNER
702. PRADO
705. REISZ ENGINEERING COMPANY
707. RESTORATION PRESERVATION ARCHITECTURE, INC.
721. RICHARD A. SCHRAMM, ARCHITECTURE, INC.
733. SLACK ASSOCIATES INC.
736. SOLARAIRE
737. SOLAR BUILDING CORPORATION
746. SOLAR DESIGNS
747. SOLAR ENERGY ASSOCIATES, LTD.
768. PAOLO SOLERI ASSOCIATES, INC.
COSANTI FOUNDATION
781. R. D. STRAYER, INC.
805. TECHNICAL VENTURES
812. JOSEPH B. THOMAS IV, ARCHITECT
815. TOTAL ENVIRONMENTAL ACTION, INC.
826. DONALD WATSON
832. RAY A. WILEY, ARCHITECT/ENGINEER
837. WORMSER SCIENTIFIC CORPORATION
842. BRATTLEBORO DESIGN GROUP
996. BIG FIVE COMMUNITY SERVICES, INC.
1002. DOMESTIC TECHNOLOGY INSTITUTE
1005. FARALLONES INSTITUTE RURAL CENTER
1012. NEW MEXICO SOLAR ENERGY ASSOCIATION
1013. NEW YORK INSTITUTE OF TECHNOLOGY
CENTER FOR ENERGY POLICY AND RESEARCH
1016. OZARK INSTITUTE
1019. SANTA CRUZ ALTERNATIVE ENERGY COOP
1030. UPLAND HILLS ECOLOGICAL AWARENESS CENTER

EDUCATION and INFORMATION

992. THE ALTERNATE ENERGY INSTITUTE
P.O. Box 3100
Estes Park, Colorado 80517

Art Anderson
(303) 586-5636

Consulting and information dissemination. The Alternate Energy Institute publishes a solar newspaper called *Solar Utilization News.* The Institute also holds solar seminars, solar courses and solar film festivals.

993. ALTERNATIVE PUBLISHERS
P.O. Box 357
Lakeside, California 92040

Jack Hedger
(714) 561-4531

Expect to issue first book on alternate energy entitled *Build Your Own Electric Plant* in May 1980. Gives complete directions for the home craftsperson to build his/her own solar electric generating plant, up to 10 kW_e. Illustrated with items that have been successfully built and tested. *A Solar Power-Tower Y-O-U Can Build* is scheduled for summer 1980.

994. ALTERNATIVE SOURCES OF ENERGY, INC.
107 South Central Avenue
Milaca, Minnesota 56353

Donald Marier
(612) 983-6892

Alternative Sources of Energy, Inc. is a nonprofit, tax-exempt scientific and educational organization for those concerned with the development and use of renewable resources. Our major thrust is to publicize and share practical applications of alternative technologies as a means of attaining some degree of energy independence. We serve as a focus for a variety of resource people.

Alternative Sources of Energy, Inc. offers the following services. 1. *Alternative Sources of Energy Magazine,* a bi-monthly publication; 2. a technical paper series which offers papers and pamphlets containing practical information, technical reports, state-of-the-art overviews, and bibliographies in all areas of renewable energy technologies; 3. programs such as lectures, courses, conferences and exhibits; and 4. the Energy Information and Referral Service (EIRS). ASE, in conjunction with a network of professional contacts, offers information search and referral as well as consulting services in all areas of renewable energy and conservation. Information for EIRS is provided from ASE's extensive energy information library and files. Bibliographic searches are available on a fee basis.

995. AMERICAN GAS ASSOCIATION
1515 Wilson Boulevard
Arlington, Virginia 22209

Mark Menzer
(703) 841-8562

Trade association involved with backup systems for domestic hot water systems.

996. BIG FIVE COMMUNITY SERVICES, INC.
215 North 16th
P.O. Box 371
Durant, Oklahoma 74701

Hub Wood, Jr.
(405) 924-5331

Big Five has conducted a workshop open to the public wherein passive solar collectors, hot water heaters, food dehydrators and a freestanding greenhouse were constructed. Workshops in solar principles are scheduled for the near future. The workshops will also result in the construction of prototype designs of solar units which can be attached to low-income homes. These units will be constructed using readily available materials and simple handtools.

997. CENTRAL UPPER PENINSULA PLANNING AND DEVELOPMENT REGIONAL COMMISSION (CUPPAD)
2415 14th Avenue South
Escanaba, Michigan 49829

Charles Kehler
(906) 786-9234

Planning/technical assistance agency involved in the performance of energy audits.

998. CLINTON COUNTY ENERGY COMMITTEE
306 Elm Street
St. Johns, Michigan 48879

Thomas P. Warstler, County Planner
(517) 224-8857 ext. 272

The Clinton County Energy Committee helps determine the appropriate activities for local government conservation programs.

999. COMMUNITY ACTION OF LARAMIE COUNTY CHEYENNE COMMUNITY SOLAR GREENHOUSE
1603 Central Avenue, No. 400
Cheyenne, Wyoming 82001

Shane Smith
(307) 635-9340

The Cheyenne Solar Greenhouse is a prototype, the first community greenhouse in the U.S. We believe the community greenhouse concept can provide simple solutions to complex economic, social and energy programs that affect every community.

The community is involved in different ways: seniors volunteer their time, many youths participate through various work experience programs, and juveniles in trouble work as an alternative to court fines. We are a training site for the handicapped utilizing horticultural therapy and increasing their employability. The greenhouse is also a solar demonstration site. There is excellent social interaction between all who are involved, which is uncommon in our stratified society.

A slide show and audio tape presentation on the history, design, maintenance and management of the greenhouse project is available for sale or rent.

1000. CONVECTION LOOPS
Box AF
Stanford, California 94305

Jim Berk, Editor
(415) 321-9953

The magazine was initially published as a free newsletter in late 1979. It "went public" with the January 1980 issue. The only public forum for the revolution in human housing.

1001. CORNERSTONES ENERGY GROUP, INC.
54 Cumberland Street
Brunswick, Maine 04011

Nancy March
(207) 729-0540

EDUCATION & INFORMATION 367

Cornerstones, an owner-builder school, emphasizes passive solar construction. It offers three-week residential courses in Energy-Efficient Housebuilding, Energy Efficient House Renovation, Passive Solar Greenhouses, and Housebuilding for Women. A Design Seminar, a Design Workshop, and a one-week Energy Audit training course are also available.

1002. DOMESTIC TECHNOLOGY INSTITUTE
Box 2043
Evergreen, Colorado 80439

Malcolm Lillywhite/Lynda Lillywhite
(303) 674-1597

The Domestic Technology Institute is a private, nonprofit research and training organization that specializes in domestic lifestyle alternatives and domestic solar energy technology. The Institute has developed a broad research and training capability utilizing small-scale, decentralized technology to encourage self-sufficiency on an individual, family and community basis.

The Institute is currently engaged in a variety of research, training and community development programs. Through seminars, conferences and workshops, DTI has trained approximately 6000 people of varied backgrounds across the nation and in six foreign countries. The Institute's training programs emphasize "hands-on" experiential learning. Programs range from designing, building, operating and maintaining simple passive solar systems to designing integrated solar-assisted food and ethanol fuel processing systems.

1003. EMMET COUNTY COOPERATIVE EXTENSION ENERGY PROJECT
441 Bay Street
Petoskey, Michgian 49770

Martha Drake
(616) 347-4645

Countywide education on solar. Promotion through media, classes, etc.

1004. ENERGY INFO
P.O. Box 98
Dana Point, California 92629

Robert Morey
(714) 496-2574

Publisher of two monthly newsletters: *Energy Info* summarizes R&D development in all areas of energy; *Advanced Battery Technology* summarizes both technical and business news of international battery industry. *Batteries Today,* a quarterly magazine, contains features on battery design, technology and applications.

1005. FARALLONES INSTITUTE RURAL CENTER
15290 Coleman Valley Road
Occidental, California 95465

Peter R. Zweig
(707) 874-3060

The Farallones Institute Rural Center is a small educational organization committed to the development of practical, conserving and biologically sound alternatives to the deteriorating resource and environmental situation. Our work covers the fields of: passive solar water and space heating, onsite waste management systems, organic horticulture, and environmentally responsive land use planning and structural design. Our aim in working with these systems is to integrate locally controlled technologies that contribute toward individual and community self-reliance. The community contains seven solar residences, greenhouses, commercial and owner-built composting toilets, grey water systems, kitchen facilities, wood and metal shops, a dairy barn, agricultural areas, and three acres of French Intensive gardens.

Workshops are held at the rural center ranging in length from one day to five weeks. Consulting services are available from our staff in the areas of horticulture, edible landscaping, food systems, solar energy, waste management, and international development.

1006. JAN JOHNSEN
SOLAR GREENHOUSE CONSULTANT
Box 393-B
High Falls, New York 12440

Jan Johnsen
(914) 687-9334

Consultant to public and private institutions and individuals. Lecturer and advisor to educational groups.

1007. MAINE OFFICE OF ENERGY RESOURCES
55 Capitol Street
State House Station 53
Augusta, Maine 04333

Rick McGinley
(207) 289-3811

Provides extensive information on solar/alternative energies via booklets, fact sheets and newsletters. Areas covered include: solar energy basics, federal and state solar legislation, consumer tips, sources of climatic data, a bibliography for children and young adults, eutectic salts, and underground houses.

1008. MID-AMERICAN SOLAR ENERGY COMPLEX (MASEC)
Alpha Business Center
8140 26th Avenue South
Minneapolis, Minnesota 55420

(612) 853-0456

MASEC is one of the four federally funded regional solar centers created by the Department of Energy, under congressional mandate, to speed the commercialization of solar products, processes and services. It serves the 13-state Midwest region: Illinois, Indiana, Iowa, Kansas, Michigan, Minnesota, Missouri, Nebraska, Ohio, North Dakota, South Dakota and Wisconsin.

Current areas of technological interest include: passive and active solar heating and cooling; wind generation systems; photovoltaics; agricultural and industrial process heat.

370 SOLAR CENSUS

1009. NATIONAL SOLAR HEATING AND COOLING
INFORMATION CENTER
P.O. Box 1607
Rockville, Maryland 20850

(800) 523-2929
(800) 462-4983 (Pennsylvania)
(800) 523-4700 (Alaska, Hawaii)

The National Solar Heating and Cooling Information Center is a complete, one-stop service facility for all information, domestic and foreign, technical and nontechnical, on any aspect of solar heating and cooling. Our aim is to make everyone aware of the practical feasibility of solar energy and to encourage the public and industry to consider solar energy systems for homes and commercial buildings. Established by HUD, in cooperation with DOE, under provisions of PL 93-409.

1010. NAVARRO COLLEGE
P.O. Box 1170
Corsicana, Texas 75110

Alan Boyd
(214) 874-6501

Navarro College, as Project Center, with four cooperating institutions (Brevard College in Cocoa Beach, FL; Cerro Cosa Community College in California; Malaspina College in British Columbia, and North Lake College in Dallas, TX) has received funding from the National Science Foundation to design, develop, implement, test, evaluate and disseminate an associate degree curriculum to train solar energy technicians. This program, with Malaspina College funded by the Provincial Government of British Columbia, constitutes both a national and international effort to train viable and marketable solar energy technicians. One- and two-year courses available.

1011. **NEW ENGLAND FUEL INSTITUTE**
20 Summer Street
Watertown, Massachusetts 02172

Joseph S. Matachinskas
(617) 924-1000

The New England Fuel Institute (NEFI) is a nonprofit business league, established in 1943. The association represents the interests of over 1100 independent heating oil and oil heating equipment distributor member companies throughout the six New England states. NEFI has now taken the initiative to establish the first comprehensive vocational level "hands-on" solar heating installation and maintenance course in the U.S. The development of this one-month (160-hr) solar program accompanies the growing trend toward use of alternative energy systems in conjunction with domestic heating equipment. Through Solar Heating Installation and Maintenance is a relatively new field in vocational education, the Technical Training Center has emerged as a recognized forerunner in this specialized area of occupational training.

1012. **NEW MEXICO SOLAR ENERGY ASSOCIATION**
P.O. Box 2004
Santa Fe, New Mexico 87501

Steven G. Meilleur/Michael Shepard
(505) 983-1006

The New Mexico Solar Energy Association (NMSEA), started in 1974, is a nonprofit R&D/educational organization dedicated to furthering the use of solar energy and related arts. NMSEA's programs center around an effort to teach and demonstrate solar energy principles in a way that equips people to think and act for themselves. In the last five years we have initiated a variety of programs designed to put solar energy into practical use in the home, business, farm or ranch. These programs emphasize low-cost do-it-yourself solar applications which encourage both individual self-reliance and a spirit of community cooperation. We are involved in passive single-family residential design, passive retrofits, and a variety of low-cost services for the northern New Mexico community.

1013. **NEW YORK INSTITUTE OF TECHNOLOGY CENTER FOR ENERGY POLICY AND RESEARCH**
Old Westbury, New York, 11568

Gale Tenen Spak, Ph.D.
(516) 686-7578

The Center for Energy Policy and Research was established in 1975 to inform public and private sector decisionmakers and professionals, students and the general public, in the complexities of efficient energy utilization and conservation.

The Center undertakes information research and dissemination activities to assist public, quasi-public and private-sector organizations in the practical utilization of current knowledge in the energy field. Among current enterprises is a contract between the Center and the U.S. Department of of Energy under which we operate a national energy information clearinghouse for Energy Extension Service.

1014. **NORTHEAST SOLAR ENERGY CENTER**
470 Atlantic Avenue
Boston, Massachusetts 02110

Steven R. Nelson, Manager,
 Information and Education Division
(617) 292-9319

The Northeast Solar Energy Center is the U.S. Department of Energy's lead institution for accelerating the commercialization of solar products, processes and services in a nine-state region: New York, New Jersey, Pennsylvania and the six New England states. It is operated under contract to the Department by the not-for-profit Northern Energy Corporation, and is one of the four regional solar energy centers across the country. Its counterparts are based in Atlanta (Southern Solar Energy Center); Minneapolis (Mid-American Solar Energy Complex); and

Portland, Oregon (Western Solar Utilization Network). NESEC provides a wide range of informational and technical assistance services in the area of solar market development, information and education, planning and systems analysis, and technology development. It publishes numerous brochures and practical materials on solar applications, runs workshops and seminars, provides direct technical assistance, and maintains an extensive collection of solar-related materials in its library.

Current technology areas of interest and commercialization activity include active and passive heating systems, wind generation systems, photovoltaics, and industrial process heat.

1015. **OREGON SOLAR INSTITUTE**
215 S.E. 9th, Room 21
Portland, Oregon 97214

James E. Molle
(503) 232-4741

The Oregon Solar Institute is a nonprofit educational corporation. Our purpose is to accelerate the transition to solar energy through workshops, a monthly newsletter, movies/slides, and solar energy equipment demonstration. Our emphasis is on technology that is resource-abundant, cost-effective, has the greatest potential for widespread use, and will provide the user with energy independence. We have accordingly chosen to promote solar electric/thermal (photovoltaic concentrator) power systems because they meet all of the above requirements as of this writing. In our workshops we instruct participants in construction of their own flat plate photovoltaic panels and provide an introduction to concentrator photovoltaics. We have a catalog of various solar energy products which show the most promise of offering a significant portion of energy savings to a user.

374 SOLAR CENSUS

1016. **OZARK INSTITUTE**
P.O. Box 549
Eureka Springs, Arkansas 72632

Barry R. Weaver
(501) 253-7384

The Ozark Institute conducts seminars and workshops on energy and the family farm, including information appropriate to low- and moderate-income poeple. Areas covered include: applying alternative technologies to farm operations; energy conservation techniques; long-range planning for maximum farm profits; alcohol and methane production; solar farm buildings; sources of financial assistance; integrated pest management; and wood heating and woodlot management.

1017. **PRACTICAL SOLAR**
Box 1067 Blair Station
Silver Spring, Maryland 20910

Allan L. Frank
(301) 587-6300

Practical Solar is a monthly newsletter directed toward contractors, installers, retailers and distributors of solar equipment and systems. Featuring information on getting into business and staying there, interviews with solar company officials on how to run a solar business. Covers items on installation, sales, licensing, certification and more.

1018. **ROARING FORK RESOURCE CENTER**
(A Branch of the Colorado Energy Extension Service)
P.O. Box 9950
Aspen, Colorado 81611

Mary Ward, Director
(303) 925-8885

The Roaring Fork Resource Center is a nonprofit energy organization which has, since 1974, provided ecuational programs and information and referral services in the areas of renewable resources, energy conservation, alternative energy and appropriate technology. RFRC facilitated Aspen Earth Week '80, presented the Aspen Energy Forums, published the *Sun Journal* and has set up numerous workshops and classes for local residents. It currently serves the Northwest Colorado Council of Governments region as a Colorado Energy Extension Office, continuing to expand services in a larger geographic area.

1019. **SANTA CRUZ ALTERNATIVE ENERGY COOP**
 No. 6
 303 Potrero
 Santa Cruz, California 95060

 P.O. Box 66959
 Scotts Valley, California 95066

 Brian Williams/Scott Roseman
 (408) 425-SOLA

 We provide information on all aspects of solar energy. We publish a semi-quarterly newspaper, the *Santa Cruz Alternative Energy News*. We sell books, solar gadgets/toys, materials and consulting services. We conduct workshops twice a month. We have monthly meetings open to the public. We go to public schools and conduct classroom presentations. We have a library open to the public, and an energy hotline (425-SOLA). We are currently researching a solar food dryer.

1020. **SHELTER INSTITUTE**
 38 Center Street
 Bath, Maine 04530

 Patsy Hennin
 (207) 442-7938

 Shelter Institute began six years ago to teach people how to design and build their own energy-efficient and cost-conscious houses. We have had over 5000 students from every state and from many foreign countries attend our program (Mexico, Canada, Peru, Iran, Turkey, Germany, Senegal, Scotland, Spain, Ireland). To date 2000 houses have been completed by our graduates—every size and shape and level of sophistication.

1021. **SOLAR AGE MAGAZINE**
Church Hill
Harrisville, New Hampshire 03450

Ken Kerber
(603) 827-3347

We are the publishers of *Solar Age* magazine. Our publication is the official publication of the prestigious American Section of the International Solar Energy Society, now in its 25th year. We also publish the *Solar Products Specifications Guide,* a continually updated manual of technical information for solar products. We also publish two comprehensive catalogs, *The Solar Age Resource Book* (1979) and the *Solar Age Catalog* (1977).

Solar Age covers a broad spectrum of solar usage and has great coverage of active and passive technology.

1022. **SOLAR ENERGY INTELLIGENCE REPORT**
Box 1067 Blair Station
Silver Spring, Maryland 20910

Allan L. Frank
(301) 587-6300

Solar Energy Intelligence Report (SEIR) is a weekly newsletter covering national, state, local and international news on all forms of solar energy. Regulation, legislation, technology, business, economics followed closely. Also: calendar of events, new publications, grants and contracts, government procurement opportunities, patents, new products and services, and more. The only weekly solar newsletter in existence. Since 1975.

EDUCATION & INFORMATION 377

1023. **SOLAR GREENHOUSE ASSOCIATION**
34 North Gore Avenue
Webster Groves, Missouri 63119

Ida Pedersen
(314) 962-4176

Contract No. MA-SC-79-MO-0021 with the Mid-American Solar Energy Complex.

1. Plan and conduct a series of seminars and workshops on greenhouse design, construction and operation.

2. Organize small greenhouse construction projects by groups or individuals. Provide technical assistance as needed and available.

3. Publicize the seminars and projects in appropriate publications and other media, with appropriate reference to MASEC sponsorship.

1024. **SOLAR LIBERATION ENGINEERING**
Division of Davis Equities Corporation
311 South Bristol Street
Los Angeles, California 90049

Martha L. Sherwood, Communications Director
(213) 451-8094

We are the publishers of the *Peoples' Solar Sourcebook*, the nation's largest catalog of solar (and other renewable energy) products and equipment. In addition to publishing and promoting the catalog and updating the public's eduation in solar power, we solicit solar product inventors to provide us with a prospectus of their projects for consideration of financial support.

1025. **SOUTHERN SOLAR ENERGY CENTER, INC. (SSEC)**
61 Perimeter Park
Atlanta, Georgia 30341

Information Services
(404) 458-8765

378 SOLAR CENSUS

SSEC is one of the four federally funded regional solar energy centers. The mission of these centers is to foster the more rapid utilization of solar technologies. Current programs emphasize technology areas which DOE has determined to be commercially viable, including: passive/hybrid heating and cooling; active hot water systems; active heating and cooling; agricultural and industrial process heat; wood combustion; small wind energy systems (windmills); and photovoltaics.

1026. **SUNRAE INC.**
SOLAR USE NOW FOR RESOURCES AND EMPLOYMENT
5679 Hollister Avenue, Room 6
Goleta, California 93017

Diane Conn/Peter Alpert
(805) 964-4483

SUNRAE is a consumer group/lobby that primarily focuses on California. We do all that we can to ensure the development of solar energy as an alternative to rapidly dwindling stocks of nonrenewable and dangerous nuclear technologies. Our income is derived from memberships, donations, t-shirt and paraphernalia sales, recycling and consulting.

We feel that education of the public is equally important as the education of our decisionmakers. Our community outreach programs include slide shows, lectures, films and organizing energy-oriented events and fairs. We have scheduled a series of hands-on solar workshops encompassing San Luis Obispo, Santa Barbara and Ventura counties. Intern programs are arranged for students interested in different aspects of renewable resources.

1027. **TEXAS A & M UNIVERSITY**
Center for Energy and Mineral Resources
College Station, Texas 77843

Nancy Hawkins
(713) 845-8025

The Texas A & M Center for Energy and Mineral Resources coordinates energy and mineral programs in teaching, research and public service throughout the Texas A & M University system. Our major function, in addition to funding energy research, is the retrieval and dissemination of energy information.

1028. **THE TEXAS SOLAR ENERGY SOCIETY, INC.**
1007 South Congress, No. 359
Austin, Texas 78704

Russel E. Smith, Executive Director
(512) 443-2528

The Texas Solar Energy Society (TX-SES), a nonprofit educational organization, was founded in 1976 (in Austin, TX) by approximately 200 Texas citizens who shared a common interest in the promotion of solar energy utilization in the state. TX-SES has subsequently become affiliated with the American Section of the International Solar Energy Society (ISES). The purpose of the TX-SES is:

"To further the development of solar energy and related arts, sciences and technologies with concern for the ecologic, social, and economic fabric of the state. This shall be accomplished through exchange of ideas and information by means of meetings, publications and information centers. The Society shall serve to inform the public, institutional and government bodies and seek to raise the level of public and governmental awareness of its purpose."

TX-SES now represents over 400 individuals directly and and many more indirectly through its local affiliate chapters.

1029. **TEXAS STATE TECHNICAL INSTITUTE (TSTI)**
Solar Energy Mechanic Program
Route 3
Sweetwater, Texas 79556

Ronnie Freeman

The Sweetwater campus of Texas State Technical Institute offers a nine-month program to train solar energy mechanics. The course is an option in the Air Conditioning and Refrigeration Technology/Mechanics curriculum. Students may take only the Solar Energy Mechanic option, or they may receive this training in addition to their Air Conditioning and Refrigeration Technology/Mechanics course of study.

The course covers solar radiation and collectors, solar collector design and installation, solar load calculation and heat storage, and solar controls. Students also study basic air conditioning and refrigeration, mathematics, drafting and basic algebra. Upon successful completion of the course, students will receive a certificate of completion.

1030. **UPLAND HILLS ECOLOGICAL AWARENESS CENTER**
2575 Indian Lake Road
Oxford, Michigan 48051

Ann Franklin
(313) 693-1021

The Upland Hills Ecological Awareness Center (UHEAC) is a nature and conference center powered by alternative energy sources. It is an entirely unique building, built largely by volunteer labor and incorporating onsite natural materials. The objectives of UHEAC are to encourage an understanding of our complex relationship to the environment and actively attempt to heighten people's awareness that positive solutions to environmental/energy problems do exist. The Center will demonstrate the practical application of these energy systems together with various approaches to energy-conscious living.

Adult workshops in energy, food production and preparation, health and environment are available, as well as high school and elementary school workshops.

1031. **URBAN OPTIONS**
135 Linden Street
East Lansing, Michigan 48823

Randy Eveleigh
(517) 351-3757

Urban Options is a nonprofit solar and weatherization demonstration project, offering seminars and workshops.

1032. **WAYNE STATE UNIVERSITY**
US/WCP
Detroit, Michigan 48202

David R. Bowen
(313) 577-4631

Offer course on Residential Solar Energy in a conference format. Course covers scientific background, active and passive systems for space heating, domestic hot water application, retrofitting, consumer issues, financing and economics, heat load management, and solar energy policy at the state and federal levels.

1033. **WESTERN SOLAR UTILIZATION NETWORK (WESTERN SUN)**
921 S.W. Washington Street
Suite 160
Portland, Oregon 97205

(503) 241-1222

Western SUN was created by the Department of Energy under congressional mandate to encourage the commercialization of solar energy in thirteen western states.

The primary focus of Western SUN is to identify programs that have particular regional importance. The basic goals are to: provide consumers with a basic understanding of solar technologies and their applicability on a local basis; assist low-income families in reducing energy consumption and expenditures; establish a network of experienced individuals and groups to provide assistance to specific audiences and assist small businesses in entering the field and providing useful services.

The technological areas that are funded by Western SUN are: passive solar heating and cooling; active solar heating and cooling; small wind energy conversion systems; wood combustion/biomass; and agricultural and industrial process heat.

1034. **WHITEFLY CONTROL COMPANY**
P.O. Box 986
Milpitas, California 95035

John A. Morris/Rose M. Morris
(408) 263-4344

Whitefly Control Company is a primary supplier of *Encarsia formosa*, a minute parasite to harmful and common plant pests—white flies. By using *Encarsia formosa*, excellent control of whiteflies is possible without using pesticide poisons.

* * *

Also see the following entries—

112. ENVIRONMENTAL ENERGY MANAGEMENT & MANU-FACTURING CORPORATION (E^2M^2)
13. ALPHA SOLARCO INC.
151. HELIOTROPE GENERAL
189. THE L.A. SOLAR MART
257. RESIDENTIAL ENERGY SYSTEMS INC.
321. SOLAR MUSIC INC.
388. SUNTREE SOLAR COMPANY
490. C. PHILLIP COLVER AND ASSOCIATES, INC.
499. CROFT & COMPANY
517. DULANEY & ASSOCIATES
528. ELSWOOD–SMITH–CARLSON, ARCHITECTS
537. ENERGY MANAGEMENT CONSULTANTS, INC.
545. ENVIRONMENTAL DESIGN ALTERNATIVES
603. UNIVERSITY OF ILLINOIS
609. INTEGRAL DESIGN
633. DR. JAN F. KREIDER, P.E.
636. LANCASTER COUNTY COMMUNITY ACTION PROGRAM
655. MAINEFORM ARCHITECTURE
666. ROBERT MITCHELL, SOLAR SYSTEMS DESIGN, INC.
678. NATURAL ENERGY WORKSHOP, INC.
691. PASSIVE SOLAR ALTERNATIVES

EDUCATION & INFORMATION

723. SEAgroup
724. SENNERGETICS
747. SOLAR ENERGY ASSOCIATES, LTD.
756. SOLAR PATHWAYS ASSOCIATES
768. PAOLO SOLERI ASSOCIATES, INC.
769. SOLSTICE DESIGNS, INC.
773. S S SOLAR INC.
785. SUNDESIGN SERVICES INC.
790. SUNSEED ENERGY
792. SUNSPACE UNLIMITED
793. SUNSTRUCTURES
794. SUNSYSTEMS
815. TOTAL ENVIRONMENTAL ACTION, INC.
827. WEST CAP
841. ZAUGG & ZAUGG ARCHITECTS, AIA
845. UNIVERSITY OF ALABAMA
853. AMITY FOUNDATION
875. COLORADO STATE UNIVERSITY
877. COPPER DEVELOPMENT ASSOCIATION, INC.
882. UNIVERSITY OF DELAWARE
884. DETROIT LAKES TECH
888. EARTHMIND
902. UNIVERSITY OF FLORIDA
903. FLORIDA SOLAR ENERGY CENTER
905. HAWAII NATURAL ENERGY INSTITUTE
908. UNIVERSITY OF IDAHO
915. LOS ALAMOS SCIENTIFIC LABORATORY
917. UNIVERSITY OF MAINE
919. UNIVERSITY OF MIAMI
920. MISSISSIPPI STATE UNIVERSITY
926. M&W ENTERPRISES
928. NATIONAL CENTER FOR APPROPRIATE TECHNOLOGY
931. NEW ALCHEMY INSTITUTE
932. NEW LIFE FARM, INC.
937. UNIVERSITY OF OKLAHOMA

939. UNIVERSITY OF OREGON
942. PASSIVE SOLAR INSTITUTE
943. THE PENNSYLVANIA STATE UNIVERSITY
952. RESEARCH INSTITUTE FOR ADVANCED TECHNOLOGY
957. SHELTON ENERGY RESEARCH
961. SOLAR ENERGY RESEARCH INSTITUTE (SERI)
966. SOLAR SURVIVAL
967. **SOLAR** SUSTENANCE TEAM
968. SOLAR THERMAL TEST FACILITIES USERS ASSOC.
985. UNIVERSITY OF UTAH
989. WESTERN ILLINOIS UNIVERSITY

* * *

ORGANIZATIONS/INFORMATION SOURCES

NATIONAL

CENTER FOR RENEWABLE
RESOURCES
1001 Connecticut Avenue N.W.
Suite 510
Washington, D.C. 20036

(202) 466-6880

CITIZENS' ENERGY PROJECT
1110 6th Street N.W.
Suite 300
Washington, D.C. 20009

(202) 387-8998

INTERNATIONAL SOLAR ENERGY
SOCIETY, AMERICAN SECTION
P.O. Box 1416
U.S. Highway 190 West
Killeen, Texas 76541

NATIONAL ASSOCIATION OF
SOLAR CONTRACTORS
910 17th Street N.W.
Suite 928
Washington, D.C. 20006

(202) 785-3244

NATIONAL CENTER FOR
APPROPRIATE TECHNOLOGY
P.O. Box 3838
Butte, Montana 59701

(406) 494-4572

NATIONAL SOLAR HEATING &
COOLING INFORMATION CENTER
P.O. Box 1607
Rockville, Maryland 20850

(800) 523-2929
(800) 462-4983 (Pennsylvania)
(800) 523-4700 (Alaska, Hawaii)

NATIONAL TECHNICAL INFOR-
MATION SERVICE (NTIS)
5285 Port Royal Road
Springfield, Virginia 22161

NEW YORK INSTITUTE OF
TECHNOLOGY
CENTER FOR ENERGY POLICY
AND RESEARCH
Old Westbury, New York 11568

(516) 686-7578

SOLAR ENERGY RESEARCH
INSTITUTE (SERI)
1617 Cole Boulevard
Golden, Colorado 80401

(303) 231-1415

SOLAR ENERGY INDUSTRIES
ASSOCIATION (SEIA)
1001 Connecticut Avenue N.W.
Suite 800
Washington, D.C. 20036

(202) 293-2981

SOLAR LOBBY
1001 Connecticut Avenue N.W.
5th Floor
Washington, D.C. 20036

(202) 466-6350

U.S. Department of Energy
TECHNICAL INFORMATION
CENTER
P.O. Box 62
Oak Ridge, Tennessee 37830

(615) 576-1301
(615) 576-6800

REGIONAL

NORTHEAST SOLAR ENERGY CENTER (NESEC)
470 Atlantic Avenue
Boston, Massachusetts 02110
(617) 292-9319

MID-AMERICAN SOLAR ENERGY COMPLEX (MASEC)
Alpha Business Center
8140 26th Avenue South
Minneapolis, Minnesota 55420
(612) 853-0456

SOUTHERN SOLAR ENERGY CENTER (SSEC)
61 Perimeter Park
Atlanta, Georgia 30341
(404) 458-8765

WESTERN SOLAR UTILIZATION NETWORK (WESTERN SUN)
715 S.W. Washington, Suite 160
Portland, Oregon 97205
(503) 241-1222

STATE

ALABAMA
Alabama Solar Energy Center
University of Alabama
P.O. Box 1247
Huntsville, Alabama 35807
(800) 572-7226

ALASKA
Department of Commerce and Economic Development
Energy and Power Development
338 Denali Street
Anchorage, Alaska 99501
(907) 276-0508

ARIZONA
Arizona Solar Energy Commission
1700 West Washington Street
Phoenix, Arizona 85007
(800) 325-5499

ARKANSAS
Department of Energy
3000 Kavanaugh Boulevard
Little Rock, Arkansas 72205
(800) 325-5499

CALIFORNIA
California Energy Commission
1111 Howe Avenue
Sacramento, California 95825
(800) 852-7516

COLORADO
Office of Energy Conservation
1600 Downing Street, 2nd Floor
Denver, Colorado 80218
(303) 839-2507

Roaring Fork Resource Center
P.O. Box 9950
Aspen, Colorado 81611
(303) 925-8885

CONNECTICUT
Office of Policy and Management
80 Washington Street
Hartford, Connecticut 06115
(203) 566-5803

EDUCATION & INFORMATION 387

DELAWARE
Governor's Energy Office
P.O. Box 1401
Dover, Delaware 19901
(800) 282-8616

FLORIDA
Florida Solar Energy Center
300 State Road 401
Cape Canaveral, Florida 32920
(305) 783-0300

GEORGIA
Office of Energy Resources
Room 615
270 Washington Street S.W.
Atlanta, Georgia 30334
(404) 656-5176

HAWAII
State Energy Office
P.O. Box 2359
Honolulu, Hawaii 96804
(808) 548-4150

IDAHO
Office of Energy
State House
Boise, Idaho 83720
(208) 334-3800

ILLINOIS
Institute of Natural Resources
325 West Adams
Springfield, Illinois 62706
(217) 785-2800

INDIANA
Department of Commerce
440 North Meridian
Indianapolis, Indiana 46204
(317) 232-8940

IOWA
Energy Policy Council
Lucas Building, 6th Floor
State Capitol Complex
Des Moines, Iowa 50319
(800) 532-1114

KANSAS
Energy Office
214 West 6th
Topeka, Kansas 66603
(800) 432-3537

KENTUCKY
Department of Energy
Bureau of Energy Management
Iron Works Pike
Lexington, Kentucky 40578
(800) 432-9014

LOUISIANA
Department of Natural Resources
Research and Development
P.O. Box 44156
Baton Rouge, Louisiana 70804
(504) 342-4594

MAINE
Office of Energy Resources
State House
Station Number 53
Augusta, Maine 04330
(800) 452-4648

MARYLAND
Energy Office
Room 1302
301 West Preston Street
Baltimore, Maryland 21201
(800) 492-5903

MASSACHUSETTS
Executive Office of Energy Resources
Renewable Energy Division
73 Tremont Street
Room 849
Boston, Massachusetts 02108
(800) 922-8265

MICHIGAN
Department of Commerce
Energy Administration
P.O. Box 30228
Lansing, Michigan 48909
(800) 292-4704

MINNESOTA
Energy Agency
980 American Center Building
150 East Kellog
St. Paul, Minnesota 55101
(800) 652-9747

MISSISSIPPI
Governor's Office of Energy
and Transportation
510 George Street
Jackson, Mississippi 39205

MISSOURI
Department of Natural Resources
Division of Energy
P.O. Box 176
Jefferson City, Missouri 65102
(800) 392-0717

MONTANA
Department of Natural Resources
and Conservation
Energy Division
32 South Ewing
Helena, Montana 59601
(406) 449-4624

NEBRASKA
Solar Office
Nebraska Hall
University of Nebraska
Lincoln, Nebraska 68588
(402) 472-3414

NEVADA
Department of Energy
400 West King Street
Carson City, Nevada 89710
(800) 992-0900

NEW HAMPSHIRE
Governor's Council on Energy
2½ Beacon Street
Concord, New Hampshire 03301
(800) 852-3311

NEW JERSEY
State Energy Office
Office of Alternative Technology
101 Commerce Street
Newark, New Jersey 07102
(800) 492-4242

NEW MEXICO
Energy and Minerals Department
P.O. Box 2770
Santa Fe, New Mexico 87501
(800) 432-6782

New Mexico Solar Energy
Association
P.O. Box 2004
Santa Fe, New Mexico 87501
(505) 932-1006

NEW YORK
State Energy Office/Solar Program
2 Rockefeller Plaza
Albany, New York 12223
(800) 342-3772

EDUCATION & INFORMATION

NORTH CAROLINA
Department of Commerce
Energy Division
430 North Salisbury Street
Raleigh, North Carolina 27611
(919) 733-2230

NORTH DAKOTA
Energy Management and Conservation
1533 North 12th Street
Bismarck, North Dakota 58501
(701) 224-2250

OHIO
Department of Energy
30 East Broad Street, 34th Floor
Columbus, Ohio 43215
(800) 282-9234

OKLAHOMA
Department of Energy
4400 North Lincoln Boulevard
Suite 251
Oklahoma City, Oklahoma 73105
(405) 521-3941

OREGON
Department of Energy
Room 111
Labor and Industries Building
Salem, Oregon 97310
(800) 452-7813

Oregon Solar Institute
215 S.E. 9th, Room 21
Portland, Oregon 97214
(503) 232-4741

PENNSYLVANIA
Governor's Energy Council
1625 North Front Street
Harrisburg, Pennsylvania 17101
(800) 882-8400

RHODE ISLAND
Governor's Energy Office
80 Dean Street
Providence, Rhode Island 02903
(401) 277-3773

SOUTH CAROLINA
Division of Energy Resources
1122 Lady Street, 11th Floor
Columbia, South Carolina 29201
(800) 922-5310

SOUTH DAKOTA
State Solar Office
Suite 102
Capitol Lake Plaza Building
Pierre, South Dakota 57501
(800) 592-1865

TENNESSEE
Tennessee Energy Authority
Suite 707
Capitol Building
226 Capitol Boulevard
Nashville, Tennessee 37219
(800) 342-1340

TEXAS
Texas Energy and Natural Resources
Advisory Council
411 West 13th Street
Room 800
Austin, Texas 78701
(512) 475-5588

UTAH
Energy Office
231 East 400 S., Suite 101
Salt Lake City, Utah 84111
(800) 662-3633

390 SOLAR CENSUS

VERMONT
State Energy Office
4 East State Street
Montpelier, Vermont 05602

(800) 642-3281

VIRGINIA
Energy Information and Services Center
310 Turner Road
Richmond, Virginia 23225

(800) 552-3831

WASHINGTON
State Energy Office
400 East Union Street, 1st Floor
Olympia, Washington 98504

(206) 754-1350

WEST VIRGINIA
Fuel and Energy Office
1262½ Greenbrier Road
Charleston, West Virginia 25311

(800) 642-9012

WISCONSIN
Division of State Energy
101 South Webster, 8th Floor
Madison, Wisconsin 53702

(608) 266-8234

WYOMING
Energy Extension Service
P.O. Box 3965
University Station
Laramie, Wyoming 82070

(800) 442-6783

ALPHABETICAL INDEX
(with entry code)

A

A.A.I. Corporation [1]
Abacus Controls Inc. [2]
Abacus Group Inc. [431]
Abernathy Solar Consultant [432]
Acorn Structures [433]
AcroSun Industries Inc. [3]
Acurex Solar Corporation [4]
Adobe Solar Ltd. [434]
Advance Air Control [435]
Advanced Energy Systems [436]
Advanced Energy Systems Inc. [5]
Advanced Energy Technology Inc. [6]
Advance Energy Technologies Inc. [7]
Advance Technology Engineering [8]
Aeolian Kinetics [9]
Aesop Institute [844]
Affiliated Engineers Inc. [437]
AFG Industries Inc. [10]
AIDCO Maine Corp. [438]
Alabama, University of—Johnson Environmental & Energy Center [845]
Aldermaston Inc. [11]
Alkar Engineering & Manufacturing Co. [439]

Allen & Sheriff [440]
Allied Industries International [12]
Alpha Solarco Inc. [13]
Altas Corporation [846]
Altenburg and Company [441]
Alternate Energy Industries Corporation (AEIC) [442]
Alternate Energy Institute [992]
Alternate Energy Research and Development [847]
Alternative Energy Concepts of California Inc. [443]
Alternative Energy Resources Inc. [14]
Alternative Energy Works Inc. [444]
Alternative Publishers [993]
Alternative Resources, Inc. [445]
Alternative Sources of Energy [994]
Alufoil Packaging Co. Inc. [15]
AMAF Industries [848]
AMCOR Group Ltd., The [16]
American Creative Engineering [152]
American Energy Savers Inc. [17]
American Gas Association [995]
American Heliothermal Corp. [18]
American Home Solar Energy Systems Inc. [19]
American Solar Heat Corp. [20]

American Solar King Corp. [21]
American Sun Corp. [22]
American Timber Homes/Solartran Division [23]
Ametek, Power Systems Group [24]
Amity Foundation [853]
Anachem Inc. [850]
Ana-Lab Corp. [851]
ANCO Engineers Inc. [446]
Andersen Corporation [25]
Anderson, John, Associates [447]
Andrews, Leroy, Architects Inc. [448]
Angel, Mull & Associates Inc. [449]
Applied Solar Energy Corp. [26]
Approtech Solar Products [27]
APTEC Corporation [28]
Aquasolar Inc. [29]
Architects Partnership, Inc. The [450]
Architectural Alliance [451]
Architectural Bureau [452]
Architectural Collective/Architronics Inc. [453]
Architectural Concepts [454]
Architectural & Industrial Marketing Inc. [455]
Architecture One [456]
Architekton Inc. [457]
Architronics Inc. [453]
ARCO Solar Inc. [30]
Argonne National Laboratory [852]
Aries Consulting Engineers Inc. [458]
Arizona, University of [853]

Arizona State University [854,855]
ARKLA Industries Inc. [31]
Armstrong, Torseth, Skold & Rydeen [459]
Associated Enterprises [856]
Astro Solar Corp. [857]
Astro Research Corporation [32]
Atek Design [858]
Atkinson/Karius/Architects [460]
Atlantic Richfield Company [225]
ATR Electronics Inc. [33]
Automatic Solar Covers Inc. [34]
Ayres, Lewis, Norris & May Inc. [461]
Aztech International Ltd. [35]

B

Bakewell Corporation [462]
Balance Associates [463]
Baldwin Solar Technologies [859]
Barber-Nichols Engineering Co. [860]
Basic Environmental Engineering Inc. [36]
Battelle Memorial Institute [861]
Beck, Mark, Associates [464]
Beckman, Blydenburgh & Associates [465]
Bell, Howard, Enterprises Inc. [466]
Bell, Robert A., Architects Ltd. [467]
Bell & Gossett [37]
Berg & Associates [468]
Berkeley Solar Group [469]
Bernheim, Kahn & Lozano [470]
Berry Solar Products [38]

ALPHABETICAL INDEX 393

Big Five Community Services [996]
Bio-Energy Systems Inc. [39]
Bio-Gas of Colorado Inc. [862]
Black, W.J., Architect [471]
Black-Huettenrauch-Schuyler [472]
Blue White Industries [40]
Bockman, J.A., Solar Engineering [473]
Bogen, Paul F., Architect [474]
Boston College [863]
Brattleboro Design Group [842]
Bregar, Robert J., Associates [475]
Brookhaven National Laboratory [864]
Brown University [865]
Brown & Brown [476]
Browning, D.W., Contracting Co. [41]
Buckmaster Industries Inc. [477]
BUDCO [42]
Burt Hill Kosar Rittlemann Associates [478]
Bushnell, Robert H. [479]

C

Caldwell Construction [480]
Cal Energy Consultants Inc. [481]
California, University of—Lawrence Livermore National Laboratory [866]
California Solar Designs [482]
Calmac Manufacturing Corp. [43]
Candrex Pacific [483]
Canfield Group [44]
Capitol Consultants Inc. [484]
Cardor Company [45]
Carolina Solar Systems [46]
Cary Arboretum of the New York Botanical Garden [867]

Cascade Solar Technics [868]
Cellular Product Services Inc. [47]
Center for Energy Policy and Research [1013]
Center for Maximum Potential Building Systems [869]
Center for Technology Assessment [681]
Central Montana Human Resources Development Council [870]
Central States Energy Research Corp. [485]
Central Upper Peninsula Planning & Development Regional Commission [997]
Century Fiberglass Products [48]
Chamberlain Manufacturing Corp. [871]
Champion Home Builders Co. [49]
Chemax Manufacturing Corp. [50]
Chemplast Inc. [51]
Cheyenne Community Solar Greenhouse [999]
Chicago, University of—Enrico Fermi Institute [872]
Chicago Solar Corporation [52]
Childs, Willard D. [873]
Cider Bluff Inc. [486]
Clairex Electronics [53]
Clark & Walter Architects [487]
Clinton Co. Energy Committee [998]
Close Associate Inc. [488]
CMI Solarglas [54]
Coating Laboratories Inc. [55]
Cole Solar Systems Inc. [56]
Collins, E.C., Associates [489]
Colorado, University of [874]

394 SOLAR CENSUS

Colorado State University [875]
Columbia Chase Corporation [57]
Colver, C.P., & Associates Inc. [490]
Commercial Plumbing Corp. [491]
Community Action of Laramie Co. County [999]
Community Action Commission for the City of Madison and the County County of Dane Inc. [492]
Community Builders [493]
Compool Corporation [58]
Comstock Construction Co., Inc. [494]
Con-Egy [59]
Conklin & Rossant [495]
Connecticut, University of [876]
Conserdyne Corporation [60]
Contemporary Systems Inc. [61]
Contextus Corporation [496]
Controlex Engineering [62]
Controls/Inc. [63]
Convection Loops [1000]
Coordinated Systems Inc. [497]
Copper Development Assoc. Inc. [877]
Coppersmith's [64]
Cornerstones Energy Group Inc. [1001]
Corporate Energy Developments Inc. [878]
Cosanti Foundation [768]
Cover Pools Inc. [65]
Creative Alternatives [498]
Creative Energy Products [66]
Creighton Solar Concepts Inc. [67]
Crescent Engineering Co., Inc. [68]
Croft & Company [499]
Crowther Solar Group [500]

Crystalite Embedments Inc. [879]
Crystal Systems Inc. [880]
CSI Solar Systems Division [69]

D

Dale and Associates Inc. [70]
Daniel Enterprises Inc. [501]
DAS/Solar Systems [502]
Davis Engineering Inc. [503]
Davis, Jacoubowsky, Hawkins Associates [504]
Dearing, L.M., Associates Inc. [71]
Decker Manufacturing [165, 881]
Deco Products [72]
Delaware, University of [882]
del Sol Control Corporation [73]
Delta H Systems [74]
Delta Solar System Company [75]
Dencor Energy Cost Controls Inc. [76]
Denver Research Institute [883]
Deposition Technology Inc. [77]
Des Lauriers, R.E., Architect [505]
De Soto Inc. [78]
Detroit Environmental Control Engineers Inc. [506]
Detroit Lakes Tech [884]
Design/Build Inc. [468]
Devices & Services Co. [79]
Devon D Inc. [991]
dh2W Inc. [507]
Dickey/Kodet/Architects Inc. [508]
Dimetrodon [509]
Dixie Royal Homes Inc. [80]
Dixon/Carter Architects [510]

ALPHABETICAL INDEX 395

DIY-Sol Inc. [81]
Dodge Products Inc. [82]
Domestic Technology Institute [1002]
Donovan Enterprises [511]
Dooling & Siegel Architects [512]
Doucette Industries [83]
Dow Corning Corporation [84]
Downing Leach [513]
Dressler Energy Consulting and Design Corporation [514]
DSET Laboratories Inc. [885]
DSS Engineers Inc. [515]
Dubin-Bloome Associates [516]
Dulaney & Associates [517]
Dunham Associates [886]
Du Pont de Nemours and Company, E.I., Inc. [85]
Dwyer Instruments Inc. [86]
Dyna Technology Inc. [87]

E

Earle Engineering [887]
Earthmind [888]
Earth Services Inc. [88]
Earth-Sheltered Housing Systems [518]
Earth-Sun-Design [519]
EASCO Aluminum [89]
Ecoenergetics Inc. [889]
Ecotope Group [520]
Ecotronic Laboratories Inc. [890]
ECS—Environmental Control Systems [521]
Edge Research [522]
Edmund Scientific Co. [90]
edSON Solar Systems [91]

Edwards Engineering Corp. [92]
EIC Corporation [891]
Eisler Engineering Company [93]
Ekose'a Inc. [523]
E&K Service Co. [94]
Eland Electric Corp. [892]
Electra Sol Labs Inc. [95]
Electrical Construction Co. [524]
Electrolab Inc. [96]
Electrostatic Consulting Associates [525]
Eley, Charles, Associates [526]
Elkhart Products Corporation [97]
Ellmore/Titus/Architects Inc. [527]
Elmwood Sensors Inc. [98]
Elswood—Smith—Carlson, Architects [528]
Emerson Electric Co. [153]
Emmet Co. Cooperative Extension Energy Project [1003]
Enerad Inc. [893]
Enercom Inc. [529]
Enercon Ltd. [530]
Energy Applications [531]
Energy Associates [532]
Energy Associates [99]
Energy Center Inc. [533]
Energy Control Systems [100]
Energy Design Consultants and Builders Inc. [534]
Energy Design Corp. [101]
Energy Designs/Architects [535]
Energy Development Co. [894]
Energy Dynamics Corporation [895]
Energy Engineering Group Inc. [536]

396 SOLAR CENSUS

Energy Equipment Sales [102]
Energy Factory, The [103]
Energy Harvester [104]
Energy Info [1004]
Energy Mgmt. Consultants Inc. [537]
Energy Management and Control Corporation (EMC^2) [538]
Energy Materials Inc. [105]
Energy Planning & Investment Corporation (EPIC) [539]
Energy Research Institute of Delaware [896]
Energy Science Corporation [897]
Energy Specialties [540]
Energy Systems Group/Division of Rockwell International [898]
Energy Systems Inc. [106]
Energy Transfer Systems Inc. [107]
Energy Saving Shop [75]
Energy Systems, Inc. [106]
Energy 2000 [335]
Enerspan, Inc. [108,789]
Enertech Corporation [109]
Ener-Tech, Inc. [541]
Engel Industries, Inc. [110]
Engineering Consulting Services [542]
Entropy Limited [111]
Engineers—Architects, P.C. (EAPC) [543]
Entec [544]
Environmental Design Alternatives, Architects [545]
Environmental Design Group [546]
Environmental Energy Management & Manufacturing Corp. (E^{2}M^2) [112]
Environmental Instrumentation [547]
Environmental Power Corp. [548]
Environmental Research Institute of Michigan (ERIM) [899]

Environment Associates [549]
Eppley Laboratory, Inc. [113]
Ergenics [114]
Erie Manufacturing Co. [115]
Erie Scientific Co. [116]
Escher: Foster Technology Associates, Inc. [900]
ESR—Environmental Systems Research [550]
E-Tech, Inc. [117]
Evanoff, John M., Architect [551]
Evjen Associates, Inc., The [552]

F

Fabrico Manufacturing Corp. [118]
Fafco, Inc. [119]
Falbel Energy Systems Corp. [120]
Farallones Institute [1005]
Fellers, Paul, Architect [553]
Fermi, Enrico, Institute [872]
Fermi National Accelerator Laboratory (Fermilab) [901]
Feuer Corporation [554]
Filon Div. of Vistron Corp. [121]
First Solar Industries, Inc. [122]
Fischer Sun Cooker [123]
Fisher/Roberts [555]
Flagala Corporation [124]
Fleming, W. S., and Associates, Inc. [556,557]
Flinn Saito Andersen Architects [558]
Florida, University of [902]
Florida Solar Energy Center [903]
Follensbee, James, & Assoc., Ltd. [559]
Fonda-Bonardi, Mario [560]
Ford Products Corporation [125]
Forest City Dillon, Inc. [561]

ALPHABETICAL INDEX

Forster-Morrell Engineering Associates, Inc. [562]
Four Seasons Solar Products Corp. [126]
Freeberg Company [563]
Free Energy Systems Inc. [127]
Freeman, Ronnie [564]
Friedman Sagar McCarthy Miller and Associates [565]
Friedrich & Dimmock, Inc. [128]
Fuad & Associates [437]
Fusion Incorporated [129]

G

Gade, Randolph Steven [566]
Gantec Corporation [130]
G.E. Associates, Inc., Architects [567]
General Energy Development Corporation [131]
General Energy Devices, Inc. [132]
General Extrusions Inc. [134]
General Instrument Corp. [133]
General Solar Corp. [568]
General Solar Systems Division/General Extrusions Inc. [134]
Genesis Architecture [569]
Geoscience Ltd. [904]
Glass-Lined Water Heater Company, The [135]
Glenn, Gary, AIA—Architect [570]
Glumac & Associates, Inc. [571]
Goodwin Controls Co. [572]
Go Solar Inc. [573]
Gove Associates Inc. [574]
Gower, Rondal, Associates [575]
Graham, W.J., Co. Inc. [136]
Graheck, Bell, Kline & Brown [576]
Granbery, Carleton, Architect [577]

Grayson Associates, Inc. [578]
Green, Peter H., & Associates [579]
Green, Wm. E., Associates [580]
Green Horizon [581]
Green Mountain Homes, Inc. [137]
Greer, George, Design & Construction [582]
Griswold Controls [138]
Groth, Glenn F., Architect [583]
Grumman Energy Systems, Inc. [139]
Grundfos Pumps Corporation [140]
Gulf Thermal Corporation [141]

H

Habitec: Architecture & Planning [584]
Haines Tatarian Ipsen & Associates [585]
Haleakala Solar Resources, Inc. [142]
Halstead & Mitchell Co. [143]
Hammel Green & Abrahamson Inc. [586]
Hammer, Siler, George Associates [587]
Hanat Sales Company [144]
Harris Architects, Inc. [588]
Haverstick & Associates [589]
Hawaii Natural Energy Institute [905]
Hawkins & Associates [590]
Hawkweed Group, Ltd. [591]
Hawthorne Industries, Inc. [145]
Hedland Products [146]
Heliodyne, Inc. (Richmond, CA) [147]
Heliodyne, Inc. (Rockford, IL) [148]
Helios International Corp. [149]
Helio Thermics Inc. [150]

398 SOLAR CENSUS

Heliotrope General [151]
Helix Solar Systems [152]
Hemphill, Vierk & Dawson [592]
HGF/Centrum Architects Inc. 593
H&H Precision Products [153]
H&H Tube & Manufacturing Co. [154]
Hitachi Chemical Co. America, Ltd. [155]
Hi-Tech, Inc. [156]
Hittman Associates, Inc. [906]
Hollis Observatory [157]
Holt + Fatter + Scott Inc. [594]
Honeywell Inc. [907]
Hooker/De Jong Associates Architects [595]
Hoster's HVAC Engineering Co. [596]
Hot Stuff [158]
Hoyem-Basso Associates, Inc. [597]
HSR Associates, Inc. [598]
Huffman, Ray, Energy Management Company [708]
Hughes Aircraft [364]
Hunter Hunter Associates Architects [599]
Hy-Cal Engineering [159]
Hydro-Flex Corporation [160]
Hylton, Joe, & Associates [600]
Hyperion, Inc. [161]
Hyperion, Incorporated [162]

I

IBE Energy [601]
Idaho, University of [908]
i e associates, inc. [602]
ILI, Inc. [163]
Illinois, University of [603]
Ilse Engineering, Inc. [164]
Imagineering Solar [604]
Immormino, M.R. [605]
Impac Corporation [165]
Inatome & Associates, Inc. [606]
Inco Ltd. [114]
Independent Energies Inc. [166]
Independent Energy, Inc. [167]
Independent Power Developers Inc. [909]
Indiana Solar Designs [607]
Infinity Energy Systems [168]
Innovative Energy Corp. [608]
Insolarator [169]
Integral Design [609]
Integrated Energy Systems [610]
Interactive Resources, Inc. [611]
International Solar Technologies, Inc. [430A]
Intertechnology/Solar Corp. [170]
Ionic Solar Inc. [612]
Ironwood, Inc. [613]
Irvine Engineering [614]
Iskander, Kamal S., and Associates [615]
Ista Energy Systems Corp. [171]
ITT Fluid Handling Division [172]

J

Jackson, W.L., Manufacturing Co. [173]
Jacobs-Del Solar Systems, Inc. [616]
Janofsky, Jack, & Associates [174]
JBF Scientific Corporation [910]
J & D Solar Contracting [617]
Jeannette, Brion S., & Assoc. [618]
Jetel Company [175]

J.F.A. Services, Inc. [619]
Jira Heating & Cooling, Inc. [620]
Johnson, Jan [1006]
Johnson Controls [176]
Johnson Engineering [621]
Johnson Environmental & Energy Center [845]
Jones, M. Dean, Architect [622]
Jordan's Solar Equipment Sales [177]

K

Kaelin Industries, Inc. [178]
Kalwall Corporation [179]
Kaman Aerospace Corp. [911]
Kansas, University of—Center for Research, Inc. [912]
Kathabar Systems [180]
Kay, Arlan, & Associates [623]
Keane Associates [624]
Kelbaugh & Lee Architects [625]
KEM Associates, Inc. [181]
Keniston, Stanley, AIA [626]
C. Kessel Co. [627]
Kessel Insolar Designs [628]
K & G Manufacturing Inc. [389]
Kiene & Bradley Partnership, The [629]
Kingscott Associates, Inc. [630]
Kingston Industries Corp. [182]
Kirkham, Michael & Associates [631]
Kirkhill Rubber Company [183]
Kohloss, Frederick H., & Associates, Inc. [632]
Komara Company [184]
Kreider, Dr. Jan F., P.E. [633]

Krumbhaar & Holt Architects [634]

L

Lafont Corporation [185]
Lamb, Wm. & Co. [186]
Lambeth, James, Architect [635]
Lamco, Inc. [187]
Lancaster Co. Community Action Program [636]
Larkin, Paul, P.E. [637]
Lasco Industries [188]
L.A. Solar Mart, The [189]
La Tour, Gene, General Contractor [638]
Lear, K.T., Associates, Inc. [639]
Le Bleu Enterprises [640]
Lee & Associates [641]
Lennox Industries Inc. [190]
Leslie, Steven M., Consultant [642]
Letro Thermometer, Inc. [191]
Levi & Co. [643]
Lewis & Associates [192]
Lewis, John R., P.E. [644]
Lewis, Malcolm, Associates [645]
Libbey-Owens-Ford Company [193]
Li-Cor [194]
Lipe-Rollway Corporation [913]
Liss Engineering Inc. [646]
Livermore, Lawrence, National Laboratory [866]
Living Systems [647]
Lobato, Rudolph B., Associates [648]
Lodi Solar Co. Inc. [195]
Londe-Parker-Michels, Inc. [649]
Long, Edward J., Consulting Engineers [650]

Long Island Lighting Co. (LILCO) [914]
Los Alamos Scientific Laboratory [915]
LSW Engineers [651]
Lundquist, Wilmar, Schultz & Martin Inc. [652]
Luther, H.D., Manufacturing Co. [196]
Lutton, Paul H. [653]

M

MacBall Industries, Inc. [197]
MacDonald, Angus, AIA [799]
MacDonald Engineering [916]
Madlin's Enterprises [654]
Maine, University of [917]
Maine Office of Energy Resources [1007]
Maineform Architecture [655]
March Manufacturing Inc. [198]
Mark Enterprises Inc. [199]
Marshall's Drafting/California Builders [656]
Martin Processing, Inc. [200]
Martis, James A., Jr. Architects [657]
Massdesign Architects & Planners, Inc. [658]
Mass Energy Systems, Inc. [430C]
Math/Tec Inc. [659]
Matrix, Inc. [201]
McConnell, Steveley, Anderson [660]
McCracken Solar Co. [661]
Mechanical Seals Corporation [202]
Mechanical Technology Inc. [918]

Megatherm [203]
Mehrkam Energy Development Co. [204]
Merdler, Stephen, Associates [662]
Merrymeeting Architects [663]
Miami, University of [919]
Microcontrol Systems, Inc. [205]
Mid-American Solar Energy Complex (MASEC) [1008]
Midland-Ross Corporation [180]
Mid Pacific Solar Corporation/Solar Disc Corporation [206]
Midwest Components Inc. [207]
Miller & Sun Enterprises, Inc. [664]
Millville Windmills, Inc. [208]
Milton International, Inc. [665]
Minnesota, University of [983]
Minnesota, Mining and Manufacturing (3M) [209]
Minnesota Tritec Inc. [210]
Mississippi State University [920]
Mitchell, Robert, Solar System Design Inc. [666]
Mitchell & Jensen, Architect & Engineer [667]
Mobil Corporation [921]
Modupac Inc. [922]
Mogavero & Unruh [668]
Monogram Industries, Inc. [211]
Monosolar, Inc. [211]
Monroe Modular Homes, Inc. [212]
Monsanto Company [213]
Monsul, James, A., & Associates [669]

ALPHABETICAL INDEX 401

Montana State University [923]
Moore Grover Harper, P.C. [670]
Moore/Weinrich Architects [671]
Morelan, Jim D., Architect & Associates [672]
Moreland Associates [673]
Mor-Flo Industries, Inc. [214]
Mormec Engineering, Inc. [674]
Mortensen, Glen H., Inc. [675]
Mosfet Microlabs Inc. [924]
Motorola Semiconductor Group [215]
Mototron Inc. [925]
MPD Technology Corp. [114]
Multi-Duti Manufacturing, Inc. [216]
Murphy, C.F., Association [676]
M&W Enterprises [926]

N

Nabozny, Ronald [677]
National Aeronautics and Space Administration (NASA) [927]
National Center for Appropriate Technology [928]
National Institute of Creativity, Inc. [929]
National Patent Development Corp. [930]
National Products, Inc. [217]
National Semiconductors Ltd. [218]
National Solar Corporation [219]
National Solar Heating & Cooling Information Center [1009]
National Solar Supply [220]
National Stove Works [221]
Natural Energy Workshop, Inc. [678]
Natural Power Inc. [222]
Navarro College [1010]
Neill & Gunter Inc. [679]
New Alchemy Institute, The [931]
New Day Builders [680]
New England Fuel Institute [1011]
New Jersey Institute of Technology [681]
New Life Farm, Inc. [932]
New Mexico Solar Energy Assoc. [1012]
New York Botanical Garden [867]
New York Institute of Technology [1013]
Nielsen Engineering & Research, Inc. [933]
Nobbe, Daniel A., & Associates, Inc. [682]
North American Solar Development Corp. [223]
North Dakota, University of [934]
Northeast Solar Energy Center [1014]
Northern Solar Power Co. [224]
Northrup Inc. [225]
North Wind Power Co., Inc. [226]
Novan Energy, Inc. [227]
NRG Company [228]
Nuclear Technology Corporation [229]
NUTEK [229]
Nydam, A. J., Co. Inc. [683]

O

Oak Ridge National Laboratory [935]
OEM Products, Inc. [230]
Oklahoma, University of [936]
Oklahoma State University [937,938]
Olin Corp./Brass Group [231,232]
Olmon & Hutchinson Architects [684]
Olson and Associates [685]
Olson and MacDonald, Inc. [686]
Olympic Solar Corporation [233]
Oregon, University of [939]
Oregon Solar Institute [1015]
Orlando Laboratories Inc. [940]
Owens-Illinois Sunpak Division [234]
Ozark Institute [1016]

P

Paas, Al, and Associates [687]
Pacific Sun Inc. [688]
Paintridge Design & Development, Inc. [689]
Panich, David, Architect [690]
Park Energy Company [235]
Parkway Windows [941]
Passive Solar Alternatives [691]
Passive Solar Institute [942]
Paulson Engineering [692]
Pedersen, Arthur Hall [693]
Pedersen, Edward [694]
Pennsylvania State University, The [943]
People's Sun Solar Construction Co. [695]
Perkins, Morris Richard [696]
Peters-Williams-Kubota, P.A. [697]
Petite, David L., & Associates [698]
Phelps Dodge Brass Company [236]
Phifer Western [237]
Philips Industries, Inc. [188,238]
Pioneer Solar Construction [699]
Piper Hydro Inc. [239]
Pipe Systems Inc. [244]
Pittman Earthworks [700]
Pittsburgh Corning Corp. [240]
Poley, Lincoln A., Architect [701]
Polytechnic Institute of New York [944]
Porter, W.H., Inc. [241]
Power-Sonic Corp., The [242]
Practical Solar [1017]
Prado [702]
Progress Industries Inc. [243]
PSI Energy Systems, Inc. [244]

Q

Quality Energy Systems [843]

R

Radco Products, Inc. [245]
Radiant Equipment Co. [945]
Radiation Research Associates, Inc. [946]
Radice, F.C., Research & Development [947]
RA Energy Systems, Inc. [246]
Ragsdale, Lee, Associates [247]

ALPHABETICAL INDEX 403

Ralos Manufacturing Co. [248]
Ramada Energy Systems, Inc. [249]
Ram Products Inc. [250]
Raypak Inc. [251]
RCA Corporation [948]
R C Solar Co. Inc. [252]
R & D Enterprise [253]
REA Associates Inc. [949]
Reaction Research Laboratories [254]
Real Gas & Electric Co., Inc. [703]
Rectech Inc. [950]
Refine Building & Construction Co. [704]
Refrigeration Research Inc. [255]
Reisz Engineering Company [705]
Remmers Engineering [706]
Research & Design Institute [951]
Research Institute for Advanced Technology [952]
Research Products Corp. [256]
Residential Energy Systems Inc. [257]
Restoration Preservation Architecture, Inc. [707]
Revere Solar and Architectural Products, Inc. [258]
Rhemco [708]
Rho Sigma [259]
Richdel, Inc. [260]
Rigidized Metals Corporation [261]
Rise Inc. [382]
R-M Products [262]
Roaring Fork Resource Center [1018]
Robertshaw Controls Company [263]

Rochester Institute of Technology [953]
Rockrise/Odermatt/Mountjoy Associates (ROMA) [709]
Rockwell International [898]
Roggensack Insulation & Solar Inc. [710]
RoKi Associates, Inc. [711]
Roll Forming Corporation [264]
ROMA [709]
Rom-Aire Solar Corporation [265]
Rosenberg, Paul, Associates [712]
Ross & Baruzzini, Inc. [713]
Roth, Eckert & Jacobson, Inc. [714]
R. P. Woodcrafters [715]
Rural Housing Research (USDA) [984]
Rusth Industries [266]
Ryniker Steel Products Co. [267]

S

SAF Energy Consultants, Inc. [954]
Saltzberg, Edward, & Associates [716]
Santa Cruz Alternative Energy Coop [1019]
Santa Cruz Solarworks Inc. [268]
Sav-On Energy Products, Inc. [269]
Sav Solar Systems, Inc. [270]
Schemmer Associates, Inc., The [717]
Schipke Engineers [718]
Schloh, E. Gary, AIA [719]
Schmidt/Claffey Architects, Inc. [720]
Schramm, Richard A., Architecture, Inc. [721]

404 SOLAR CENSUS

Schriever, Lee—Architect [722]

Schwinghamers [955]

Scientific Building and Energy Consultants, Inc. [956]

J. Scotts, Inc. [271]

SEAgroup [723]

Seahorse Plastics Corporation [336]

Sealed Air Corporation [272]

Semcor Energy Systems [273]

Sennergetics [724]

Sepe, William R. [725]

Sera Solar Corporation [274]

SETSCO [275]

Shannon, Robert Foote, Architect [726]

Shelley Radiant Ceiling Co., Inc. [276]

Shelter Institute [1020]

Shelton Energy Research [957]

Sheridan Solar [727]

Shock Hydrodynamics Division/ Whittaker Corporation [958]

Siegel, E. F., and Associates, Ltd. [728]

Sierra Solar Systems, Inc. [729]

Silicon Sensors, Inc. [277]

Simmons & Sun, Inc. [730]

SIRI Corporation [168]

Skillestad Engineering, Inc. [278]

Skujins, John, AIA [731]

Skytherm Processes & Engineering [732]

Slack Associates Inc. [733]

Slemmons Associates Architects, P.A. [734]

Smith, A. O. [279]

Smith, Dan—Architects [735]

Smith, Harold A., & Associates [280]

Solag [281]

Solahart California [282]

Solar Age Designers [959]

Solar Age Magazine [1021]

Solar-Air by Jira [620]

Solaraire [736]

Solara Associates Inc. [283]

Solarado Inc. [284]

Solaray Corporation, The [285]

Solar Building Corporation [737]

Sol-Arc [738]

Solar City [286]

Solar Clime Designs [739]

Solar Collector Sales Co., Inc. [287]

Solar Comfort Systems Manufacturing, Inc. [288]

Solarcon, Inc. [740]

Solar Construction and Design Inc. [289]

Solar Contact Systems, Inc. [290]

Solar Control Corporation [291]

Solar Cooker Parts [292]

Solarcrete of Southern California [741]

Solar Dakota [742]

Solar Data Systems International [293]

ALPHABETICAL INDEX

Solar Design Associates Inc. [743]
Solar Design Consultants Inc. [744]
Solar Design Group, Ltd. [745]
Solar Designs (Placerville, CA) [746]
Solar Designs, (San Francisco, CA) [294]
Solar Development Co. [960]
Solar Development Inc. (of Denver) [295]
Solar Development Inc. [296]
Solar Disc Corporation [206]
Solar Dynamics of Arizona [297]
Solar Earth Consultants, Inc. [730]
Solar Electric Inc. [298]
Solar Energies of California [299]
Solar Energy Associates, Ltd. [747]
Solar Energy of Colorado, Inc. (SECO) [300]
Solar Energy Design Corporation of America [748]
Solar Energy Design, Inc. [749]
Solar Energy Engineering [751]
Solar Energy Engineering, Inc. [301]
Solar Energy Engineering/Solar Energy Systems [750]
Solar Energy Equipment Corp. [752]
Solar Energy Intelligence Report [1022]
Solar Energymaster [302]
Solar Energy Research Corp. [303]
Solar Energy Research Institute (SERI) [961]
Solar Energy Thermal System Company [275]

Solar Engineering and Manufacturing [304]
Solar Enterprises Hawaii Inc. [962]
Solar Environmental Engineering Company, Inc. [963]
Solar Equipment Corporation [305]
Solarex Corporation [306]
Solar-Eye Products Inc. [307]
Solar Fab, Inc. [308]
Solargizer International Inc. [309]
Solarglas [54]
Solar Greenhouse Assoc. [1023]
Solargy Corporation [310]
Solar Heat Co. [311]
Solar Heater Manufacturing Co. [312]
Solar Homes Inc./Suntrol [313]
Solar Home Systems Inc. [753]
Solar, Incorporated [314]
Solar Industries, Inc. [315]
Solar Industries, Incorporated [316]
Solar Industries of Florida [317]
Solar International [754]
Solar Kinetics, Inc. [318]
Solar King International, Inc. [319]
Solar Liberation Engineering [1024]
Solar Magnetic Labs [320]
Solar Music Inc. [321]
Solarnetics Corporation [322]
Solaron Corporation [323]
Solarpak, Inc. [755]
Solar Patents [964]

Solar Pathways, Inc. [324]
Solar Pathways Associates [756]
Solar P.I.E. [757]
Solar Power Co., Inc. [325]
Solar Power Corporation [326]
Solar Products, Inc. [327]
Solar Products Information & Engineering (Solar P.I.E.) [757]
Solar Products Manufacturing Corp. [328]
Solar Products Sun Tank Inc. [329]
Solar Radiation Industries [330]
Solar Research Systems [331]
Solar Resources [332]
Solar Resources, Inc. [333]
Solar Sauna [758]
Solar Search Corporation [334]
Solar Services, Inc. [759]
Solar Shelter, Inc. [335]
Solar Site Selector [192]
Solar Spectrum Inc. [336]
Solar Structures Inc. [760]
Solar Sunstill [965]
Solar Supply, Inc. [337]
Solar Survival [966]
Solar Sustenance Team [967]
Solar Systems Design Inc. [666]
Solar Technologies Inc. [761]
Solar Technology Associates [762]
Solar Technology Corporation [338]
Solartec Inc. [339]
Solartek [763]

Solar Thermal Systems Inc. [340]
Solartherm, Inc. [341]
Solar Thermal Test Facilities Users Association [968]
Solar-Therm-El [764]
Solartherm Manufacturing Corp. [342]
Solartran [23]
Solar-Trol [343]
Solartronics Inc. [344]
Solar Unlimited Inc. (Huntsville, AL) [345]
Solar Unlimited, Inc. (Sun Prairie, WI) [346]
Solar Usage Now, Inc. [347]
Solar Water Heaters of N.P.R. Inc. [348]
Solar/Wind Energy Systems Inc. [765]
Solar Works Inc. [766]
Solar World, Inc. [349]
Solation Products Inc. [350]
Sol Dev Co Inc. [767]
Solec International, Inc. [351]
Solectro-Thermo, Inc. [352]
Sol Energy Corporation [969]
Solergy Company [353]
Soleri, Paolo, Associates, Inc. [768]
Solex Corporation [354]
Solex Solar Energy Systems, Inc. [355]
Sol-Lector, Inc. [356]
Sollos Incorporated [357]

ALPHABETICAL INDEX

Solpower Industries, Inc. [358]
Solstice Designs, Inc. [769]
Sol-Temp Inc. [359]
Sonnewald Service [360]
Southeastern Solar Systems, Inc. [361]
Southern Solar Energy Center, Inc. (SSEC) [1025]
Southwest Ener-Tech, Inc. [362]
Southwest Solar Corporation [363]
Space/Time Designs, Inc. [770]
Specialty Manufacturing [169]
Spectrolab [364]
Spitznagel Partners, Inc., The [771]
Spring Mountain Energy Systems, Inc. [772]
SRI International [970]
S S Solar Inc. [773]
Stanford Resources, Inc. [971]
Stanford University [972]
Stanley Consultants, Inc. [774]
Stanley, E. J., Enterprises [775]
State Industries, Inc. [365]
Steed/Hammond/Paul [776]
Steel, Gerald B. [777]
Stein Partnership, The [778]
Stella, Edward V. [779]
Stern Research Corporation [973]
Straub, Van Dine & Dziurman Associates, Inc. [780]
Strayer, R.D., Inc. [781]
Structural Composites Industries, Inc. (SCI) [366]
Stutzman, James R., Architects [782]

Subjinske, R.S., & Assoc. Inc. [783]
Sunburst Solar Energy, Inc. [367]
Sun Cycle Co./Solar R&D [974]
Sundance Building Contractors [784]
Sundesign Services Inc. [785]
Sundex Solar Equipment Sales [368]
Sunduit, Inc. [369]
Sunearth of California [370]
Sunearth Solar Products Corp. [371]
Sunflower Energy Works [372]
Sun Harvesters of Oregon [786]
Sun Industries, Inc. [373]
Sun Life Systems Inc. [374]
Sunlight Energy Systems Inc. [787]
Sunmaster Corporation [375]
Sun Mizer Inc. [788]
Sun of Man Solar Systems [376]
Sun-Pac, Inc. [377]
Sun Pak [234]
Sunpower Inc. [975]
Sunpower Research & Development Corp. [976]
Sunpower Systems Corp. [378]
SUNRAE Inc. [1026]
Sun Ray of California [379]
Sun-Ray Solar Equipment Co., Inc. [380]
Sun-Ray Solar Heaters [381]
Sunrise Solar Products [789]
Sunsearch, Inc. [977]
Sunseed Energy [790]
Sunshelter Design [791]

408 SOLAR CENSUS

Sunshine Development Corp. Inc. [382]
Sun Solector Corp. [390]
Sunsource of Arizona [383]
Sunspace Unlimited [792]
Sunspool Corporation [384]
Sunstar Group [149]
Sun Stone [5]
Sun Stone Company, Inc. [385]
Sunstructures [793]
Sunstructures Construction [793]
Sunsystems [794]
Suntank Inc. [329]
Suntappers [795]
Suntek Research Associates [978]
Suntek Solar Products Ltd. [386]
Suntime Inc. [387]
Suntrak [317]
Suntree Solar Company [388]
Suntrol [313]
Suntroller Products [796]
Suntron [389]
Sunway Heating Systems, Inc. [797]
Sun-West Solar Energy Systems, Inc. [798]
Sunwind Ltd. [844]
Sunworks Division of Sun Solector Corp. [390]
Survival Consultants [799]
Swaney, Jack W. [800]
Sybron Corporation [116]
SYenergy Methods, Inc. [391]
SYMCOM INC. [886]

Synago Corporation [801]
Synergy, Inc. [802]

T

Talley Industries [890]
Taliman USA Inc. [754]
Tambling, Philip S., Architect [803]
Tarbell, Eaton W., & Associates, Inc. [804]
TA-SOLAR [208]
Technical Ventures [805]
Technitrek Corporation [392]
Teledyne Mono-Thane [393]
Telluride Designworks [806]
Tennessee, University of [979]
Tennessee Valley Authority [807]
Terra-Dome Corporation [808]
Terra-Light Inc. [394]
Terra-Sol [809]
Terrien Architects [810]
Tetra Corporation [980]
Texas A&M University [1027]
Texas Solar Energy Society, Inc., The [1028]
Texas State Technical Institute (TSTI) [1029]
Texxor Corporation [395]
Theorem Solar Energy [811]
Thermaco [396]
Thermal Energy Storage, Inc. [397]
Thermal Engineering & Design Co. [398]
Thermatool Corporation [399]

ALPHABETICAL INDEX

Thermex Solar Corporation [400]
Thermodular Designs, Inc. [401]
Thermo Electron Corporation [981]
Thermograte, Inc. [402]
Thomas, Joseph B., IV, Architect [812]
Thomason Solar Homes, Inc. [403]
Timberman, Paul, Architect [813]
Tios Corporation [814]
Tjernagel, Sven, Solar Systems [404]
Tona-Midsouth, Ltd. [405]
Total Energy [406]
Total Environmental Action, Inc. [815]
Tri-Ex Tower Corporation [407]
Trolex Corporation [343]
Turbonics Inc. [408]
Turner Greenhouses [409]
Twin Pane Division/Philips Industries [238]

U

UCE Inc. [982]
Unarco-Rohn [410]
Underground Space Center [983]
Unisol Energy Corporation [816]
United Energy Corporation [411]
United Standard Management Corp. [332]
United States Solar Industries [412]
Universal Design and Construction [817]
Universal Solar Development, Inc. [413]
Upland Hills Ecological Awareness Center [1030]
Urban Options [1031]
Urethane Molding [414]
USDA-SEA-AR [984]
U.S. Solar Corporation [415]
Utah, University of [985]

V

Vachon, Nix & Associates [818]
Valley Pump Company [416]
Valour Fiberglass Manufacturing, Inc. [430B]
Vanguard Energy Systems [417]
Vaughn Corporation [418]
Vegetable Factory, Inc. [419]
Viking Energy Corporation [819]
Viracon/Dial [420]
Virginia Polytechnic Institute and State University [986]
Vistron Corporation [121]
Von Arx, Joseph, Passive Solar Custom Homes [820]
Vulcan Solar Industries, Inc. [421]

W

Wagner, Seiss, Architect [821]
Walgamuth & Associates [822]
Ward, James,A., P. E. [823]
Ware, C. Edward, Associates, Inc. [824]
Warren, André R., Consultant & Design Inc. [825]
Washington, University of [987]

Watermeister [422]
Watson, Donald [826]
Wayne State University [1032]
Weather Energy Systems, Inc. [423]
Wenning Associates [760]
West CAP [827]
West Central Wisconsin Community Action Agency [827]
Western Illinois University [988]
Western Solar Development, Inc. [424]
Western Solar Utilizatoin Network [1033]
Western SUN [1033]
Westinghouse Electric Corp. [828]
Whatworks [829]
Whipple, John, Architect [830]
Whitefly Control Co. [1034]
Whittaker Corporation [958]
Wiel, S., & Associates [831]
Wiley, Ray A., Architect/Engineer [832]
Wilson & Company [833]
WINCO [87]
Windependence Electric Power Co. [989]

Winzler and Kelley [834]
Wisconsin, University of—Madison [990]
Wittman, Lawrence, & Co., Inc. [425]
Wolverine Solar Industries, Inc. [426]
Woodford/Sloan, AIA, Architect [835]
Woods and Associates, Inc. [836]
Wormser Scientific Corp. [837]

X

Xerxes Fiberglass, Inc. [48]

Y

Yang, William J., & Associates [838]
Yellott, John, Engineering Associates Inc. [839]
Ying Manufacturing Corp. [427]
Young, R.M., Company [428]
Younger, Herbert C., P.E. [840]

Z

Zaugg & Zaugg Architects, AIA [841]
Zephyr Wind Dynamo Company [429]
Zia Associates, Inc. [430]

TRADENAME INDEX
(by entry code)

AMERICAN   214

AQUARIUS   426

ARCTIC SOLAR HEATER   320

CALQFLO   38

CHILL CHASER   408

CHROMONYX BLACK CHROME   233

CLEARLITE   10

COE   45

CONSERVATIONIST, THE   279

COOLECTOR   300

DAYSTAR   340

DELTA   120

DELTA NATIONAL   298

DELTA-T   151

DYNA-STIL   297

EAGLE SUN   415

ECONOSOL   13

ECON-O-SUN   278

EFFICIENCY   116

ELECTRA   322

ELECTRA-SOLARIUS   322

ENERGY MACHINE   342

ENERGY SAVER   214

ENERSORB   78

EQUINOX   269

FOAMEDGE   393

FOAMGLAS   240

GREENROOM   338

HEAT-STICK   253

HELIO-MATIC   151

HOTSTREAM   214

HELIO-PAK   147

HELIOPASS   166

HELIOPHASE   28

HELIOSTILL   27

HI-CAL   152

HOMESIDE   7

INSULJAC   414

ISIS   165

KATHENE   180

KGL   152

KING-LUX   182

KOPPER KING   400

LASCOLITE   188

412 SOLAR CENSUS

LOF SUNPANEL 193
LOK RING 202
MAXORB 114
MICROL 205
MIROBLACK 18
NOVA 19
NOVAMET 114
OPTIMA 226
PHIFERGLAS 237
PLASTICOOL 55
PLEXUS 423
POOLSAVER 34
PRESTIGE 214
REDI-MOUNT 57
REFLEC-TEC 130
RIGID-TEX 261
ROTA FLOW 74
SAF 954
SANDS 214
SCOTCHCAL 209
SELECTIVE VEE 161
SKYSORB 114
SLIMLINE 316
SOLAIRE 31
SOL-A-METER 201
SOLAQUA 13
SOLAR-AIRE 314
SOLAR APPLIANCE 18

SOLAR AWNING 320
SOLAR BANK 7
SOLAR-BOND 231
SOL-ARC 380
SOLARCAP 71
SOLARCRAFT 365
SOLAR DISC 363
SOLAR DYNE 297
SOLARGLAS 54
SOLARGY 42
SOLARIS 403
SOLARISTOCRAT 27
SOLAR-KAL 179
SOLAR MODULE 88
SOLAROASTER 27
SOLAROLL 39
SOLAROOF 120
SOLAR PAK 52
SOLAR PATHFINDER 324
SOLAR POWER CALCULATOR 293
SOLAR ROOM 333
SOLAR SITE SELECTOR 192
SOLAR-SLAB 137
SOLARSTREAM 214
SOLARSTRIP 38
SOLARSYPHON DIODE 376
SOLARVENT 90

TRADENAME INDEX

SOLA-SENTRY 297	SUNPAK 234
SOLATEX 10	SUNPUMP/SUNCYCLE 111
SOLATRON 132	SUNSCREEN 237
SOLECTOR PAK 390	SUNSHINE 370
SOL-HEATER 331	SUN-SPOT 328
SOLO 27	SUN STONE 5
SOL-R 94	SUNSTREAM 139
SOLSTAR 295	SUNSYM 390
SOLTEC 338	SUNTAINER 314
SPRAY CEL 180	SUNTANK 430B
STRATOSOL 438	SUN-TEMP 328
STRATOTHERM 266	SUNTREK 13
SUNADEX 10	SUNTROL 313
SUN-BANK 328	SUNVERTER 2
SUNBOX 224	SUNWALL 179
SUNCATCHER 345	SUNWAVE 433
SUNCEIVER 143	TA-SOLAR 208
SUNCELL 256	TECHNITROL 392
SUN COOKER 123	TEDLAR 85
SUN DOME 118	TEFLON 85
SUN HAUS 423	TEMPSTAT 263
SUNJAMMER 24	TEMP TRAK 151
SUNKIT 430A	THERMAFIN 399
SUN-LITE 179	THERMALATOR 81
SUN MACHINE, THE 248	THERMINOL 213
SUNMAT 43	THERMOL 244
SUNMATE 299	THRUFLO 235
SUNMETER 20	TIMBERLINE 166

414 SOLAR CENSUS

TROL-A-TEMP 343
VERTAFIN 49
WINCHARGER 87
WINDOW WARMERS 66

XSOLTHERM 81
ZEROENERGY 7
ZEROTHERM 7
ZODIAC 426

SUBJECT INDEX
(by entry code)

absorber plates 10, 13, 21, 22, 27, 39, 43, 64, 67, 68, 75, 81, 89, 114, 150, 155. 161, 164, 169, 184, 219, 235, 269, 272, 276, 278, 284, 290, 298 303, 308, 319, 320, 323, 334, 345, 359, 362, 374, 380, 403, 413, 427

all-copper 3, 13, 19, 20, 56, 64, 141, 147, 177, 193, 195, 220, 227, 231, 232, 236, 245, 246, 258, 266, 268, 294, 297, 301, 312, 339, 358, 363, 369, 377, 388, 392, 394, 399, 400, 424

aluminum 15, 49, 89, 106, 120, 126, 139, 187, 214, 309, 382

plastic 119, 188, 249

with all-copper waterways 6, 56, 64, 91, 139, 141, 143, 161, 195, 214, 251, 269, 309, 311, 341, 367, 371, 415

absorbers, R&D 861, 863, 866, 867, 870, 901, 902, 908, 927, 930, 952, 953, 954, 956, 961, 977, 984, 987, 1010

agricultural applications 1, 4, 13, 39, 52, 55, 101, 104, 109, 112, 148, 152, 155, 166, 170, 181, 184, 186, 187, 188, 194, 208, 210, 227, 228, 234, 241, 253, 254, 260, 281, 295, 304, 308, 309, 314, 320, 323, 326, 334, 335, 336, 346, 359, 369, 372, 373, 375, 382, 385, 389, 403, 413, 426, 437, 438, 451, 498, 506, 509, 514, 516, 538, 540, 542, 549, 55 , 552, 602, 610, 611, 620, 630, 639, 641, 649, 664, 683, 688, 690, 705, 706, 715, 720, 732, 733, 735, 756, 766, 768, 797, 805, 807, 818, 819, 827, 833

R&D 845, 849, 861, 862, 868, 875, 884, 889, 899, 901, 902, 905, 908, 911, 920, 931, 934, 935, 939, 952, 953, 954, 959, 960, 961, 964, 966, 969, 970, 973, 975, 976, 979, 986, 988, 999, 1002, 1005, 1007, 1012, 1016, 1019, 1025, 1033

Also see animal confinement shelters, bioconversion, grain drying systems, irrigation systems, methane digestion systems

air handling systems 52, 70, 74, 100, 150, 156, 158, 180, 262, 278, 291, 323, 343, 423, 507

aluminum collector frame manufacturing 89

animal confinement shelters 188, 334, 335, 875, 906, 935, 986, 989, 1016

backup systems, design applications 437, 445, 451, 461, 468, 477, 478, 486, 492, 494, 499, 506, 507, 509, 514, 516, 521, 525, 528, 534, 539, 540, 542, 545, 550, 552, 556, 565, 566, 573, 574, 602, 604, 610, 613, 614,

416 SOLAR CENSUS

619, 632, 637, 640, 645, 647, 648, 662, 664, 668, 670, 683, 687, 690, 694, 706, 708, 711, 714, 716, 720, 722, 725, 741, 745, 749, 750, 765, 768, 772, 779, 788, 798, 800, 804, 805, 812, 816, 817, 828, 832, 833

backup systems, manufacture 5, 6, 10, 13, 16, 35, 42, 70, 75, 112, 125, 150, 152, 170, 174, 181, 203, 207, 214, 225, 227, 242, 251, 254, 270, 289, 295, 298, 304, 315, 323, 328, 336, 359, 362, 408, 417, 421, 426, 427

backup systems, R&D 846, 847, 968, 890, 892, 901, 902, 914, 917, 927, 932, 934, 942, 957, 959, 984, 995, 1000, 1010

battery chargers 87, 127, 175, 887, 890, 975, 1005

bioconversion 27, 170, 281, 304, 309, 438, 516, 542, 544, 552, 602, 745, 765, 805, 833, 844, 850, 852, 861, 862, 866, 875, 883, 889, 894, 901, 902, 905, 908, 919, 928, 934, 935, 942, 950, 961, 970, 975, 976, 988, 999, 1002, 1005, 1010

Btu meters 60, 171, 222, 259

building performance monitoring 469, 520, 523, 526, 530, 531, 536, 537, 538, 544, 556, 561, 562, 570, 597, 603, 610, 616, 621, 649, 659, 666, 667, 674, 676, 688, 702, 723, 738, 779, 815, 818, 823, 828, 831, 837, 935

chillers, design 437, 438, 451, 459, 478, 481, 494, 497, 505, 507, 514, 516, 521, 544, 552, 556, 597, 604, 606, 632, 645, 651, 668, 716, 765, 807, 833, 843

chillers, manufacture 21, 31, 42, 92, 134, 159, 187, 230, 304, 355, 405, 427

chillers, R&D 860, 867, 902, 915, 918, 927, 975

coatings 3, 13, 18, 27, 38, 56, 67, 77, 78, 79, 107, 114, 116, 119, 152, 161, 170, 174, 229, 255, 276, 355, 356, 362, 374, 381, 390, 403, 404, 424, 451, 481

coatings, R&D 79, 863, 866, 867, 877, 883, 901, 915, 927, 928, 930, 952, 953, 956, 961, 965, 970, 984, 1010, 1012

collectors, air

   design applications 432, 435, 436, 437, 438, 444, 446, 451, 452, 459, 461, 467, 473, 474, 478, 486, 488, 490, 494, 498, 505, 506, 516, 523, 526, 532, 534, 535, 539, 542, 547, 549, 550, 552, 555, 556, 561, 563, 565, 566, 570, 572, 577, 593, 597, 598, 601, 604, 607, 613, 620, 626, 627, 632, 633, 634, 636, 637, 639, 640, 644, 645, 648, 651, 652, 671, 677, 678, 682, 683, 687, 694, 706, 707, 710, 712, 714, 716, 717, 720, 722, 725, 733, 735, 741, 742, 743, 759, 765, 766, 767, 768, 769, 772, 773, 774, 781, 782, 784, 786, 792, 797, 798, 802, 805, 806, 809, 814, 815, 817, 827, 828, 833, 837, 842, 996 1002

   manufacture 7, 12, 14, 15, 27, 35, 52, 54, 61, 70, 75, 81, 100, 103, 110, 120, 122, 124, 148, 150, 161, 165, 166, 178, 189, 195,

SUBJECT INDEX 417

196, 210, 221, 223, 234, 235,
249, 252, 254, 256, 262, 265,
267, 278, 287, 288, 295, 298,
299, 304, 307, 308, 309, 313,
314, 315, 320, 323, 334, 335,
336, 341, 346, 352, 359, 362,
368, 372, 379, 382, 385, 389,
424, 425

R&D 859, 865, 866, 870, 875,
876, 878, 901, 904, 908, 915,
920, 923, 927, 928, 930, 932,
934, 952, 954, 956, 958, 959,
960, 961, 977, 979, 981, 984,
986, 991, 1000, 1010, 1012

collectors, concentrating

design applications 438, 451, 459,
461, 492, 494, 505, 514, 516,
542, 550, 552, 556, 566, 612,
613, 616, 620, 632, 633, 644,
651, 652, 676, 677, 678, 706,
712, 716, 717, 720, 745, 751,
773, 788, 795, 805, 814, 817,
833

manufacture 1, 4, 13, 27, 59, 101,
104, 111, 116, 120, 123, 130,
134, 149, 170, 181, 195, 213,
243, 270, 283, 292, 304, 305,
315, 317, 318, 332, 336, 338,
344, 346, 350, 368, 378, 405,
427

R&D 844, 852, 853, 861, 866
868, 872, 876, 900, 901, 904,
905, 908, 927, 928, 930, 952,
958, 960, 961, 964, 981, 985,
1010, 1012

collectors, liquid

design applications 432, 436, 437,
438, 442, 443, 444, 445, 446,
451, 452, 459, 461, 473, 474,
477, 478, 480, 490, 494, 497,
498, 499, 502, 503, 505, 506,
507, 514, 516, 520, 521, 526,
533, 539, 540, 542, 547, 550,
552, 555, 556, 561, 563, 565,
568, 573, 577, 597, 598, 601,
604, 606, 607, 609, 612, 613,
620, 626, 627, 630, 632, 633,
636, 637, 639, 640, 644, 645,
648, 651, 652, 659, 661, 664,
667, 670, 677, 678, 682, 694,
703, 706, 708, 710, 712, 714,
716, 717, 720, 722, 724, 725,
729, 732, 733, 735, 741, 742,
743, 752, 754, 759, 760, 765,
766, 767, 772, 773, 774, 781,
784, 786, 787, 788, 790, 796,
798, 802, 805, 814, 817, 827,
828, 833, 837, 842, 996

manufacture 1, 3, 6, 7, 12, 13,
14, 16, 18, 19, 20, 21, 22, 24,
27, 39, 41, 43, 45, 46, 54, 56,
57, 59, 64, 67, 69, 91, 92, 94,
95, 99, 101, 104, 106, 107,
108, 109, 110, 112, 114, 119,
120, 124, 132, 141, 142, 143,
149, 152, 155, 161, 164, 166,
169, 170, 181, 184, 187, 189,
190, 193, 195, 206, 208, 213,
214, 219, 220, 221, 225, 227,
231, 241, 245, 246, 249, 251,
253, 255, 258, 262, 269, 272,
275, 283, 286, 287, 288, 290,
294, 296, 297, 298, 299, 300,
301, 302, 303, 304, 307, 308,
309, 310, 311, 312, 313, 315,
316, 317, 319, 320, 323, 328,
331, 335, 336, 339, 340, 341,
344, 345, 346, 347, 348, 353,
355, 356, 358, 361, 362, 363,
368, 370, 371, 373, 374, 375,
377, 379, 381, 388, 390, 392,
396, 400, 403, 404, 405, 412,
413, 415, 421, 424, 425, 427

R&D 846, 847, 852, 855, 859,
861, 862, 866, 867, 868, 870,

872, 875, 876, 896, 900, 901,
902, 904, 908, 920, 923, 927,
928, 930, 936, 953, 956, 958,
959, 960, 961, 964, 974, 977,
979, 981, 986, 991, 1010

collectors, vacuum tube
  design applications 442, 459, 478,
  494, 514, 516, 542, 552, 556,
  566, 626, 632, 633, 644, 645,
  651, 652, 682, 706, 716, 717,
  720, 735, 795, 817, 833, 837
  manufacture 13, 21, 29, 59, 101,
  104, 122, 149, 206, 234, 254,
  375
  R&D 852, 872, 875, 876, 900,
  901, 908, 927, 928, 930, 961,
  977, 1010

commercialization 845, 889, 903,
914, 944, 961, 996, 1008, 1009,
1012, 1014, 1015, 1025, 1028,
1033

computer analysis 189, 227, 256, 262,
293, 353, 379, 423, 469, 480, 481,
482, 483, 499, 501, 502, 514, 520,
524, 530, 536, 538, 544, 556, 562,
574, 587, 603, 631, 649, 654, 659,
667, 670, 688, 692, 703, 706, 738,
747, 750, 756, 773, 794, 815, 833,
845, 846, 861, 874, 915, 944, 977

computer software 9, 168, 170, 189,
193, 227, 390, 423, 431, 463, 469,
481, 501, 536, 556, 603, 631, 649,
658, 659, 670, 688, 706, 740, 747,
748, 756, 763, 815, 828, 833, 840,
845, 846, 861, 915, 944, 958, 961,
963, 977

construction management 513, 515,
516, 520, 523, 530, 555, 559, 561,
570, 582, 596, 610, 611, 612, 616,

621, 630, 657, 665, 667, 669,
674, 689, 691, 695, 697, 704,
723, 741, 746, 764, 767, 779,
785, 793, 794, 798, 804, 807,
818, 837

consulting engineers 175, 228, 262,
360, 374, 438, 442, 443, 458,
460, 461, 465, 467, 468, 469,
470, 472, 478, 479, 481, 482,
483, 484, 485, 490, 497, 500,
501, 502, 503, 506, 511, 512,
513, 514, 515, 516, 520, 524,
526, 530, 531, 536, 537, 538,
544, 548, 549, 550, 552, 555,
556, 557, 559, 562, 566, 570,
574, 579, 580, 587, 597, 608,
609, 610, 611, 613, 614, 615,
616, 619, 621, 625, 626, 630,
632, 633, 637, 642, 644, 646,
649, 650, 651, 658, 665, 667,
670, 673, 674, 676, 679, 682,
685, 688, 691, 692, 693, 697,
698, 703, 705, 706, 708, 712,
713, 716, 717, 718, 728, 729,
737, 738, 744, 746, 747, 749,
750, 755, 756, 764, 767, 769,
771, 773, 777, 779, 781, 783,
794, 798, 804, 807, 814, 815,
818, 819, 823, 828, 833, 837,
838, 906, 960

controls/instrumentation, design/
applications 437, 445, 458, 459,
461, 474, 477, 478, 481, 494,
497, 499, 502, 505, 506, 514,
516, 521, 524, 526, 533, 540,
542, 550, 556, 572, 597, 598,
606, 612, 632, 633, 640, 644,
645, 646, 652, 656, 666, 668,
671, 677, 682, 683, 702, 706,
708, 716, 717, 722, 724, 729,
733, 741, 742, 743, 747, 759,
761, 766, 767, 796, 798, 817,
833, 840, 846

SUBJECT INDEX 419

controls/instruments, manufacture 1, 2, 9, 13, 16, 20, 28, 30, 37, 40, 42, 58, 60, 61, 62, 63, 67, 70, 73, 74, 76, 79, 81, 82, 86, 92, 96, 98, 99, 100, 106, 107, 112, 113, 115, 133, 138, 146, 147, 148, 149, 150, 151, 153, 156, 157, 158, 159, 167, 168, 170, 171, 175, 176, 181, 191, 192, 194, 195, 205, 206, 210, 214, 222, 227, 235, 243, 254, 258, 259, 260, 263, 286, 290, 291, 293, 295, 296, 297, 303, 304, 307, 310, 314, 315, 316, 319, 323, 324, 328, 330, 335, 337, 343, 344, 346, 349, 355, 359, 374, 376, 383, 385, 398, 400, 403, 404, 427, 428, 430, 640, 646

controls/instruments, R&D 846, 862, 866, 867, 868, 874, 875, 876, 886, 887, 890, 901, 904, 907, 908, 918, 922, 927, 928, 938, 956, 959, 961, 981, 984, 1010

Also see Btu meters, flow measurement devices, heliostats, insolation monitors, pyranometers, pyrheliometers, radiometers, solar controllers, temperature sensors, thermometers, thermostats, tracker controls, wind survey instruments and controls

convective loop design 523, 628, 655, 739, 785

Also see envelope concept

cooling systems, design applications 437, 438, 451, 452, 461, 468, 478, 481, 494, 497, 503, 504, 505, 506, 514, 516, 526, 531, 539, 542, 550, 556, 565, 570, 594, 597, 598, 604, 606, 632, 640, 645, 647, 654, 683, 706, 709, 716, 720, 723, 732, 734, 739, 741, 765, 768, 781, 816, 818, 832

cooling systems, manufacture 1, 13, 21, 22, 31, 42, 52, 54, 67, 92, 100, 103, 170, 180, 187, 190, 206, 225, 228, 230, 237, 303, 318, 334, 336, 342, 354, 362, 375, 377, 403, 411

cooling systems, R&D 860, 866, 878, 902, 906, 908, 927, 938, 942, 944, 952, 956, 959, 975, 979, 1000, 1009, 1010

copper tubing and sheet 154, 877

corrosion inhibiting systems 18, 41, 78, 251

corrosion research 854, 861, 883, 885, 930

data acquisition systems 9, 649, 840, 845, 846, 876, 944, 961

desalination 515, 818, 896, 906

distillation, R&D 296, 333, 334, 639, 705, 844, 845, 861, 862, 869, 896, 902, 905, 907, 908, 918, 928, 932, 934, 935, 939, 952, 959, 961, 970, 976, 979, 996, 1002, 1010, 1016

distillation unit manufacture 13, 21, 27, 101, 112, 166, 168, 170, 195, 253, 260, 265, 270, 278, 283, 295, 297, 298, 304, 305, 309, 320

distillation systems, design 438, 439, 515, 542, 544, 552, 557, 566, 568, 593, 602, 604, 616, 626, 632, 661, 667, 708, 732, 733, 766, 829, 833

domestic hot water systems,
  design applications 432, 433, 434,
    435, 436, 437, 438, 439, 442, 443,
    444, 445, 446, 451, 452, 454, 455,
    459, 461, 468, 473, 477, 478, 479,
    480, 481, 482, 486, 488, 490, 492,
    494, 496, 497, 498, 499, 500, 501,
    502, 503, 504, 505, 509, 510, 511,
    514, 516, 518, 521, 526, 533, 534,
    535, 539, 540, 542, 543, 544, 545,
    546, 547, 548, 549, 550, 552, 555,
    556, 561, 563, 564, 566, 568, 569,
    570, 572, 573, 575, 576, 577, 583,
    588, 591, 592, 593, 595, 596, 597,
    598, 601, 602, 604, 605, 609, 611,
    612, 613, 623, 624, 625, 626, 627,
    630, 631, 632, 633, 635, 636, 640,
    641, 642, 643, 644, 645, 647, 648,
    650, 651, 652, 654, 659, 661, 662,
    667, 668, 670, 677, 678, 686, 687,
    692, 695, 698, 699, 702, 703, 705,
    706, 708, 709, 710, 711, 712, 716,
    717, 719, 720, 722, 723, 724, 726,
    728, 729, 734, 735, 739, 741, 745,
    749, 750, 751, 752, 753, 754, 757,
    759, 760, 765, 766, 767, 768, 769,
    772, 773, 774, 775, 781, 783, 784,
    786, 787, 788, 790, 792, 793, 795,
    798, 802, 803, 809, 811, 814, 815,
    816, 817, 818, 819, 827, 828, 829,
    831, 832, 833, 834, 837, 839, 843,
    845, 996, 1002

domestic hot water systems,
  manufacture 6, 7, 12, 13, 14, 16,
    18, 20, 21, 22, 24, 27, 28, 31, 29,
    41, 42, 45, 46, 49, 54, 56, 57, 59,
    60, 67, 68, 88, 91, 92, 94, 99, 100,
    101, 106, 107, 108, 109, 111, 114,
    117, 120, 125, 132, 134, 142, 143,
    147, 149, 150, 152, 155, 161, 162,
    163, 164, 166, 169, 170, 174, 177,
    181, 187, 190, 203, 206, 208, 210,
    214, 219, 221, 225, 227, 230, 231,
    235, 239, 246, 247, 251, 255, 256,
    258, 260, 262, 263, 265, 269, 270,
    278, 279, 282, 286, 288, 290, 295,
    297, 299, 300, 302, 303, 307, 308,
    309, 311, 312, 313, 315, 316, 317,
    320, 322, 323, 327, 328, 329, 335,
    338, 339, 340, 342, 344, 345, 347,
    348, 350, 353, 355, 356, 358, 359,
    361, 362, 363, 365, 368, 369, 370,
    371, 372, 373, 374, 375, 377, 378,
    379, 381, 382, 383, 385, 386, 388,
    390, 396, 400, 403, 404, 412, 413,
    418, 421, 424, 426, 427

domestic hot water systems, R&D
    847, 852, 853, 861, 866, 867, 868,
    869, 870, 874, 875, 876, 900, 901,
    902, 903, 904, 905, 906, 908, 914,
    920, 927, 928, 931, 936, 938, 939,
    944, 953, 956, 959, 961, 963, 964,
    966, 970, 974, 977, 979, 984, 991,
    1000, 1009, 1010, 1012

earth-sheltered design 446, 454, 468,
    470, 487, 517, 518, 519, 528, 549,
    551, 558, 589, 599, 600, 607, 617,
    628, 639, 655, 673, 690, 699, 700,
    702, 720, 730, 747, 768, 771, 780,
    782, 785, 800, 808, 817, 822, 937,
    983

  Also see subterranean structures

economic research 587, 613, 845,
    846, 915, 936, 961, 970

education 13, 112, 151, 189, 257,
    321, 323, 388, 490, 499, 517,
    528, 537, 545, 587, 603, 609,
    633, 636, 655, 666, 678, 691,
    723, 724, 747, 756, 768, 769,
    773, 785, 790, 792, 793, 794,
    815, 827, 841, 845, 849, 875,
    877, 882, 884, 888, 902, 903,
    905, 908, 915, 917, 919, 920,
    926, 928, 931, 932, 937, 939,

SUBJECT INDEX 421

942, 943, 952, 957, 961, 966, 967, 968, 985, 988, 992, 994, 996, 999, 1001, 1002, 1003, 1004, 1005, 1006, 1010, 1011, 1012, 1014, 1015, 1016, 1018, 1019, 1020, 1023, 1024, 1027, 1028, 1029, 1030, 1031, 1032

Also see technical training

electric vehicles 866, 888, 905

electroplating 38, 233

energy audits 112, 189, 192, 283, 379, 439, 446, 457, 461, 472, 481, 499, 500, 506, 513, 515, 529, 530, 537, 538, 545, 548, 557, 562, 587, 589, 608, 610, 611, 616, 631, 637, 649, 665, 667, 669, 674, 679, 682, 688, 711, 713, 723, 728, 729, 738, 747, 749, 750, 755, 756, 764, 771, 793, 794, 804, 809, 814, 818, 823, 828, 831, 837, 838, 843, 997, 1001

energy management planning 205, 497, 499, 506, 513, 514, 515, 516, 530, 537, 538, 544, 548, 579, 582, 583, 584, 585, 586, 587, 591, 597, 610, 611, 615, 626, 630, 631, 665, 674, 676, 678, 679, 681, 682, 685, 686, 688, 696, 702, 713, 723, 728, 738, 744, 747, 749, 755, 756, 771, 804, 814, 818, 819, 828, 831, 837, 883, 906, 952, 961, 962, 997

engineering design
See consulting engineers

engines 13, 101, 130, 206, 304, 305, 438, 439, 461, 514, 516, 525, 542, 544, 602, 632, 641, 650, 705, 716, 733, 818

Rankine cycle systems 4, 818, 860, 907, 918

R&D 844, 860, 861, 862, 866, 883,

887, 900, 902, 904, 907, 908, 918, 927, 960, 961, 975, 981

Stirling 861, 918, 975

envelope concept 496, 523, 546, 566, 645, 655, 701, 738, 745, 809

eutectic salts
See storage (phase-change)

fans 70. 90, 186, 266, 423

fan coil units 408

feasibility studies 103, 256, 446, 457, 461, 467, 469, 472, 501, 508, 513, 514, 515, 520, 526, 530, 531, 536, 556, 559, 562, 574, 579, 587, 610, 611, 612, 616, 626, 633, 644, 658, 665, 674, 679, 681, 688, 694, 696, 703, 706, 728, 738, 749, 771, 793, 794, 804, 805, 815, 818, 819, 823, 837, 838, 861, 906

fireplace inserts/heat extraction systems 45, 185, 361, 402, 404, 412

flow measurement devices 40, 86, 138, 146, 153, 171

Fresnel lenses 981

furnaces, design applications 438, 445, 461, 505, 509, 514, 516, 521, 542, 544, 552, 565, 596, 597, 602, 613, 640, 641, 645, 650, 654, 655, 683, 714, 716, 765, 766, 812, 817

furnaces, manufacture 13, 27, 36, 49, 67, 170, 181, 190, 221, 252, 262, 295, 304, 355, 381

422 SOLAR CENSUS

furnaces, R&D 866, 868, 870, 878, 896, 902, 908, 917, 927, 928, 952, 958, 961, 987

geothermal systems 67, 170, 335, 336, 373, 442, 515, 604, 631, 860, 883, 886, 891, 905

glass manufacture 10, 25, 54, 116, 128, 401

   bending 93, 430

glazing, collector

   design applications 451, 452, 459, 461, 467, 474, 506, 516, 539, 548, 549, 550, 552, 566, 574, 579, 607, 613, 626, 632, 645, 650, 677, 678, 687, 705, 720, 725, 733, 742, 767, 768, 786, 817, 827, 833, 837

   manufacture 10, 27, 46, 51, 54, 85, 99, 103, 106, 116, 119, 121, 128, 141, 150, 161, 170, 179, 188, 206, 227, 304, 307, 311, 323, 336, 347, 355, 359, 362, 374, 381, 403, 419, 436

glazing, passive systems 10, 25, 46, 51, 80, 121, 150, 188, 212, 235, 249, 270, 271, 298, 304, 308, 323, 327, 333, 336, 338, 347, 359, 380, 381, 401, 419, 423, 444, 451, 452, 459, 460, 461, 474, 506, 508, 516, 518, 523, 534, 539, 545, 548, 549, 552, 555, 563, 565, 566, 567, 570, 578, 579, 595, 598, 603, 604, 613, 618, 625, 626, 627, 633, 635, 636, 643, 645, 647, 659, 662, 666, 668, 677, 678, 686, 687, 691, 692, 699, 701, 705, 707, 708, 710, 719, 720, 722, 725, 727, 730, 733, 742, 744, 745, 767, 768, 782, 785, 786, 792, 807, 809, 817, 822, 826, 827, 829, 832, 833, 837, 849, 1000, 1005, 1010

glazing, for photovoltaics 116, 127, 206, 270, 304, 359, 502, 516, 545, 566, 574, 650, 668, 725, 829

glazing, R&D 866, 867, 901, 902, 908, 928, 939, 941, 942, 956, 959, 961, 965, 984, 985, 986, 1010

grain drying systems 52, 74, 273, 281, 295, 335, 369, 385, 549, 604, 705, 818, 819, 875, 935, 986, 988, 996, 999, 1005, 1019

greenhouses, design applications 434, 437, 438, 444, 445, 451, 452, 454, 459, 460, 461, 467, 474, 478, 481, 490, 492, 493, 496, 498, 506, 508, 509, 516, 520, 522, 523, 526, 528, 534, 535, 539, 540, 545, 546, 548, 549, 550, 552, 553, 555, 556, 563, 565, 566, 569, 570, 573, 574, 575, 578, 579, 581, 582, 591, 593, 597, 598, 600, 601, 602, 604, 610, 612, 617, 618, 622, 624, 625, 626, 627, 628, 636, 639, 640, 645, 648, 649, 652, 656, 663, 666, 667, 668, 678, 685, 686, 687, 691, 692, 695, 699, 701, 702, 705, 707, 710, 711, 715, 717, 719, 720, 727, 729, 730, 735, 739, 741, 743, 744, 745, 746, 753, 757, 758, 759, 765, 766, 767, 768, 769, 773, 774, 775, 777, 781, 782, 784, 785, 788, 790, 792, 793, 794, 795, 797, 799, 802, 803, 804, 805, 806, 807, 809, 812, 813, 815, 817, 820, 822, 826, 827, 829, 833, 842, 849, 858, 926, 999, 1001, 1002, 1012, 1023

SUBJECT INDEX 423

greenhouses, manufacture 13, 27, 46, 59, 67, 74, 80, 95, 118, 121, 125, 150, 152, 155, 170, 179, 181, 184, 188, 195, 212, 228, 268, 271, 281, 313, 317, 327, 333, 336, 338, 346, 359, 369, 380, 381, 385, 401, 403, 409, 411, 419, 423

Also see kits

greenhouses, R&D 874, 875, 884, 889, 901, 908, 926, 928, 931, 932, 939, 952, 959, 961, 963, 964, 966, 967, 984, 988, 996, 999, 1004, 1006, 1010, 1016, 1023, 1034

hardware (mounting, etc.) 7, 23, 52, 61, 72, 104, 112, 131, 297, 307, 351

heat exchangers, design applications 451, 452, 459, 461, 474, 478, 490, 497, 498, 499, 509, 511, 515, 516, 521, 525, 539, 542, 545, 549, 550, 552, 556, 574, 597, 598, 606, 607, 630, 632, 633, 640, 644, 645, 649, 668, 683, 692, 705, 706, 708, 716, 717, 720, 722, 724, 742, 743, 744, 745, 765, 766, 774, 788, 798, 802, 807, 809, 817, 832, 833, 837

heat exchangers, manufacture 7, 18, 21, 37, 39, 45, 64, 67, 74, 81, 83, 91, 92, 99, 111, 131, 137, 138, 142, 152, 160, 161, 163, 164, 166, 170, 172, 173, 177, 181, 184, 185, 199, 206, 208, 210, 214, 225, 227, 232, 253, 254, 270, 287, 288, 301, 303, 304, 313, 328, 329, 330, 331, 334, 335, 344, 345, 346, 347, 355, 359, 362, 369, 371, 374, 381, 383, 385, 396, 402, 403, 426, 439

heat exchangers, R&D 846, 852, 853, 861, 862, 866, 867, 868, 869, 880, 901, 902, 904, 905, 918, 919, 928, 933, 934, 939, 941, 949, 956, 961, 976, 984, 986, 1010

heat pumps 5, 7, 21, 127, 140, 167, 172, 177, 186, 198, 216, 225, 230, 254, 260, 265, 288, 312, 313, 335, 337, 353, 383, 385, 388, 392, 416, 458, 473, 502, 507, 525, 594, 596, 601, 610, 644, 706, 724, 734, 754, 783, 818

R&D 846, 848, 861, 864, 896, 934, 953, 973, 975, 979, 980, 986

water source 119, 388, 417, 604, 783, 973

heat recovery systems 5, 45, 105, 140, 142, 189, 303, 334, 379, 395, 408, 443, 446, 461, 514, 556, 631, 644, 664, 667, 705, 749, 788, 832, 833, 837, 862, 901

heat transfer cement 50

heat transfer fluids 67, 187, 199, 213, 327, 328, 342, 353, 706, 896, 898, 905, 922

heliostats 67, 706

historic site preservation/restoration 667, 684, 707, 804

horticulture/aquaculture 849, 867, 889, 926, 931, 959, 999, 1004, 1005, 1006 industrial/institutional specialization 1, 4, 211, 437, 438, 442, 443, 444, 446, 447, 449, 451, 455, 457, 458, 459, 465, 469, 472, 491, 494, 497, 499, 512, 513, 514, 516, 526, 529, 530, 531, 532, 537, 538, 539, 546, 554, 561, 562, 565, 567, 571, 574, 579, 580, 584, 585, 586, 597, 598, 616, 630, 651, 676, 681, 688, 696, 697, 705, 709, 713, 717, 718, 738, 755, 774, 804, 816, 818, 833, 838, 945

Also see solar community development

industrial process heating systems, design/engineering 442, 446, 501, 502, 514, 515, 531, 584, 616, 688, 705, 749, 751, 773, 802, 818, 828, 846, 852

industrial process heating systems, manufacture 1, 4, 13, 18, 21, 24, 31, 48, 101, 114, 134, 147, 170, 172, 234, 265, 295, 318, 319, 323, 351, 390, 400, 421

industrial process heating systems, R&D 861, 866, 900, 901, 904, 906, 940, 945, 961, 976, 1008, 1014, 1025, 1033

information dissemination 603, 691, 723, 769, 770, 815, 877, 882, 884, 888, 903, 919, 939, 952, 957, 961, 968, 992, 993, 994, 1000, 1003, 1004, 1006, 1007, 1009, 1012, 1013, 1014, 1015, 1016, 1017, 1018, 1019, 1021, 1022, 1023, 1024, 1025, 1026, 1027, 1028

insolation monitors 4, 82, 86, 157, 192, 194, 222, 248, 259, 324

insulated window coverings 66, 174, 212, 237, 241, 327, 338, 379, 393, 442, 518, 620, 636, 645, 653, 666, 670, 782, 935, 1031

insulation, collector 5, 12, 38, 47, 84, 94, 99, 110, 141, 142, 152, 155, 166, 170, 195, 206, 227, 240, 241, 284, 286, 288, 298, 311, 355, 359, 369, 374, 381, 403, 414, 437, 445, 452, 459, 460, 461, 474, 490, 494, 516, 539, 540, 542, 548, 550, 552, 566, 573, 597, 604, 607, 613, 632, 633, 640, 650, 652, 666, 705, 716, 720, 730, 742, 798, 805, 809, 827, 829, 833

insulation, movable 338, 451, 549, 570, 647, 653, 662, 668, 710, 732, 782, 842

insulation, pipe 21, 46, 99, 136, 142, 155, 166, 170, 195, 206, 227, 240, 286, 288, 290, 311, 335, 369, 374, 381, 393, 414, 426, 437, 445, 446, 452, 459, 460, 461, 474, 477, 490, 494, 516, 534, 539, 540, 542, 548, 550, 552, 566, 567, 573, 596, 597, 598, 606, 607, 613, 632, 633, 640, 645, 650, 652, 666, 705, 716, 720, 724, 725, 742, 766, 774, 787, 788, 798, 805, 809, 822, 827, 829, 833

insulation, R&D 853, 862, 867, 868, 901, 902, 904, 905, 928, 935, 936, 939, 942, 983, 984, 1010, 1012

insulation, super- 7, 462, 532, 560, 630, 636, 647, 699, 710, 720, 738, 782, 785, 832, 837, 942

SUBJECT INDEX 425

insulation, tank 5, 12, 21, 46, 47, 84, 99, 142, 155, 166, 170, 181, 195, 206, 227, 240, 241, 286, 288, 290, 311, 336, 355, 369, 381, 396, 437, 445, 446, 452, 459, 460, 461, 474, 477, 490, 494, 516, 534, 539, 540, 542, 548, 550, 552, 566, 573, 596, 597, 598, 606, 607, 613, 632, 633, 640, 645, 650, 652, 666, 705, 716, 720, 724, 742, 788, 805, 809, 822, 827, 829, 833

kits, do-it-yourself 7, 8, 12, 16, 52, 61, 81, 108, 126, 145, 278, 288, 304, 305, 340, 347, 383, 409, 430A, 603, 609, 769, 797, 799, 993

  controls 145, 383

  greenhouse 7, 67, 103, 126, 338, 409, 419, 423, 522, 581, 799, 967

  insulation 66

  site evaluation 757

  swimming pool 8, 108, 846, 991

  testing 743

inverters 2, 33, 175

irrigation systems 4, 228, 260, 819, 860, 861, 1016

land use planning 509, 513, 515, 539, 546, 574, 579, 582, 587, 610, 612, 647, 667, 676, 688, 689, 709, 738, 768, 777, 794, 807, 815, 818, 819, 837, 1004

laser technology 858, 935

life cycle costing 103, 469, 481, 530, 537, 556, 587, 631, 662, 667, 688, 696, 738, 747, 748, 749, 771, 804, 815, 818, 828, 837, 838, 906

meteorological/climatological data 256, 428, 845, 855, 952, 1007

methane digestion systems 170, 281, 304, 520, 544, 593, 602, 610, 616, 668, 694, 805, 819, 827, 829, 833, 834, 845, 862, 866, 889, 894, 902, 905, 919, 928, 932, 953, 961, 969, 970, 998, 1005, 1016

microprocessor systems 167, 168, 205, 349, 446, 499, 536, 890, 963

mobile homes 460, 462, 496, 797

modular/prefabricated solar homes/buildings 7, 23, 80, 137, 212, 336, 460, 498, 581, 782, 808, 979

  Also see kits, greenhouse

modular systems 13, 16, 42, 54, 61, 71, 101, 103, 104, 107, 108, 111, 130, 172, 225, 227, 245, 251, 350, 357, 359, 430, 845

  controls 88, 92, 147, 158, 167, 172, 263, 337, 349, 383, 392, 404, 423, 952

  heat transfer 37, 92, 199, 303, 330

  pumps 296, 337

  solar-electric 357

movable insulation, See insulation

natural lighting 188, 447, 588, 631, 647, 676, 686, 709, 723, 738, 747, 782, 821

ocean thermal conversion 27, 67, 266, 515, 557, 844, 845, 852, 866, 873, 904, 905, 910, 919, 935, 960, 961, 970, 971

passive design emphasis 55, 125, 137, 142, 166, 179, 243, 289, 387, 434, 435, 445, 446, 447, 448, 450, 451, 452, 462, 465, 472, 476, 487, 493, 496, 508, 512, 517, 518, 520, 523, 527, 536, 541, 549, 550, 553, 555, 558, 560, 563, 564, 565, 569, 570, 576, 581, 588, 590, 594, 595, 599, 609, 610, 617, 622, 623, 625, 627, 628, 634, 635, 638, 643, 647, 649, 653, 655, 660, 662, 673, 686, 687, 689, 690, 691, 699, 700, 701, 702, 711, 715, 719, 722, 723, 726, 727, 730, 732, 745, 747, 768, 769, 777, 778, 780, 782, 785, 790, 794, 809, 812, 820, 822, 826, 829, 832, 842, 1001, 1012

passive systems, R&D 875, 881, 915, 923, 928, 931, 935, 942, 945, 961, 978, 979, 980, 1000, 1004, 1033

pest control 392, 926, 931, 1014, 1034

photovoltaic array support structures 29, 131, 214, 218, 351

photovoltaic cells 26, 30, 53, 67, 77, 90, 162, 206, 211, 215, 218, 274, 277, 306, 317, 323, 326, 341, 347, 351, 357, 359, 364, 442, 451, 460, 462, 478, 496, 514, 516, 531, 540, 556, 566, 632, 668, 699, 712, 720, 742, 805, 807, 818, 819, 829, 861

photovoltaic panels 13, 22, 26, 30, 69, 90, 96, 104, 127, 131, 162, 170, 186, 206, 211, 215, 218, 243, 254, 270, 274, 277, 298, 306, 308, 315, 317, 323, 326, 337, 341, 347, 351, 352, 357, 359, 364, 442, 445, 451, 460, 462, 478, 496, 502, 514, 516, 531, 540, 544, 556, 566, 632, 668, 699, 703, 712, 720, 729, 741, 742, 772, 787, 805, 807, 818, 819, 861

photovoltaics, R&D 297, 411, 478, 545, 658, 815, 819, 844, 852, 853, 861, 863, 865, 866, 868, 875, 880, 881, 882, 885, 890, 892, 897, 898, 900, 902, 903, 905, 906, 924, 927, 930, 935, 939, 943, 948, 952, 958, 961, 969, 970, 971, 972, 973, 977, 980, 981, 982, 988, 1008, 1010, 1014, 1015, 1025

pyranometers 79, 194

pyrheliometers 113, 159

radiometers 113, 201, 259, 961

recreational vehicles 127, 214

reflective mirror manufacture 77, 130, 217, 250
    aluminized sheet 182, 209, 261

reflectors, concentrating 1, 12, 13, 27, 116, 123, 130, 182, 209, 211, 250, 270, 292, 318, 368, 460, 516, 544, 556, 566, 570, 616, 632, 633, 635, 668, 712, 716, 720, 817

reflectors, flat sheet 5, 12, 15, 54, 77, 116, 127, 130, 163, 174, 182, 209, 211, 217, 250, 270, 298, 311, 347, 359, 460, 490, 516, 544, 545, 556, 566, 570, 607, 632, 633, 635, 668, 687, 712, 716, 720, 722, 745, 790, 805, 817, 829, 837, 843

reflectors, R&D 852, 866, 872, 900, 905, 961, 1010, 1015

refrigerant-charged solar systems, 28, 255, 564, 725, 759

SUBJECT INDEX 427

remote power systems 26, 127, 215, 373

roll-forming 110, 196, 264, 430C

rubber (sealants, etc.) 84, 114, 183, 184

sheet metal 61, 261

silicon cells 26, 30, 82, 218, 357

silicon cell, **R&D** 861, 863, 880, 882, 961, 982

site planning 9, 192, 194, 222, 248, 324, 495, 512, 515, 544, 546, 559, 570, 574, 579, 587, 591, 610, 612, 626, 631, 635, 655, 670, 676, 678, 688, 694, 711, 723, 738, 748, 756, 770, 771, 794, 800, 804, 805, 815, 818, 821, 837

solar central receiver systems 898, 933, 936, 968, 987

solar community development 103, 574, 578, 610, 647, 649, 686, 695, 702, 762, 768, 900, 935, 999, 1005, 1012

solar controllers 4, 62, 63, 73, 74, 86, 151, 176, 640, 666

solar-electric conversion systems 2, 4, 22, 26, 69, 96, 104, 127, 170, 206, 211, 215, 218, 254, 305, 306, 326, 351, 352, 353, 357, 359, 364, 373, 462, 514, 706, 818, 819

solar-electric conversion R&D 334, 844, 847, 852, 860, 861, 882, 883, 893, 897, 898, 906, 918, 936, 960, 961, 971, 1015

solar film suppliers 51, 77, 85, 114, 200

solar-hydrogen systems 850, 900

solar novelties 12, 127, 321, 1019

solar ovens 27, 123, 292

solar roofs 1, 18, 20, 54, 120, 174, 179, 320, 336, 369, 391, 457, 732, 801

solar ponds 187, 438, 478, 499, 555, 556, 616, 618, 630, 640, 716, 732, 767, 785, 790, 805, 817, 822, 829, 843

solar pond, **R&D** 846, 851, 852, 866, 902, 905, 915, 920, 931, 961, 985, 986, 988

solar-powered satellites 852, 987

solar-powered vehicles 844, 974

solar thermal energy conversion system, **R&D** 101, 706, 852, 860, 861, 873, 875, 891, 961, 968, 982, 990

solder manufacture 44, 129

space heating/cooling systems, design applications 433, 435, 436, 437, 438, 439, 442, 443, 444, 445, 446, 451, 452, 454, 455, 459, 461, 468, 477, 478, 479, 480, 486, 488, 490, 492, 494, 496, 497, 498, 499, 500, 501, 502, 503, 504, 505, 509, 510, 514, 516, 518, 521, 533, 539, 540, 542, 543, 544, 545, 546, 547, 548, 549, 552, 553, 555, 556, 560, 561, 564, 566, 568, 570, 572, 573, 575, 576, 577, 583, 588, 590

428 SOLAR CENSUS

592, 593, 596, 597, 598, 601, 602, 604, 605, 612, 613, 623, 624, 625, 626, 627, 630, 631, 632, 635, 636, 640, 641, 642, 643, 644, 645, 647, 650, 651, 652, 654, 661, 667, 668, 670, 676, 677, 678, 686, 687, 692, 695, 698, 702, 703, 705, 706, 708, 709, 710, 711, 716, 717, 719, 720, 722, 723, 724, 728, 729, 732, 734, 735, 739, 741, 745, 749, 750, 751, 755, 757, 759, 760, 765, 766, 767, 768, 772, 774, 775, 781, 782, 784, 787, 788, 790, 791, 792, 793, 794, 795, 802, 803, 811, 814, 815, 816, 817, 818, 828, 829, 832, 833, 834, 837, 839, 842

space heating/cooling systems, manufacture 13, 20, 21, 22, 24, 27, 31, 39, 42, 57, 60, 61, 94, 100, 101, 103, 107, 109, 119, 120, 132, 134, 147, 163, 164, 169, 177, 181, 187, 190, 203, 206, 221, 223, 227, 230, 231, 239, 247, 251, 255, 256, 265, 282, 286, 288, 295, 300, 303, 309, 312, 315, 318, 320, 321, 323, 327, 335, 340, 342, 345, 350, 353, 355, 358, 359, 361, 369, 373, 375, 378, 379, 385, 390, 403, 404, 412, 421, 426, 427

space heating/cooling systems, R&D 852, 861, 870, 874, 875, 878, 900, 902, 904, 905, 906, 908, 920, 927, 928, 931, 936, 938, 939, 959, 961, 966, 970, 977, 979, 1009

spas/hot tubs 6, 12, 58, 108, 169, 177, 187, 206, 227, 246, 284, 288, 290, 300, 304, 316, 358, 363, 370, 386, 422, 427, 445, 477, 511, 533, 546, 632, 843

storage batteries 127, 186, 242, 254, 351, 478

storage batteries, R&D 854, 866, 986, 1004

storage bins/tanks, design application 432, 433, 446, 460, 467, 490, 505, 507, 509, 514, 516, 545, 566, 613, 618, 622, 632, 633, 640, 645, 652, 668, 670, 683, 706, 716, 719, 733, 741, 742, 767, 805, 809, 817, 829

storage bins/tanks, manufacture 12, 48, 52, 61, 81, 117, 150, 173, 179, 214, 228, 235, 258, 289, 336, 359, 385, 403, 404, 406, 430B, 436

storage bins/tanks, R&D 867, 875, 905, 934, 938, 961, 988, 1010

storage, fuel cell 69, 478, 632, 668, 716, 866, 905, 927, 961

storage, liquid—design applications 432, 436, 437, 445, 460, 490, 494, 499, 514, 516, 521, 526, 532, 533, 539, 540, 565, 566, 604, 606, 607, 612, 613, 618, 622, 627, 628, 630, 632, 633, 635, 640, 644, 645, 647, 652, 659, 668, 687, 706, 708, 711, 716, 717, 720, 729, 741, 742, 745, 766, 767, 774, 781, 798, 805, 812, 817, 822, 837

storage, liquid—unit manufacture 13, 16, 18, 21, 22, 46, 52, 99, 109, 112, 124, 125, 135, 149, 163, 164, 170, 177, 203, 206, 213, 214, 227, 241, 381, 385, 396, 403, 404, 412, 426

rods 244

water in plastic packs 5

SUBJECT INDEX 429

storage, liquid—R&D 837, 859, 862, 866, 867, 868, 875, 901, 902, 905, 920, 936, 942, 953, 960, 961, 977, 979, 986, 1010

storage, phase-change 5, 105, 149, 166, 170, 206, 227, 230, 243, 278, 281, 314, 327, 346, 347, 373, 374, 388, 395, 397, 403, 419, 432, 435, 436, 451, 460, 473, 506, 514, 516, 526, 545, 566, 570, 612, 613, 620, 622, 632, 644, 645, 662, 668, 703, 706, 711, 716, 717, 722, 733, 741, 742, 767, 781, 786, 798, 817, 829, 837

storage, phase-change—R&D 105, 846, 852, 882, 902, 905, 907, 934, 952, 961, 977, 987, 1010

   calcium chloride hexahydrate 105

   eutectic salts 27, 315, 435, 506, 711, 959

   freon 28

storage, solid 5, 81, 103, 346, 359, 432, 435, 446, 460, 475, 510, 514, 516, 518, 526, 528, 534, 539, 540, 545, 563, 565, 566, 570, 607, 613, 618, 622, 625, 630, 632, 635, 636, 730, 741, 745, 767, 772, 777, 781, 805, 809, 817, 822, 829

storage, solid—R&D 852, 905, 935, 961, 991, 1000

storage tubes 179

subterranean structures 528, 589, 599, 600, 607, 673, 690, 699, 701, 785, 808, 817, 989, 1007

   Also see earth-sheltered design

sunspaces 327, 401, 423, 451, 498, 636, 690, 711, 782, 792

superinsulation, See insulation

swimming pool blankets/enclosures 34, 65, 71, 118, 197, 272, 323, 425

swimming pool heating systems, design applications 435, 436, 437, 443, 445, 454, 459, 460, 477, 479, 481, 490, 496, 501, 503, 505, 508, 511, 514, 516, 521, 526, 528, 533, 540, 546, 547, 548, 550, 552, 555, 556, 566, 568, 570, 572, 596, 597, 598, 604, 612, 613, 618, 622, 627, 628, 630, 632, 635, 639, 640, 643, 645, 647, 651, 656, 659, 661, 668, 685, 695, 699, 701, 703, 708, 716, 717, 720, 729, 742, 752, 754, 757, 759, 765, 774, 775, 786, 787, 788, 790, 793, 795, 802, 809, 811, 817, 821, 829, 833, 837, 839, 843

swimming pool heating systems, manufacture 5, 6, 7, 8, 12, 18, 21, 22, 24, 29, 39, 43, 46, 52, 54, 56, 57, 60, 68, 81, 91, 94, 95, 99, 107, 108, 111, 114, 118, 119, 132, 142, 143, 147, 155, 162, 163, 169, 170, 181, 187, 195, 206, 227, 231, 232, 246, 251, 260, 262, 269, 272, 286, 287, 290, 295, 298, 300, 308, 309, 311, 313, 316, 317, 320, 322, 323, 327, 329, 331, 335, 344, 346, 355, 358, 362, 368, 369, 370, 374, 381, 386, 392, 403, 412, 421, 425, 426, 427, 991

technical training programs 21, 112, 323, 412, 1002, 1010, 1011, 1020, 1029

temperature sensors 62, 73, 74, 82, 88, 145, 159, 176, 207, 222, 259, 260, 263, 270, 307, 428 testing facilities 1, 333, 502, 526, 538, 541, 542, 544, 562, 749, 806, 831, 833, 837, 845, 846, 871, 876, 885, 903, 904, 915, 921, 923, 928, 935, 944, 945, 946, 958, 968, 970, 989

thermal energy storage system R&D 105, 713, 837, 852, 859, 861, 862, 882, 883, 898, 900, 906, 939, 961, 970, 979, 990

thermosiphonic systems 16, 64, 81, 225, 251, 376, 655, 797

thermometers 88, 133, 151, 191, 285, 890

thermostats 74, 76, 98, 291, 342, 984

differential 20, 37, 88, 107, 145, 151, 176, 222, 259, 328, 507

tracker controls 4, 13

tracking system development 4, 13, 134, 168, 208, 243, 344, 352, 907, 961, 977

trombe walls 27, 39, 114, 161, 195, 206, 249, 268, 333, 434, 438, 444, 451, 452, 474, 478, 498, 499, 506, 508, 516, 526, 534, 536, 545, 549, 550, 552, 555, 556, 565, 570, 572, 576, 593, 595, 598, 601, 610, 612, 613, 625, 626, 630, 633, 634, 635, 636, 639, 640, 643, 645, 647, 648, 651, 652, 662, 671, 677, 678, 682, 687, 690, 691, 695, 706, 708, 710, 714, 716, 717, 720, 722, 725, 730, 735, 741, 742, 745, 753, 765, 767, 781, 782, 785, 792, 805, 809, 815, 817, 821, 833, 842, 996

trombe walls, R&D 875, 876, 901, 904, 908, 915, 920, 928, 935, 942, 952, 961, 980, 1005, 1010, 1012

tubing 60, 64, 97, 154, 179, 184, 228, 877

tube joints 202

wind generation systems 12, 17, 27, 32, 59, 67, 69, 87, 108, 109, 142, 163, 204, 208, 226, 227, 254, 292, 304, 310, 323, 366, 407, 410, 411, 416, 428, 429, 427, 438, 441, 442, 445, 451, 468, 475, 478, 508, 509, 516, 540, 550, 552, 557, 566, 572, 583, 596, 597, 604, 610, 622, 624, 632, 640, 643, 652, 655, 668, 694, 699, 702, 703, 712, 717, 720, 737, 741, 742, 745, 765, 766, 771, 788, 799, 805, 817, 827, 829, 843

wind generation systems, R&D 858, 861, 866, 869, 873, 875, 886, 887, 888, 890, 894, 898, 905, 909, 910, 911, 913, 916, 925, 927, 928, 929, 931, 934, 939, 945, 947, 951, 953, 959, 961, 970, 979, 986, 988, 989, 1005, 1008, 1014, 1025, 1033

wind generation systems, support structures/towers 32, 87, 310, 407, 410

wind survey instruments 9, 82, 222, 226, 428, 911

wood burning systems 17, 59, 75, 125, 150, 166, 181, 185, 221, 257, 361, 404, 528, 613, 617, 634, 647, 648, 655, 749, 804, 812, 817, 827, 847, 917, 928, 957, 1005, 1016, 1025, 1033

GEOGRAPHICAL INDEX

ALABAMA

University of Alabama [845]
Johnson Environmental & Energy
  Center
P.O. Box 1247
Huntsville, Alabama 35807

Alabama Solar Energy Center
University of Alabama
P.O. Box 1247
Huntsville, Alabama 35807

Halstead & Mitchell Co. [143]
P.O. Box 1110
Scottsboro, Alabama 35768

Reisz Engineering Company [705]
2607 Leeman Ferry Road
Huntsville, Alabama 35805

Solar Unlimited Inc. [345]
37 Traylor Island
Huntsville, Alabama 35801

ALASKA

Alaska Department of Commerce
  and Economic Development
Energy and Power Development
338 Denali Street
Anchorage, Alaska 99501

ARIZONA

University of Arizona [853]
Solar Energy Research Facility
Civil Engineering Building, Rm. 206
Tucson, Arizona 85721

Arizona Solar Energy Commission
1700 West Washington Street
Phoenix, Arizona 85007

Arizona State University [854]
Center for Solid State Science
Tempe, Arizona 85281

Arizona State University [855]
Laboratory of Climatology
Tempe, Arizona 85281

Brown & Brown [476]
726 North Country Club Road
Tucson, Arizona 85716

DSET Laboratories Inc. [885]
Box 1850, Black Canyon Stage
Phoenix, Arizona 85029

Ecotronic Laboratories Inc. [890]
Talley Industries
7745 E. Redfield Road
Scottsdale, Arizona 85260

432 SOLAR CENSUS

Enercom Inc. [529]
2323 South Hardy Drive
Tempe, Arizona 85282

Energy Planning & Investment
  Corporation (EPIC) [539]
833 N. 4th Avenue
Tucson, Arizona 85705

Jetel Company [175]
2811 North 24th Street
Phoenix, Arizona 85008

Matrix Inc. [201]
537 South 31st Street
Mesa, Arizona 85204

Motorola Semiconductor Group [215]
P.O. Box 2953
Phoenix, Arizona 85062

Ramada Energy Systems, Inc. [249]
1421 S. McClintock
Tempe, Arizona 85281

Schwinghamers [955]
3310 North 27th Avenue
Phoenix, Arizona 85017

Solar Dynamics of Arizona [297]
1100 North Lake Havasu Avenue
Lake Havasu City, Arizona 86403

Solar Resources [332]
A Division of United Standard
  Management Corporation
1735 East Indian School Road
Phoenix, Arizona 85016

Solar World, Inc. [349]
4449 North 12th Street, Suite 7
Phoenix, Arizona 85014

Sol Dev Co Inc. [767]
P.O. Box 43125
Tucson, Arizona 85733

Paolo Soleri Associates, Inc. [768]
Cosanti Foundation
6433 Doubletree Road
Scottsdale, Arizona 85253

Sunpower Systems Corp. [378]
510 S. 52nd Street
Tempe, Arizona 85281

Sunsource of Arizona [383]
3441 N. 29th Avenue
Phoenix, Arizona 85017

John Yellott Engineering
  Associates Inc. [839]
901 West El Caminito Drive
Phoenix, Arizona 85021

ARKANSAS

Arkansas Department of Energy
3000 Kavanaugh Boulevard
Little Rock, Arkansas 72205

James Lambeth, Architect [635]
1591 Clark
Fayetteville, Arkansas 72701

Ozark Institute [1016]
P.O. Box 549
Eureka Springs, Arkansas 72632

Valley Pump Company [416]
Division of Valley Industries, Inc.
P.O. Box 1364
Commerce & Exchange Streets
Conway, Arkansas 72032

CALIFORNIA

Abernathy Solar Consultant [432]
1250 North Walden Lane
Anaheim, California 92807

AcroSun Industries, Inc. [3]
1024 W. Maude Avenue
Sunnyvale, California 94086

GEOGRAPHICAL INDEX 433

Acurex Solar Corporation [4]
485 Clyde Avenue
Mountain View, California 94042

Advanced Energy Technology Inc. [6]
121 C Albright Way
Los Gatos, California 95030

Advance Technology Engineering [8]
P.O. Box 176
Canoga Park, California 91305

Aesop Institute/Sunwind Ltd. [844]
P.O. Box 880
Sebastopol, California 95472

Allen & Sheriff [440]
3020 South Robertson, BL S4
Los Angeles, California 90034

Altas Corporation [846]
500 Chestnut Street
Santa Cruz, California 95060

Alternative Energy Concepts of
 California Inc. [443]
P.O. Box 564
Palm Springs, California 92060

Alternative Publishers [993]
P.O. Box 357
Lakeside, California 92040

Alternative Resources, Inc. [445]
4 East LaVerne
Mill Valley, California 94941

American Home Solar Energy
 Systems, Inc. [19]
23142 Alcalde Drive
Laguna Hills, California 92653

ANCO Engineers Inc. [446]
1701 Colorado Avenue
Santa Monica, California 90404

Leroy Andrews Architects Inc. [448]
2284 South Victoria Avenue
Suite 2A
Ventura, California 93003

Applied Solar Energy Corporation [26]
P.O. Box 1212
Industry, California 91746

Approtech Solar Products [27]
770 Chestnut Street
San Jose, California 95110

APTEC Corporation [28]
1637 Pontius Avenue
Los Angeles, California 90025

ARCO Solar Inc. [30]
20554 Plummer Street
Chatsworth, California 91311

Astro Research Corporation [32]
6390 Cindy Lane
Carpinteria, California 93013

Automatic Solar Covers, Inc. [34]
1970 Gladwick
Compton, California 90220

Baldwin Solar Technologies [859]
5088 Laurel Drive
P.O. Box 5104
Concord, California 94524

Berkeley Solar Group [469]
3026 Shattuck Avenue
Berkeley, California 94705

Blue White Industries [40]
14931 Chestnut Street
Westminster, California 92683

Buckmaster Industries, Inc. [477]
P.O. Box 730
23846 Sunnymead Boulevard
Sunnymead, California 92388

434 SOLAR CENSUS

Caldwell Construction [480]
240 Commercial Street
Nevada City, California 95959

Cal Energy Consultants Inc. [481]
711 E. Walnut, Suite 205
Pasadena, California 91101

University of California [866]
Lawrence Livermore National
　Laboratory
P.O. Box 5500
Livermore, California 94550

California Energy Commission
1111 Howe Avenue
Sacramento, California 95825

California Solar Designs [482]
383 Union Street
Encinitas, California 92024

Candrex Pacific [483]
693 Veterans Boulevard
Redwood City, California 94063

Century Fiberglass Products [48]
Xerxes Fiberglass, Inc.
1210 North Tustin Avenue
Anaheim, California 92807

Willard D. Childs [873]
582 Rancho Santa Fe Road
Encinitas, California 92024

CMI Solarglas [54]
11015 Cumpston Street
North Hollywood, California 91601

Compool Corporation [58]
333 Fairchild Drive
Mountain View, California 94043

Conserdyne Corporation [60]
4437 San Fernando Road
Glendale, California 91204

Contextus Corporation [496]
110 Orange Street
Chico, California 95927

Convection Loops [1000]
Box AF
Stanford, California 94305

Coppersmith's [64]
P.O. Box 907
Cypress, California 90630

Crescent Engineering Co., Inc. [68]
12118 S. Western Avenue
Gardena, California 90249

Daniel Enterprises, Inc. [501]
Solar Technology Division
P.O. Box 2370
1291 S. Brass Lantern Drive
La Habra, California 90631

Davis Engineering, Inc. [503]
20976 Currier Road
Walnut, California 91789

Davis, Jacoubowsky, Hawkins
　Associates, Inc. [504]
299 Cannery Row
Monterey, California 93940

L. M. Dearing Associates, Inc. [71]
12324 Ventura Boulevard
P.O. Box 1744
Studio City, California 91604

Deposition Technology, Inc. [77]
7670 Trade Street
San Diego, California 92121

Robert E. Des Lauriers, Architect
　[505]
9349 El Cajon Boulevard
La Mesa, California 92041

Donovan Enterprises [511]
3642 Eagle Street
San Diego, California 92103

Earle Engineering [887]
P.O. Box 850
Alpine, California 92001

Earthmind [888]
4844 Hirsch Road
Mariposa, California 95338

Ecoenergetics Inc. [889]
180 Viewmont Avenue
Vallejo, California 94590

edSON Solar Systems [91]
7600 Capricorn Drive
Citrus Heights, California 95610

Ekose'a, Inc. [523]
573 Mission Street
San Francisco, California 95610

Electrical Construction Company [524]
P.O. Box 1028
Palmdale, California 93550

Charles Eley Associates [526]
342 Green Street
San Francisco, California 94133

Ellmore/Titus/Architects/Inc. [527]
736 Chestnut Street
Santa Cruz, California 95060

Energy Associates [99]
101 Townsend
San Francisco, California 94107

Energy Center, Inc. [533]
1726 W. Mission, P.O. Box 2086
Escondido, California 92025

The Energy Factory [103]
1550 N. Clark
Fresno, California 93703

Energy Harvester [104]
11807 Bernardo Terrace
San Diego, California 92128

Energy Info Newsletter [1004]
P.O. Box 98
Dana Point, California 92629

Energy Management Consultants,
   Inc. [537]
1180 South Beverly Drive, Suite 315
Los Angeles, California 90035

Energy Science Corporation [897]
6211 Covington Way
Goleta, California 93017

Energy Specialties [540]
9312 Greenback Lane
Orangevale, California 95662

Energy Systems Group [898]
Division of Rockwell International
8900 DeSoto Avenue
Canoga Park, California 91304

Enerspan, Inc. [108.789]
14168 Poway Road
Poway, California 92064

Energy Systems, Inc. [106]
4570 Alvarado Canyon Road
San Diego, California 92120

Engineering Consulting Services [542]
8 South Montgomery Street
P.O. Box 1809
San Jose, California 95109

ENTEC [544]
1900 Point West Way, Suite 171
Sacramento, California 95815

436 SOLAR CENSUS

Environmental Design Group [546]
269 West Second Street
Claremont, California 91711

Environmental Energy Management &
  Manufacturing Corp. [112]
2727 Temple Avenue
Signal Hill, California 90806

Environmental Instrumentation [547]
922 S. Barrington Avenue, Suite 202
Los Angeles, California 90049

Environmental Power Corporation
  [548]
12750-146 Centralia Street
Lakewood, California 90715

ESR—Environmental Systems
  Research [550]
1102 Coloma Way
Roseville, California 95678

FAFCO, Inc. [119]
235 Constitution Drive
Menlo Park, California 94025

Farallones Institute Rural Center
  [1005]
15290 Coleman Valley Road
Occidental, California 95465

Paul Fellers, Architect [553]
244-C Commercial Street
Nevada City, California 95959

Feuer Corporation [554]
2601 Ocean Park Boulevard
Santa Monica, California 90405

Filon Division of Vistron Corp. [121]
12333 Van Ness Avenue
Hawthorne, California 90250

Fischer Sun Cooker [123]
302 Center Street
Redwood City, California 94061

Fisher/Roberts [555]
220 Pier Avenue
Santa Monica, California 90405

Mario Fonda-Bonardi [560]
3111 3rd Street, No. 19
Santa Monica, California 90405

Friedman Sagar McCarthy Miller
  and Associates [565]
353 Folsom Street
San Francisco, California 94105

Randolph Steven Gade [566]
Energy Consultant
P.O. Box 248
Riverside, California 92501

Gantec Corporation [130]
P.O. Box 88447
Emeryville, California 94662

Geoscience Ltd. [904]
410 South Cedros Avenue
Solana Beach, California 92075

Glumac & Associates, Inc. [571]
1 Embarcadero Center
San Francisco, California 94111

Go Solar Inc. [573]
835 Grand Avenue
Grover City, California 93433

W. J. Graham Co. Inc. [136]
1759 E. Borchard
Santa Ana, California 92705

Wm. E. Green Associates [580]
Consulting Engineers
P.O. Box 38674
Sacramento, California 95838

GEOGRAPHICAL INDEX 437

George Greer Design & Construction [582]
5411 Colodny Drive
Agoura, California 91301

Griswold Controls [138]
2803 Barranca
Irvine, California 92714

Grundfos Pumps Corporation [140]
2555 Clovis Avenue
Clovis, California 93612

Habitec: Architecture & Planning [584]
445 Washington Street
Santa Clara, California 93110

Haines Tatarian Ipsen & Associates [585]
121 Second Street - 4th Floor
San Francisco, California 94105

Hawkins & Associates [590]
375 S. Main Street, Box 161
Lakeport, California 95453

Heliodyne, Inc. [147]
770 South 16th Street
Richmond, California 94804

Heliotrope General [151]
3733 Kenora Drive
Spring Valley, California 92077

Helix Solar Systems [152]
Division of American Creative Engineering
P.O. Box 2038
City of Industry, California 91746

HY-CAL Engineering [159]
12105 Los Nietos Road
Santa Fe Springs, California 90670

Imagineering Solar [604]
296 Vista Conejo
Newbury Park, California 91320

Insolarator [169]
A Product of Specialty Manufacturing Inc.
7926 Convoy Court
San Diego, California 92111

Integral Design [609]
3825 Sebastopol Road
Santa Rosa, California 95401

Interactive Resources, Inc. [611]
Comprehensive Professional Services Group
117 Park Place
Point Richmond, California 94801

Kamal S. Iskander and Associates [615]
350 North Acaso Drive
Walnut, California 91789

Jacobs-Del Solar Systems, Inc. [616]
A Member of the Jacobs Engineering Group
251 South Lake Avenue
Pasadena, California 91101

Jack Janofsky & Associates [174]
dba Energy Conservation Rep
148 South Crescent Heights Blvd.
P.O. Box 48405
Los Angeles, California 90048

Brion S. Jeannette & Associates [618]
Architects & Planners
470 Old Newport Boulevard
Newport Beach, California 92663

Johnson Engineering [621]
340 North Riverside Avenue
Rialto, California 92376

438 SOLAR CENSUS

M. Dean Jones, Architect [622]
45 Sycamore
Mill Valley, California 94941

Jordan's Solar Equipment Sales [177]
671 Newcastle Road
P.O. Box 758
Newcastle, California 95658

Stanley Keniston, AIA [626]
666 State Street
San Diego, California 92101

C. Kessel Co. [627]
17940 Laramie Lane
Twain Harte, California 95383

Kessel Insolar Designs [628]
19550 Cordelia Avenue
Sonora, California 95370

Kirkhill Rubber Company [183]
300 East Cypress
Brea, California 92621

Komara Company [184]
205 South M Street
Dinuba, California 93618

Wm. Lamb & Company [186]
10615 Chandler Boulevard
North Hollywood, California 91605

Paul Larkin, P.E. [637]
7202 Bodega Avenue
Sebastopol, California 95472

Lasco Industries [188]
Division of Philips Industries Inc.
3255 East Miraloma Avenue
Anaheim, California 92806

The L.A. Solar Mart [189]
3145 Glendale Boulevard
Los Angeles, California 92806

Gene La Tour [638]
1414 Ashland Avenue
Santa Monica, California 90405

Le Bleu Enterprises [640]
844 17th Avenue
Santa Cruz, California 95062

Lee & Associates [641]
499 Van Buren
Monterey, California 93940

Steven M. Leslie, Consultant [642]
2561 Gondar Avenue
Long Beach, California 90815

Letro Thermometer, Inc. [191]
8311 Airport Road
Redding, California 96002

Lewis & Associates [192]
105 Rockwood Drive
Grass Valley, California 95945

John R. Lewis, P.E. [644]
9 First Street, Suite 819
San Francisco, California 94105

Malcolm Lewis Associates [645]
220 Park Avenue
Laguna Beach, California 92651

Liss Engineering, Inc. [646]
2862-A Walnut Avenue
Tustin, California 92680

Living Systems [647]
Route 1, Box 170
Winters, California 95694

Lodi Solar Co. Inc. [195]
16 North Cherokee Lane
Lodi, California 95240

GEOGRAPHICAL INDEX 439

LSW Engineers [651]
9455 Ridgehaven Court
San Diego, California 92123

Paul H. Lutton [653]
50 St. James Place
Piedmont, California 94611

MacBall Industries, Inc. [197]
1820 Embarcadero
Oakland, California 94606

Madlin's Enterprises [654]
P.O. Box 1443
Palm Springs, California 92263

Marshall's Drafting/California
  Builders [656]
P.O. Box 471
Hesperia, California 92345

Math/Tec Inc. [659]
118 S. Catalina
Redondo Beach, California 90277

McCracken Solar Co. [661]
329 W. Carlos
Alturas, California 96101

Mechanical Seals Corporation [202]
2540 West 237th Street
Torrance, California 90505

Mid Pacific Solar Corporation [206]
Solar Disc Corporation
14202 Ventura Boulevard
Sherman Oaks, California 91403

Millville Windmills, Inc. [208]
10335 Old 44 Drive
Millville, California 96062

Mogavero & Unruh [668]
811 J. Street
Sacramento, California 95814

Monosolar, Inc. [211]
A Subsidiary of Monogram
  Industries, Inc.
100 Wilshire Boulevard, Suite 600
Santa Monica, California 90401

Jim D. Morelan & Associates [672]
2242 Camden Avenue, Suite 6
San Jose, California 95124

Mormec Engineering, Inc. [674]
115 West Main Street, C
Visalia, California 93277

Glen H. Mortensen, Inc. [675]
1036 W. Robinhood Drive, No. 201
Stockton, California 95207

Multi-Duti Manufacturing, Inc. [216]
2736 East Walnut
Pasadena, California 91107

Nielsen Engineering & Research
  Inc. [933]
510 Clyde Avenue
Mountain View, California 94022

Olin Corporation/Brass Group [231]
1717 Kettner Boulevard, Suite 250
San Diego, California 92101

Olson and Associates Engineering,
  Inc. [685]
P.O. Box 1006
Yreka, California 96097

Olson and MacDonald, Inc. [686]
1025 North Dutton Avenue
Santa Rosa, California 95401

Pacific Sun Incorporated [688]
439 Tasso Street
Palo Alto, California 94301

440 SOLAR CENSUS

Paintridge Design & Development, Inc. [689]
803 Russell Boulevard, No. 6
Davis, California 95616

Passive Solar Institute [942]
P.O. Box 722
Davis, California 95616

Paulson Engineering [692]
110 W. Bennett
Glendora, California 91740

People's Sun Solar Construction Co. [695]
P.O. Box 5187
Santa Rosa, California 95402

David L. Petite & Associates [698]
Route 1, Box 884
Shingle Springs, California 95682

Phelps Dodge Brass Company [236]
Solar Division
2665 Woodland Drive
Anaheim, California 92801

Phifer Western [237]
14408 E. Nelson
City of Industry, California 91744

Pioneer Solar Construction [699]
P.O. Box 213
Cloverdale, California 95425

Piper Hydro, Inc. [239]
3031 E. Coronado
Anaheim, California 92806

The Power-Sonic Corporation [242]
3106 Spring Street, P.O. Box 5242
Redwood City, California 94063

Progress Industries Inc. [243]
7290 Murdy Circle
Huntington Beach, California 92647

Quality Energy Systems [843]
7411 Broadway
Lemon Grove, California 92045

RADCO Products, Inc. [245]
2877 Industrial Parkway
Santa Maria, California 93454

Radiant Equipment Company [945]
1798 Panda Way
Hayward, California 94541

RA Energy Systems, Inc. [246]
11459-A Woodside Avenue, South
Lakeside, California 92040

Raypak Inc. [251]
31111 Agoura Road
Westlake Village, California 91359

Real Gas & Electric Company, Inc. [703]
Santa Rosa, California 95402

RHEMCO—Ray Huffman Energy Management Company [708]
4688 Oregon Street
San Diego, California 92116

Rho Sigma [259]
11922 Valerio Street
North Hollywood, California 91335

Robertshaw Controls Company [263]
100 W. Victoria Street
Long Beach, California 90805

Rockrise/Odermatt/Mountjoy Associates (ROMA) [709]
405 Sansome Street
San Francisco, California 94111

Edward Saltzberg & Associates [716]
14733 Oxnard Street
Van Nuys, California 91411

GEOGRAPHICAL INDEX 441

Santa Cruz Solarworks Inc. [268]
2715 Porter Street
Soquel, California 95073

Sav-On Energy Products, Inc. [269]
1080 North 11th Street
San Jose, California 95112

Sav Solar Systems, Inc. [270]
550 W. Patrice Place, Suite A
Gardena, California 90248

E. Gary Schloh, AIA [719]
213 Bean Avenue
Los Gatos, California 95030

J. Scotts, Inc. [271]
P.O. Box 563
El Toro, California 92630

SEAgroup [723]
Solar Environmental Architecture
418 Broad Street
Nevada City, California 95959

Sealed Air Corporation [272]
2015 Saybrook Avenue
Commerce, California 90040

SEMCOR Energy Systems [273]
25651 Taladro Circle
Mission Viejo, California 92691

Sennergetics [724]
18621 Parthenia Street
Northridge, California 91324

Sera Solar Corporation [274]
3151 Jay Street
Santa Clara, California 95050

SETSCO (Solar Energy Thermal
   System Company) [275]
40 Blaine Circle
Moraga, California 94556

Shock Hydrodynamics Division
Whittaker Corporation [958]
4716 Vineland Avenue
North Hollywood, California 91602

Sierra Solar Systems, Inc. [729]
12050 Charles Drive
Grass Valley, California 95945

Skytherm Processes & Engineering
   [732]
2424 Wilshire Boulevard
Los Angeles, California 90057

Solahart California [282]
3560 Dunhill Street
Sorrento Valley West
San Diego, California 92121

Sol-Arc [738]
2040 Addison Street
Berkeley, California 94704

Solar Clime Designs [739]
Box 9955
Stanford, California 94305

Solar Contact Systems, Inc. [290]
1415 Vernon Street
Anaheim, California 92805

Solarcrete of Southern California
   [741]
P.O. Box 476
Wrightwood, California 92397

Solar Designs [746]
2864 Ray Lawyer Drive, Suite 209
Placerville, California 95667

Solar Designs [294]
470 3rd Street
San Francisco, California 94107

442 SOLAR CENSUS

Solar Development Co. [960]
Solar Energy Research & Development Project
P.O. Box 208
Herald, California 95638

Solar Development Co. [960]
5860 Callister Avenue
Sacramento, California 95638

Solar Energies of California [299]
11421 Woodside Avenue
Lakeside, California 92040

Solar Energy Associates [747]
1226 Villanova Drive
Davis, California 95616

Solar Energy Engineering, Inc. [301]
31 Maxwell Court
Santa Rosa, California 95401

Solar Energy Equipment Corporation [752]
18368 Bandilier Circle
Fountain Valley, California 92708

Solar Equipment Corporation [305]
P.O. Box 357
Lakeside, California 92040

Solar Heat Co. [311]
10347 3rd Avenue, P.O. Box 42
Hesperia, California 92345

Solar International [754]
A Division of Taliman USA Inc.
366 Bel Marin Keys
Novato, California 94947

Solar International [754]
7246 Bellaire Avenue
North Hollywood, California 91605

Solar King International, Inc. [319]
8577 Canoga Avenue
Canoga Park, California 91304

Solar Liberation Engineering [1024]
Division of Davis Equities Corp.
311 South Bristol Street
Los Angeles, California 90049

Solar Music Inc. [321]
10615 Chandler Boulevard
North Hollywood, California 91601

Solarnetics Corporation [322]
1654 Pioneer Way
El Cajon, California 92020

Solar Research Systems [331]
3001 Red Hill Avenue, 1-105
Costa Mesa, California 92626

Solar Supply, Inc. [337]
6709 Convoy Court
San Diego, California 92111

Solar Technology Associates [762]
7348 Bonita Way
Citrus Heights, California 95610

Solartherm Manufacturing Corp. [342]
768 Vella Road
Palm Springs, California 92264

Solec International, Inc. [351]
12533 Chadron Avenue
Hawthorne, California 90250

Sollos Incorporated [357]
2231 South Carmelina Avenue
Los Angeles, California 90064

Solpower Industries, Inc. [358]
10211 C Bubb Road
Cupertino, California 95014

GEOGRAPHICAL INDEX 443

Sol-Temp Inc. [359]
1505 42nd Avenue, Suite 4
Capitola, California 95010

Southwest Solar Corporation [363]
441 North Oak Street
Inglewood, California 90301

Spectrolab [364]
A Subsidiary of Hughes Aircraft
12500 Gladstone Avenue
Sylmar, California 91342

Spring Mountain Energy Systems, Inc. [772]
2617 San Pablo Avenue
Berkeley, California 94702

SRI International [970]
333 Ravenswood Avenue
Menlo Park, California 94025

Stanford Resources, Inc. [971]
1095 Branham Lane, Suite 201
P.O. Box 20324
San Jose, California 95160

Stanford University [972]
Department of Materials Science and Engineering
Stanford, California 94305

E. J. Stanley Enterprises [775]
2730 E. Broadway
Long Beach, California 90803

Gerald B. Steel [777]
Engineering Consultant
P.O. Box 2369
Rancho Santa Fe, California 92067

Stern Research Corporation [973]
Route 3, Box 274, Orcutt Road
San Luis Obispo, California 93401

Structural Composites Industries Inc. (SCI) [366]
6344 North Irwindale Avenue
Azusa, California 91702

Sunburst Solar Energy, Inc. [367]
P.O. Box 2799
Menlo Park, California 94025

Sun Cycle Co./Solar R&D [974]
P.O. Box 2111
El Centro, California 92243

Sunearth of California [370]
P.O. Box 1729
Rohnert Park, California 94928

Sun Industries, Inc. [373]
P.O. Box 1547
Carlsbad, California 92008

Sunlight Energy Systems Inc. [787]
17961 Cowan Street
Irvine, California 92714

Sun Mizer Inc. [788]
275 East Pleasent Valley Road
P.O. Box 146
Camarillo, California 93010

Sun of Man Solar Systems [376]
Box 1066
Guerneville, California 95446

Sunrae Inc. [1026]
Solar Use Now For Resources and Employment
5679 Hollister Avenue, Room 6
Goleta, California 95446

Sun Ray of California [379]
P.O. Box 46033
Los Angeles, California 90046

444 SOLAR CENSUS

Sun-Ray Solar Heaters [381]
4898 Ronson Court
San Diego, California 92111

Sunrise Solar Products [789]
A Division of Enerspan, Inc.
14168 Poway Road
Poway, California 92064

Sunseed Energy [790]
G-4 Koshland
Santa Cruz, California 95064

Sunspool Corporation [384]
439 Tasso Street
Palo Alto, California 94301

Suntappers [795]
17501 Irvine Boulevard, No. 3
Tustin, California 92680

Suntek Research Associates [978]
506 Tamal Vista Plaza
Corte Madera, California 94925

Suntroller Products [796]
6384 Rockhurst Drive
San Diego, California 92120

Sun-West Solar Energy Systems, Inc. [798]
20445 Walnut, California 91789

Technitrek Corporation [392]
1999 Pike Avenue
San Leandro, California 94577

Theorem Solar Energy [811]
697 Brokaw Road
San Jose, California 95112

Thermaco [396]
6828 7th Avenue
Rio Linda, California 95673

Thermal Energy Storage, Inc. [397]
10637 Roselle Street
San Diego, California 92121

Thermex Solar Corporation [400]
1050 N. Kraemer Place
Anaheim, California 92806

Paul Timberman, Architect [813]
Box 130
Occidental, California 95465

Tri-Ex Tower Corporation [407]
7182 Rasmussen Avenue
Visalia, California 92377

Valour Fiberglass Manufacturing, Inc. [430B]
3703 East Melville Way
Anaheim, California 92806

Vanguard Energy Systems [417]
9133 Chesapeake Drive
San Diego, California 92123

Joseph Von Arx Passive Solar
 Custom Homes [820]
5104 Saddlewood Street
Sacramento, California 95841

Seiss Wagner, Architect [821]
1309 25th Street
Sacramento, California 95816

James A. Ward, P.E. [823]
10731 Treena Street, Suite 100
San Diego, California 92131

Western Solar Development, Inc.[424]
1236 Callen Street
Vacaville, California 95688

Whatworks [829]
P.O. Box 588
Calistoga, California 94515

GEOGRAPHICAL INDEX 445

Whitefly Control Company [1034]
P.O. Box 986
Milpitas, California 95035

Winzler and Kelly [834]
P.O. Box 1345
Eureka, California 95501

Woodford/Sloan, Architects [835]
150 Green Street, Penthouse B
San Francisco, California 94111

William J. Yang & Associates [838]
1258 North Highland Avenue
Suite 302
Los Angeles, California 90038

Ying Manufacturing Corp. [427]
1957 West 144th Street
Gardena, California 90249

Herbert C. Younger, P.E. [840]
5782 Turquoise Avenue
Alta Loma, California 91701

COLORADO

Abacus Group Incorporated [431]
P.O. Box 8006
Boulder, Colorado 80302

The Alternate Energy Institute [992]
P.O. Box 3100
Estes Park, Colorado 80517

American Heliothermal Corporation
[18]
720 South Colorado Boulevard
Suite 450
Denver, Colorado 80222

John Anderson Associates [447]
1522 Blake Street
Denver, Colorado 80202

Atkinson/Karius/Architects [460]
1738 Wynkoop Street
Denver, Colorado 80202

Barber-Nichols Engineering Co. [860]
6325 W. 55th Avenue
Arvada, Colorado 90002

Bio-Gas of Colorado, Inc. [862]
5620 Kendell Court
Arvada, Colorado 80002

Robert H. Bushnell [479]
Consulting Engineer
502 Ord Drive
Boulder, Colorado 80303

Cellular Product Services, Inc.
(CPS, Inc.) [47]
3125A North El Paso
Colorado Springs, Colorado 80907

University of Colorado [874]
Civil, Environmental and Architectural Engineering
Boulder, Colorado 80309

Colorado Office of Energy Conservation
1600 Downing Street, 2nd Floor
Denver, Colorado 80218

Colorado State University [875]
Solar Energy Applications Lab
Fort Collins, Colorado 80523

C. Phillip Colver & Associates [490]
0855 Mountain Laurel Drive
Aspen, Colorado 81611

Crowther Solar Group [500]
310 Steele Street
Denver, Colorado 80206

446 SOLAR CENSUS

Delta H Systems [74]
Route 3
Sterling, Colorado 80751

Dencor Energy Cost Controls, Inc. [76]
2750 South Shoshone
Englewood, Colorado 80110

Denver Research Institute [883]
University Park
Denver, Colorado 80208

Dixon/Carter Architects [510]
P.O. Box 797
Granby, Colorado 80446

Domestic Technology Institute [1002]
Box 2043
Evergreen, Colorado 80439

Downing Leach Architects &
  Engineers [513]
3985 Wonderland Hill Avenue
Boulder, Colorado 80302

Energy Dynamics Corporation [895]
6062 E. 49th Avenue
Commerce City, Colorado 80022

Energy Engineering Group, Inc. [536]
P.O. Box 130,
1115 Washington Avenue
Golden, Colorado 80401

Energy Materials, Inc. [105]
2622 S. Zuni Street
Englewood, Colorado 80465

Entropy Limited [111]
5735 Arapahoe
Boulder, Colorado 80303

Forster-Morrell Engineering
  Associates, Inc. [562]
P.O. Box 9881
Colorado Springs, Colorado 80932

Hot Stuff [158]
406 Walnut
La Jara, Colorado 81140

Hyperion, Inc. [161]
4860 Riverbend Road
Boulder, Colorado 80303

Dr. Jan F. Kreider, P.E. [633]
Consulting Engineers
1455 Oak Circle
Boulder, Colorado 80302

LAMCO, Inc. [187]
5923 North Nevada
Colorado Springs, Colorado 80907

Rudolph B. Lobato Associates [648]
10075 East County Line Road
Longmont, Colorado 80501

Novan Energy, Inc. [227]
1630 N. 63rd
Boulder, Colorado 80301

Remmers Engineering [706]
175 Pawnee Drive
Boulder, Colorado 80303

Residential Energy Systems Inc.
  [257]
5475 East Evans Avenue
Denver, Colorado 80222

R-M Products [262]
5010 Cook Street
Denver, Colorado 80222

Roaring Fork Resource Center [1018]
A Branch of the Colorado Energy
  Extension Service
P.O. Box 9950
Aspen, Colorado 81611

GEOGRAPHICAL INDEX 447

Solarado Inc. [284]
2625 South Santa Fe Drive
Denver, Colorado 80223

Solar Control Corporation [291]
5721 Arapahoe
Boulder, Colorado 80303

Solar Design Group. Ltd. [745]
821 5th Street
Lyons, Colorado 80540

Solar Development Inc. of Denver
 [295]
11799 East 39th Avenue
Aurora, Colorado 80010

Solar Energy Design Corporation
 of America [748]
400 Remington Street
P.O. Box 1943
Fort Collins, Colorado 80522

Solar Energy of Colorado, Inc. (SECO)
 [300]
4230 Fox Street
Denver, Colorado 80216

Solar Energy Research Corporation
 [303]
10075 East County Line Road
Longmont, Colorado 80501

Solar Energy Research Institute (SERI)
 [961]
1617 Cole Boulevard
Golden, Colorado 80401

Solar Environmental Engineering
 Company, Inc. [963]
2524 E. Vine Drive
Fort Collins, Colorado 80524

Solar Industries, Inc. [315]
2525 West 6th Avenue
Denver, Colorado 80204

Solaron Corporation [323]
1885 West Dartmouth Avenue
Englewood, Colorado 80110

Solar Pathways, Inc. [324]
3710 Highway 82
Glenwood Springs, Colorado 81601

Solar Pathways Associates [756]
3710 Highway 82
Glenwood Springs, Colorado 81601

Solar Technology Corporation [338]
2160 Clay Street
Denver, Colorado 80211

Solstice Designs, Inc. [769]
Box 2043
Evergreen, Colorado 80439

Telluride Designworks [806]
222 W. Colorado Avenue
P.O. Box 1248
Telluride, Colorado 81435

Terra-Sol [809]
S. R. 2242 A
Woodland Park, Colorado 80863

Thermodular Designs, Inc. [401]
5095 Paris Street
Denver, Colorado 80239

Zia Associates, Inc. [430]
1830 N. 55th Street
Boulder, Colorado 80302

CONNECTICUT

American Solar Heat Corporation
 [20]
7 National Place
Danbury, Connecticut 06810

448 SOLAR CENSUS

BUDCO [42]
6 Cadwell Road
Bloomfield, Connecticut 06002

University of Connecticut [876]
Energy Center
Box U-139
Storrs, Connecticut 06268

Connecticut Office of Policy and
 Management
80 Washington Street
Hartford, Connecticut 06115

Coordinated Systems, Inc. [497]
1007 Farmington Avenue
West Hartford, Connecticut 06107

Copper Development Association, Inc.
 [877]
1011 High Ridge Road
Stamford, Connecticut 06905

Falbel Energy Systems Corp. [120]
114 Manhattan Street
Stamford, Connecticut 06902

First Solar Industries, Inc. [122]
P.O. Box 303
Plymouth, Connecticut 06782

Carleton Granbery, Architect [577]
111 Old Quarry
Guilford, Connecticut 06437

Kaman Aerospace Corporation [911]
Old Windsor Road
Bloomfield, Connecticut 06002

KEM Associates, Inc. [181]
153 East Street
New Haven, Connecticut 06507

K. T. Lear Associates, Inc. [639]
53 Lyness Street
Manchester, Connecticut 06040

Mark Enterprises Inc. [199]
30 Hazel Terrace
P.O. Box 3659
Woodbridge, Connecticut 06525

Moore Grover Harper, P.C. [670]
Architects and Planners
P.O. Box 235
Essex, Connecticut 06426

National Solar Corporation [219]
Main Street
Centerbrook, Connecticut 06409

NUTEK [229]
Nuclear Technology Corporation
P.O. Box 1
Amston, Connecticut 06231

Reaction Research Laboratories
 [254]
P.O. Box 356
Groton, Connecticut 06340

Solar Products Manufacturing
 Corp. [328]
1 Alcap Ridge
Cromwell, Connecticut 06416

Sun-Ray Solar Equipment Co., Inc.
 [380]
4 Pines Bridge Road
Beacon Falls, Connecticut 06403

Sunsearch, Inc. [977]
669 Boston Post Road
P.O. Box 275
Guilford, Connecticut 06437

Thermatool Corporation [399]
Solar Products Division
280 Fairfield Avenue
Stamford, Connecticut 06902

UCE Inc. [982]
24 Fitch Street
Norwalk, Connecticut 06855

Donald Watson [826]
P.O. Box 401
Guilford, Connecticut 06437

Wormser Scientific Corp. [837]
88 Foxwood Road
Stamford, Connecticut 06903

DELAWARE

Chemax Manufacturing Corp. [50]
211 River Road
New Castle, Delaware 19720

University of Delaware [882]
Institute of Energy Conversion
One Pike Creek Center
Wilmington, Delaware 19808

Delaware Governor's Energy Office
P.O. Box 1401
Dover, Delaware 19901

E. I. du Pont de Nemours and
 Company, Inc. [85]
10th and Market Streets
Wilmington, Delaware 19898

Energy Research Institute of
 Delaware [896]
500 Homestead Road, G-1
Wilmington, Delaware 19805

Hoster's HVAC Engineering Co. [596]
1102 Brickyard Road
Seaford, Delaware 19973

Solar Construction and Design Inc.
 [289]
6 Lyndhurst Avenue
Wilmington, Delaware 19803

Edward V. Stella [779]
208 Chapel Drive
Camden, Delaware 19934

DISTRICT OF COLUMBIA

Hammer, Siler, George
 Associates [587]
1140 Connecticut Avenue, NW
Washington, D.C. 20036

National Aeronautics and Space
 Administration (NASA) [927]
600 Maryland Avenue, SW
Washington, D.C. 20546

Solara Associates Inc. [283]
1001 Connecticut Avenue, NW
Suite 632
Washington, D.C. 20036

Solar Energy Assoc., Ltd. [747]
2316 39th Street
Washington, D.C. 20007

FLORIDA

American Sun Corporation [22]
2913 Ponce de Leon Boulevard
Coral Gables, Florida 33143

Aquasolar, Inc. [29]
1232 Zacchini Avenue
Sarasota, Florida 33578

Aries Consulting Engineers, Inc.
 [458]
2021 S.W. 43rd Avenue
Gainesville, Florida 32608

Astro Solar Corp. [857]
1602 Clare Avenue
West Palm Beach, Florida 33401

450 SOLAR CENSUS

D. W. Browning Contracting Co. [41]
475 Carswell Avenue
Holly Hill, Florida 32017

CSI Solar Systems Division [69]
12400 49th Street
Clearwater, Florida 33520

del Sol Control Corporation [73]
11914 U.S. 1
Juno, Florida 33408

DSS Engineers, Inc. [515]
1850 N.W. 69th Avenue
Fort Lauderdale, Florida 33313

Electra Sol Labs Inc. [95]
2326 Fieldingwood Road
Maitland, Florida 32751

Energy Equipment Sales [102]
412 Longfellow Boulevard
Lakeland, Florida 33801

Energy Transfer Systems Inc. [107]
5001 W. Waters Avenue
Tampa, Florida 33614

Flagala Corporation [124]
9700 W. Alt. 98
Panama City, Florida 32407

Florida Solar Energy Center [903]
300 State Road 401
Cape Canaveral, Florida 32920

University of Florida [902]
Solar Energy and Energy Conversion
    Laboratory
Room 325 MEB
Gainesville, Florida 32611

General Energy Devices, Inc. [132]
P.O. Box 5679
Clearwater, Florida 33515

General Instrument Corp. [133]
3811 University Boulevard, W26
Jacksonville, Florida 32217

Gulf Thermal Corporation [141]
645 Central Avenue, P.O. Box 1273
Sarasota, Florida 33578

Hawthorne Industries, Inc. [145]
Solar Energy Division
3114 Tuxedo Avenue
West Palm Beach, Florida 33405

MacDonald Engineering [916]
1235 Ashland Avenue
Wilmette, Florida 60091

University of Miami [919]
Clean Energy Research Institute
School of Architecture and
    Engineering
P.O. Box 248294
Coral Gables, Florida 33124

OEM Products, Inc. [230]
Route 3, Box 295
Dover, Florida 33527

Orlando Laboratories, Inc. [940]
Box 8008
Orlando, Florida 32856

Solar City [286]
3519 Henderson Boulevard
Tampa, Florida 33609

Solar Development Inc. [296]
3630 Reese Avenue
Riviera Beach, Florida 33404

Solar-Eye Products Inc. [307]
1300 N.W. McNab Road
Ft. Lauderdale, Florida 33309

GEOGRAPHICAL INDEX 451

Solar Heater Manufacturing Co. [312]
1011 Sixth Avenue South
Lake Worth, Florida 33460

Solar Industries of Florida [317]
Suntrak, Inc.
P.O. Box 9013
Jacksonville, Florida 32208

Solar Magnetic Labs [320]
121 Riley Avenue
Lake Worth, Florida 33461

Solar Products Sun Tank Inc. [329]
4291 N.W. 7th Avenue
Miami, Florida 33127

Solar Water Heaters of N.P.R. Inc.
[348]
1214 U.S. 19 North
New Port Richey, Florida 33552

Universal Solar Development, Inc.
[413]
1505 Sligh Boulevard
Orlando, Florida 32806

U.S. Solar Corporation [415]
P.O. Drawer K
Hampton, Florida 32044

GEORGIA

E-Tech, Inc. [117]
3570 American Drive
Atlanta, Georgia 30341

Georgia Office of Energy Resources
Room 615
270 Washington Street S.W.
Atlanta, Georgia 30334

ILI, Inc. [163]
5965 Peachtree Corners East
Norcross, Georgia 30071

National Solar Supply [220]
2331 Adams Drive, N.W.
Atlanta, Georgia 30318

Solar Data Systems International
[293]
Box 8000
Atlanta, Georgia 30357

Southeastern Solar Systems, Inc.
[361]
4705 Bakers Ferry Road, J
Atlanta, Georgia 30336

Southern Solar Energy Center, Inc.
(SSEC) [1025]
61 Perimeter Park
Atlanta, Georgia 30341

United States Solar Industries [412]
5600 Roswell Road N.E., Suite 350
Atlanta, Georgia 30342

Vachon, Nix & Associates [818]
6855 Jimmy Carter Boulevard
Norcross (Atlanta), Georgia 30071

HAWAII

Haleakala Solar Resources, Inc.
[142]
865-A Ahua Street
Honolulu, Hawaii 96819

Hawaii Natural Energy Institute
[905]
2540 Dole Street
Honolulu, Hawaii 96822

Hawaii State Energy Office
P.O. Box 2359
Honolulu, Hawaii 96804

Frederick H. Kohloss & Associates,
Inc. [632]
345 Queen Street, Suite 401
Honolulu, Hawaii 96813

452 SOLAR CENSUS

Mid Pacific Solar Corporation/
Solar Disc Corporation [206]
540 Lagoon
Honolulu, Hawaii 96819

The Solaray Corporation [285]
2414 Makiki Heights Drive
Honolulu, Hawaii 96822

Solar Enterprises Hawaii Inc.
P.O. Box 27031
Honolulu, Hawaii 96827

United Energy Corporation [411]
666 Mapunapuna Street
Honolulu, Hawaii 96819

IDAHO

University of Idaho [908]
College of Engineering
Moscow, Idaho 83843

Idaho Office of Energy
State House
Boise, Idaho 83720

Solar Power Co. Inc. [325]
556 South Main, P.O. Box 4043
Pocatello, Idaho 83201

ILLINOIS

Advanced Energy Systems [436]
3744 W. 63rd Street
Chicago, Illinois 60629

Argonne National Laboratory [852]
Solar Energy - Building 362
9700 South Cass Avenue
Argonne, Illinois 60439

Basic Environmental Engineering Inc.
 [36]
21 W. 161 Hill Avenue
Glen Ellyn, Illinois 60137

Robert A. Bell Architects Ltd. [467]
115 N. Marion
Oak Park, Illinois 60301

Bell & Gossett [37]
8234 North Austin Avenue
Morton Grove, Illinois 60053

Bernheim, Kahn & Lozano [470]
One North Wacker Drive
Suite 205
Chicago, Illinois 60606

J. A. Bockman Solar Engineering
 [473]
1912 Northfield Court
Naperville, Illinois 60540

Chamberlain Manufacturing
 Corporation [871]
845 Larch Avenue
Elmhurst, Illinois 60126

University of Chicago [872]
Enrico Fermi Institute
5630 S. Ellis Avenue
Chicago, Illinois 60422

Comstock Construction Co., Inc.
 [494]
5848 W. Higgins Road
Chicago, Illinois 60630

De Soto Inc. [78]
1700 South Mount Pleasant Road
Des Plaines, Illinois 60018

Earth—Sun—Design [519]
P.O. Box 2473
Springfield, Illinois 62705

Enercon, Ltd. [530]
500 Davis Street
Evanston, Illinois 60201

GEOGRAPHICAL INDEX 453

Energy Design Consultants
and Builders [534]
118 Brook Bridge Drive
Cary, Illinois 60013

Fabrico Manufacturing Corp. [118]
4222 South Pulaski Road
Chicago, Illinois 60632

Fermi National Accelerator Laboratory
(FERMILAB) [901]
P.O. Box 500, MS-115
Batavia, Illinois 60510

James Follensbee & Associates Ltd.
[559]
311 West Hubbard Street
Chicago, Illinois 60610

Rondal Gower Associates [575]
2404 Windsor Place
Champaign, Illinois 61820

Hanat Sales Company [144]
2220 West Hassell Road, Suite 304
Hoffman Estates, Illinois 60195

Hawkweed Group, Ltd. [591]
4643 North Clark
Chicago, Illinois 60640

Heliodyne, Inc. [148]
2629 Charles Street
Rockford, Illinois 61108

Hi-Tech, Inc. [156]
3600 16th Street
Zion, Illinois 60099

Hyperion Incorporated [162]
1300 South Maybrook Drive
Maywood, Illinois 60153

University of Illinois [603]
Small Homes Council/Building Research
1 East St. Mary's Road
Champaign, Illinois 61820

Illinois Institute of Natural
Resources
325 West Adams
Springfield, Illinois 62706

Infinity Energy Systems [168]
Division of the SIRI Corporation
7015 North Sheridan Road
Chicago, Illinois 60626

ITT Fluid Handling Division [172]
4711 Golf Road
Skokie, Illinois 60076

Johnson Controls [176]
Control Products Division
2221 Camden Court
Oakbrook, Illinois 60521

Edward J. Long Consulting
Engineers [650]
516 East Monroe Street
Springfield, Illinois 62701

March Manufacturing Inc. [198]
1819 Pickwick Avenue
P.O. Box 87
Glenview, Illinois 60025

Mobil Corporation [921]
Montgomery Ward Testing Laboratories
RD 5-5 Montgomery Ward Plaza
Chicago, Illinois 60671

C. F. Murphy Association [676]
224 South Michigan Avenue
Chicago, Illinois 60604

Olin Corporation/Brass Group [232]
East Alton, Illinois 62024

SAF Energy Consultants, Inc. [954]
P.O. Box 3052
Peoria, Illinois 61614

454 SOLAR CENSUS

Shelley Radiant Ceiling Co., Inc. [276]
456 West Frontage Road
Northfield, Illinois 60093

Sheridan Solar [727]
8429 North Monticello
Skokie, Illinois 60076

A. O. Smith [279]
P.O. Box 28
Kankakee, Illinois 60901

Solag [281]
P.O. Box 676
Roseville, Illinois 61473

Solar Design Associates Inc. [743]
205 West John
Champaign, Illinois 61820

Solarpak, Inc. [755]
6516 West Higgins Avenue
Chicago, Illinois 60656

Solar Products, Inc. [327]
2419 20th Street
Rockford, Illinois 61101

Solar Search Corporation [334]
Rural Route 1, Box 42
Rankin, Illinois 60960

Sol Energy Corporation [969]
761 Rohde
Hillside, Illinois 60162

S S Solar Inc. [773]
16 Keystone Avenue
River Forest, Illinois 60305

Sundesign Services Inc. [785]
Route 2
Cobden, Illinois 62920

Sunduit, Inc. [369]
281 E. Jackson
Virden, Illinois 62690

Unarco-Rohn [410]
6718 West Plank Road
P.O. Box 2000
Peoria, Illinois 61601

Viracon/Dial [420]
1315 North Branch
Chicago, Illinois 60622

C. Edward Ware Associates, Inc. [824]
415 Y Boulevard
Rockford, Illinois 61107

Watermeister [422]
P.O. Box 87433
Chicago, Illinois 61107

Western Illinois University [988]
Department of Agriculture
Macomb, Illinois 61455

Woods and Associates, Inc. [836]
112 Water Street
Naperville, Illinois 60540

INDIANA

Arkla Industries Inc. [31]
P.O. Box 534
Evansville, Indiana 47704

Champion Home Builders Co. [49]
Route 2, Box 321
Plainfield, Indiana 46168

Controls/Inc. [63]
1509 Woodlawn Avenue
Logansport, Indiana 46947

GEOGRAPHICAL INDEX 455

dh2W, Inc. [507]
705 Franklin Square
Michigan City, Indiana 46360

Dwyer Instruments Inc. [86]
P.O. Box 373
Michigan City, Indiana 46360

ECS—Environmental Control Systems [521]
525 East 10th Street
Jeffersonville, Indiana 47130

Elkhart Products Corporation [97]
1255 Oak Street, P.O. Box 1008
Elkhart, Indiana 46515

Indiana Department of Commerce
440 North Meridian
Indianapolis, Indiana 46204

Indiana Solar Designs [607]
R. R. No. 2, Box 59B
Paoli, Indiana 47454

International Solar Technologies, Inc. [430A]
Route 2, Box 321
Plainfield, Indiana 46168

Kaelin Industries, Inc. [178]
2210 North Grand Avenue
Evansville, Indiana 47711

Levi & Co. — Concept Builders, Inc. [643]
P.O. Box 2254
Evansville, Indiana 47714

Restoration Preservation Architecture, Inc. [707]
51 South Ritter
Indianapolis, Indiana 46219

Schmidt/Claffey Architects, Inc. [720]
333 North Pennsylvania Street
Indianapolis, Indiana 46204

Solar Shelter, Inc. [335]
Energy 2000, Marketing Division
800 South Council Street
Muncie, Indiana 47302

James R. Stutzman Architects [782]
P.O. Box 55111
Indianapolis, Indiana 46205

Suntron [389]
A Division of K&G Manufacturing
2302-10 Pennsylvania Street
Fort Wayne, Indiana 46802

Synergy, Inc. [802]
515 W. Maplewood Drive
Ossian, Indiana 46777

Universal Design and Construction [817]
Rural Route 9, Box 160
Columbus, Indiana 47201

Walgamuth & Associates [822]
608-½ Columbia Street
Lafayette, Indiana 47901

IOWA

Alufoil Packaging Co., Inc. [15]
Route 3, Box 15
1800 33rd Street
Fort Madison, Iowa 52627

Central States Energy Research Corp. [485]
128-½ E. Washington Street
P.O. Box 2623
Iowa City, Iowa 52240

Decker Manufacturing [881]
Impac Corporation
312 Blondeau
Keokuk, Iowa 52632

Deco Products [72]
506 Sanford Street
Decorah, Iowa 52101

Dulaney & Associates [517]
P.O. Box 346
Ames, Iowa 50010

Earth Sheltered Housing Systems [518]
Route 1
Marshalltown, Iowa 50158

Flinn Saito Andersen Architects [558]
604 Mulberry
Waterloo, Iowa 50703

Impac Corporation [167]
Division of Decker Manufacturing Co.
P.O. Box 365
Keokuk, Iowa 52632

McConnell, Steveley, Anderson [660]
Architects and Planners
860 17th Street, S.E.
Cedar Rapids, Iowa 52403

Solar Electric Inc. [298]
403 South Maple
West Branch, Iowa 52358

Stanley Consultants, Inc. [774]
Stanley Building
Muscatine, Iowa 52761

KANSAS

Alternative Energy Works, Inc. [444]
310 West Fourth Street, P.O. Box 271
Newton, Kansas 67114

Con-Egy [59]
P.O. Box 740, 2nd & Main
Maple Hill, Kansas 66507

Dressler Energy Consulting
 and Design Corporation [514]
4550 West 109th, Suite 170
Overland Park, Kansas 66211

Elswood—Smith—Carlson,
Architects [528]
5700 Broadmoor, Suite 703
Mission, Kansas 66202

Energy Management and Control
 Corporation [538]
634 Harrison, Suite B
Topeka, Kansas 66603

Hydro-Flex Corporation [160]
2101 N.W. Brickyard Road
Topeka, Kansas 66618

University of Kansas [912]
Center for Research, Inc.
Applied Energy Research Program
2291 Irving Hill Road
Lawrence, Kansas 66045

Kansas Energy Office
214 West 6th
Topeka, Kansas 66603

The Kiene & Bradley Partnership
 [629]
1st National Bank Building
Suite 925
Topeka, Kansas 66603

Morris Richard Perkins [696]
Architect/P.A.
245 Hillside, North
Wichita, Kansas 67214

Peters—Williams—Kubota, P.A. [697]
2500 W. 6th
Lawrence, Kansas 66044

GEOGRAPHICAL INDEX 457

Slemmons Associates [734]
1 Townsite Plaza, Suite 1515
Topeka, Kansas 66603

Solar Services, Inc. [759]
1029 North Wichita, Suite 5
Wichita, Kansas 67203

Sunflower Energy Works [372]
101 Main Street
Lehigh, Kansas 67073

Wilson & Company [833]
Engineers and Architects
P.O. Box 1648
Salina, Kansas 67401

KENTUCKY

Kentucky Department of Energy
Bureau of Energy Management
Iron Works Pike
Lexington, Kentucky 40578

National Products, Inc. [217]
900 Baxter Avenue
P.O. Box 4174
Louisville, Kentucky 40204

Roll Forming Corporation [264]
Industrial Park
Shelbyville, Kentucky 40065

Solaraire [736]
2459 Nicholasville Road
Lexington, Kentucky 40503

Solar Radiation Industries [330]
2531 Plantside Drive
Louisville, Kentucky 40299

LOUISIANA

Innovative Energy Corporation [608]
216 Conrad Street
New Orleans, Louisiana 70124

Louisiana Department of Natural
  Resources
Research and Development
P.O. Box 44156
Baton Rouge, Louisiana 70804

Solar Technologies Inc. [761]
P.O. Box 8175
Shreveport, Louisiana 71108

Sun-Pac, Inc. [377]
P.O. Box 8169
Alexandria, Louisiana 71306

MAINE

AIDCO Maine Corporation [438]
Orr's Island, Maine 04066

Altenburg and Company [441]
587 Spring Street
Westbrook, Maine 04092

Berg & Associates [468]
Design/Build Inc.
3140 Harbor Lane
Plymouth, Maine 55441

Cornerstones Energy Group, Inc.
  [1001]
54 Cumberland Street
Brunswick, Maine 04011

Keane Associates [624]
22 Monument Square
Portland, Maine 04101

Krumbhaar & Holt Architects [634]
66 Main Street
Ellsworth, Maine 04605

University of Maine [917]
Department of Industrial Cooperation
109 Boardman Hall
Orono, Maine 04469

458 SOLAR CENSUS

Maineform Architecture [655]
295 Water Street
Augusta, Maine 04330

Maine Office of Energy Resources [1007]
55 Capitol Street
State House Station 53
Augusta, Maine 04333

Merrymeeting Architects [663]
Lincoln Building
Brunswick, Maine 04011

Moore/Weinrich Architects [671]
49 Pleasant Street
Brunswick, Maine 04011

Neill & Gunter Inc. [679]
P.O. Box 1959
Portland, Maine 04104

RoKi Associates, Inc. [711]
P.O. Box 301, Route 113
Standish, Maine 04084

William R. Sepe [725]
Architect and Planner
One Maple Street
Camden, Maine 04843

Shelter Institute [1020]
38 Center Street
Bath, Maine 04530

Sunsystems [794]
Weld Road
Dryden, Maine 04225

Suntime Inc. [387]
101 High Street
Belfast, Maine 04915

Eaton W. Tarbell & Assoc. Inc. [804]
84 Harlow Street
Bangor, Maine 04401

Terrien Architects [810]
5 Moulton Street
Portland, Maine 04101

Joseph B. Thomas IV, Architect [812]
Box 428
N.E. Harbor, Maine 04662

André R. Warren [825]
Consultant & Design Inc.
P.O. Box 718
Brunswick, Maine 04011

John Whipple, Architect [830]
386 Fore Street
Portland, Maine 04101

Zephyr Wind Dynamo Company [429]
P.O. Box 241
Brunswick, Maine 04011

MARYLAND

A.A.I. Corporation [1]
P.O. Box 6767
Baltimore, Maryland 21204

AMAF Industries [848]
Box 1100
Columbia, Maryland 21044

Associated Enterprises [856]
2832 Montclair Drive
Ellicott City, Maryland 21043

Mark Beck Associates [464]
762 Fairmount Avenue
Towson, Maryland 21204

Energy Applications [531]
Long Reach Village Center
Suite 227
Columbia, Maryland 21045

Hittman Associates, Inc. [906]
9190 Red Branch Road
Columbia, Maryland 21045

Maryland Energy Office
301 West Preston Street, Rm. 1302
Baltimore, Maryland 21201

National Patent Development Corp. [930]
Energy Systems Inc.
1455 Research Boulevard
Rockville, Maryland 20850

National Solar Heating and Cooling Information Center [1009]
P.O. Box 1607
Rockville, Maryland 20850

Practical Solar [1017]
Box 1067 Blair Station
Silver Spring, Maryland 20910

E. F. Siegel and Associates Ltd. [728]
Consulting Engineers
7104 Milford Industrial Road
Baltimore, Maryland 21208

Slack Associates Inc. [733]
540 South Longwood Street
Baltimore, Maryland 21223

Solar Comfort Systems Manufacturing Co. [288]
311 Mount Vernon Place
Rockville, Maryland 20852

Solar Energy Intelligence Report [1022]
Box 1067 Blair Station
Silver Spring, Maryland 20910

Solarex Corporation [306]
1335 Piccard Drive
Rockville, Maryland 20850

Solartherm, Inc. [341]
1110 Fidler Lane, Suite 1215
Silver Spring, Maryland 20910

Thomason Solar Homes, Inc. [403]
609 Cedar Avenue
Fort Washington, Maryland 20022

MASSACHUSETTS

Acorn Structures [433]
Box 250
Concord, Massachusetts 01742

Atek Design [858]
3 Crawford Street, No. 3
Cambridge, Massachusetts 02139

Boston College [863]
Chestnut Hill, Massachusetts 02167

E. C. Collins II Associates [489]
40 Lowell Road
Concord, Massachusetts 01742

Columbia Chase Corporation [57]
Solar Energy Division
55 High Street
Holbrook, Massachusetts 02343

Commercial Plumbing Corp. [491]
732 Washington Street
Weymouth, Massachusetts 02188

Crystal Systems, Inc. [880]
Shetland Industrial Park
35 Congress Street
Salem, Massachusetts 01970

DIY-Sol, Inc. [81]
29 Highgate Road
Marlboro, Massachusetts 01752

460 SOLAR CENSUS

Dooling & Siegel Architects [512]
84 Bowers Street
Newtonville, Massachusetts 02160

EIC Corporation [891]
55 Chapel Street
Newton, Massachusetts 02158

Electrostatic Consulting Associates [525]
P.O. Box 552
Groton, Massachusetts 01450

General Energy Development Corp. [131]
661 Highland Avenue
Needham Heights, Massachusetts 92194

Grayson Associates, Inc. [578]
68 Leonard Street
Belmont, Massachusetts 02178

JBF Scientific Corporation [910]
2 Jewell Drive
Wilmington, Delaware 01887

Massachusetts Executive Office of Energy Resources
Renewable Energy Division
73 Tremont Street, Room 849
Boston, Massachusetts 02108

Massdesign Architects & Planners, Inc. [658]
138 Mt. Auburn Street
Cambridge, Massachusetts 02138

The New Alchemy Institute [931]
237 Hatchville Road
East Falmouth, Massachusetts 02536

New England Fuel Institute [1011]
20 Summer Street
Watertown, Massachusetts 02172

Northeast Solar Energy Center [1014]
470 Atlantic Avenue
Boston, Massachusetts 02110

Ralos Manufacturing [248]
331 Harvard Street
Cambridge, Massachusetts 02139

REA Associates Inc. [949]
Faulkner Street
North Billerica, Massachusetts 01821

Robert Foote Shannon, Architect [726]
49 Garden Street
Boston, Massachusetts 02114

Solar Power Corporation [326]
20 Cabot Road
Woburn, Massachusetts 01801

Solar Thermal Systems Inc. [340]
90 Cambridge Street
Burlington, Massachusetts 01803

Solectro-Thermo, Inc. [352]
1934 Lakeview Avenue
Dracut, Massachusetts 01826

Solex Corporation [354]
187 Billerica Road
Chelmsford, Massachusetts 02184

Sunspace Unlimited [792]
34 Silver Hill Road
Weston, Massachusetts 02193

Terra-Light Inc. [394]
30 Manning Road, P.O. Box 493
Billerica, Massachusetts 01821

Thermo Electron Corp. [981]
101 First Avenue
Waltham, Massachusetts 02154

Vaughn Corporation [418]
386 Elm Street
Salisbury, Massachusetts 01950

Weather Energy Systems, Inc. [423]
39 Barlows Landing Road
P.O. Box 968
Pocasset, Massachusetts 02559

MICHIGAN

Alkar Engineering & Manufacturing
  Co. [439]
25520 Ingleside
Southfield, Michigan 48034

Alternate Energy Research and
  Development [847]
P.O. Box 77
Atlanta, Michigan 49709

American Timber Homes [23]
Solartran Division
Escanaba, Michigan 49829

Architectural & Industrial Marketing,
  Inc. [455]
1203 Salzburg, P.O. Box 308
Bay City, Michigan 48706

Ayres, Lewis, Norris & May, Inc. [461]
3983 Research Park Drive
Ann Arbor, Michigan 48104

Capitol Consultants Inc. [484]
1627 Lake Lansing Road
Lansing, Michigan 48912

Cardor Company [45]
743 Bay East Drive
Traverse City, Michiga 49684

Central Upper Peninsular Planning &
  Development Regional Commission
  (CUPPAD) [997]
2415 14th Avenue South
Escanaba, Michigan 49829

Clark & Walter Architects [487]
513 S. Union
Traverse City, Michigan 49684

Clinton County Energy Committee
  [998]
306 Elms Street
St. Johns, Michigan 48879

Controlex Engineering [62]
P.O. Box 473
Birmingham, Michigan 48012

Delta Solar System Company [75]
Energy Saving Shop
2930 S. Creyts Road
Lansing, Michigan 48917

Detroit Environmental Control
  Engineers Inc. [506]
39108 Charbeneau Street
Mt. Clemens, Michigan 48043

Dow Corning Corporation [84]
2200 Salzburg Road
P.O. Box 1767
Midland, Michigan 48640

Emmet County Cooperative
  Extension Energy Project [1003]
441 Bay Street
Petoskey, Michigan 49770

Environmental Research Institute
  of Michigan (ERIM) [899]
P.O. Box 8618
Ann Arbor, Michigan 48107

Escher: Foster Technology
  Associates, Inc. [900]
P.O. Box 189
St. Johns, Michigan 48879

G. E. Associates, Inc. [567]
G-6235 Corunna Road
Flint, Michigan 48504

462 SOLAR CENSUS

Gove Associates, Inc. [574]
1601 Portage Street
Kalamazoo, Michigan 49001

Graheck, Bell, Kline & Brown [576]
Architects & Engineers, Inc.
220 West Washington Street, Suite 110
P.O. Box 789
Marquette, Michigan 49855

H&H Tube and Manufacturing Co.[154]
4000 Town Center,Suite 485
Southfield, Michigan 48075

Hooker/De Jong Associates [595]
409 Frauenthal Building
Muskegon, Michigan 49440

Hoyem-Basso Associates, Inc. [597]
25 West Long Lake Road
Bloomfield Hills, Michigan 48013

Inatome & Associates, Inc. [606]
10140 West Nine Mile Road
Oak Park, Michigan 48237

Independent Energies Inc. [166]
Route 131, P.O. Box 398
Schoolcraft, Michigan 49087

J & D Solar Contracting [617]
11400 Bacon Road
Plainwell, Michigan 49080

Kingscott Associates, Inc. [630]
Architects & Engineers
511 Monroe Street, P.O. Box 671
Kalamazoo, Michigan 49005

Michigan Department of Commerce
Energy Administration
P.O. Box 30228
Lansing, Michigan 48909

Midwest Components [207]
Port City Industrial Park
Box 787
Muskegon, Michigan 49443

Milton International, Inc. [665]
808 River Acres Drive
Tecumseh, Michigan 49286

M&W Enterprises [926]
720 W. Jefferson
Ann Arbor, Michigan 48103

Ronald Nabozny [677]
710 Randolph
Jackson, Michigan 49203

Daniel A. Nobbe & Associates
Inc. [682]
1408 Arborview
Ann Arbor, Michigan 48103

A. J. Nydam Co. Inc. [683]
2634 S. Division Avenue
Grand Rapids, Michigan 49507

Al Paas and Associates [687]
2378 E. Stadium Boulevard
Ann Arbor, Michigan 48104

Lincoln A. Poley, Architect [701]
118 West Washington Street
Marquette, Michigan 49855

W. H. Porter, Inc. [241]
P.O. Box 1112
Holland, Michigan 49423

RAM Products Inc. [250]
1111 North Centreville Road
Sturgis, Michigan 49091

Refine Building & Construction Co.
[704]
24607 W. Warren
Dearborn Heights, Michigan 48127

Refrigeration Research Inc. [255]
Solar Research Division
525 North 5th Street
Brighton, Michigan 48116

Richard A. Schramm, Architecture,
    Inc. [721]
2001 South 4th Street
Kalamazoo, Michigan 49009

Solar Age Designers [959]
39108 Charbeneau Street
Mt. Clemens, Michigan 48043

Solar Collector Sales Co., Inc. [287]
7766 Auburn Road
Utica, Michigan 48087

Solarcon, Inc. [740]
607 Church Street
Ann Arbor, Michigan 48104

Solar Energy Engineering/Solar
    Energy Systems [750]
13450 Northland Drive
Big Rapids, Michigan 49307

Solargy Corporation [310]
17914 East Warren Avenue
Detroit, Michigan 48224

Solar-Therm-El [764]
Energy Savers, Inc.
1829 Northwood Boulevard
Royal Oak, Michigan 48073

Solar/Wind Energy Systems Inc. [765]
10833 Farmington Road
Livonia, Michigan 48150

Solar Works Inc. [766]
3296 McConnell
Charlotte, Michigan 48813

Sol-Lector, Inc. [356]
1529 South Division Avenue
Grand Rapids, Michigan 49507

Straub, Van Dine & Dziurman
    Associates, Inc. [780]
1441 East Maple
Troy, Michigan 48084

Sundance Building Contractors [784]
724 Main Street
Norway, Michigan 49870

Sunstructures [793]
Sunstructures Construction
201 East Liberty
Ann Arbor, Michigan 48104

Sunway Heating Systems, Inc. [797]
409 Lawrence Street
Petoskey, Michigan 49770

Twin Pane Division [238]
Philips Industries
31251 Industrial Road
Livonia, Michigan 48150

Upland Hills Ecological Awareness
    Center [1030]
2575 Indian Lake Road
Oxford, Michigan 48051

Urban Options [1031]
135 Linden Street
East Lansing, Michigan 48823

Wayne State University [1032]
US/WCP
Detroit, Michigan 48202

Windependence Electric Power Co.
    [989]
9080 Beeman Road
Waterloo, Michigan 48118

464 SOLAR CENSUS

Wolverine Solar Industries Inc. [426]
13450 Northland Drive
Big Rapids, Michigan 49307

R. M. Young Company [428]
2801 Aero-Park Drive
Traverse City, Michigan 49684

MINNESOTA

Alternative Sources of Energy, Inc.
 [994]
107 South Central Avenue
Milaca, Minnesota 56353

Andersen Corporation [25]
Bayport, Minnesota 55003

Architectural Alliance [451]
400 Clifton Avenue South
Minneapolis, Minnesota 55403

Architecture One [456]
8 First Federal Center
Brainerd, Minnesota 56401

Armstrong, Torseth, Skold & Rydeen
 [459]
4901 Olson Highway
Minneapolis, Minnesota 55422

ATR Electronics, Inc. [33]
300 E. 4th Street
St. Paul, Minnesota 55101

Chicago Solar Corporation [52]
2001 Highway 3 North, P.O. Box 839
Faribault, Minnesota 55021

Close Associates, Inc. [488]
3101 E. Franklin Avenue
Minneapolis, Minnesota 55406

Creative Alternatives [498]
Route 1
Long Prairie, Minnesota 56347

Detroit Lakes Tech [884]
Area Vocational-Technical Institute
Highway 34 East
Detroit Lakes, Minnesota 56501

Dickey/Kodet/Architects/Inc. [508]
4930 France Avenue South
Minneapolis, Minnesota 55410

Dyna Technology, Inc. [87]
Winco Division
7850 Metro Parkway
Minneapolis, Minnesota 55420

Freeberg Company [563]
8624 Pine Hill Road
Bloomington, Minnesota 55438

Genesis Architecture [569]
417-A West Litchfield Avenue
Box 107
Willmar, Minnesota 56201

Goodwin Controls Co. [572]
1430 Ranier Lane
Plymouth, Minnesota 55447

Hammel Green & Abrahamson Inc.
 [586]
2675 University Avenue
St. Paul, Minnesota 55114

HGF/Centrum Architects Inc. [593]
6311 Wayzata Boulevard
Minneapolis, Minnesota 55416

Honeywell Inc. [907]
Rosedale Towers
1700 W. Highway 36
St. Paul, Minnesota 55113

i e associates, inc. [602]
3704 11th Avenue South
Minneapolis, Minnesota 55407

Ilse Engineering, Inc. [164]
7177 Arrowhead Road
Duluth, Minnesota 55811

Ironwood, Inc. [613]
115 North First Street
Minneapolis, Minnesota 55401

Kirkham, Michael & Associates [631]
Architects, Engineers, Planners
7601 Kentucky Avenue North
Minneapolis, Minnesota 55410

Lundquist, Wilmar, Schultz & Martin, Inc. [652]
614 Endicott-on-Fourth Building
St. Paul, Minnesota 55101

Mid-American Solar Energy Complex (MASEC) [1008]
Alpha Business Center
8140 26th Avenue South
Minneapolis, Minnesota 55420

Minnesota Energy Agency
980 American Center Building
150 East Kellog
St. Paul, Minnesota 55101

Minnesota, Mining and Manufacturing (3M) [209]
Decorative Products Division
223-1 South 3M Center
St. Paul, Minnesota 55144

Minnesota Tritec Inc. [210]
Route No. 3, Box 524
Delano, Minnesota 55328

Northern Solar Power Co. [224]
311 S. Elm Street
Moorhead, Minnesota 56560

Schipke Engineers [718]
3101 West 69th Street
Minneapolis, Minnesota 55435

Skillestad Engineering, Inc. [278]
Old Highway 52 South
P.O. Box 17
Cannon Falls, Minnesota 55009

John Skujins, AIA [731]
2300 East 22nd Street
Minneapolis, Minnesota 55406

Solar Cooker Parts [292]
3548 Grand Avenue South
Minneapolis, Minnesota 55408

Solargizer International Inc. [309]
2000 West 98th Street
Bloomington, Minnesota 55431

Solergy Company [353]
7216 Boone Avenue North
Minneapolis, Minnesota 55428

Thermograte, Inc. [402]
2785 N. Fairview Avenue
P.O. Box 43566
St. Paul, Minnesota 55164

Underground Space Center [983]
221 Church Street
University of Minnesota
Minneapolis, Minnesota 55455

MISSISSIPPI

Mississippi Governor's Office of Energy and Transportation
510 George Street
Jackson, Mississippi 39205

Mississippi State University [920]
Drawer ME
Mississippi State, Mississippi 39762

MISSOURI

The Architects Partnership, Inc. [450]
717 Cherry Street
Columbia, Missouri 65201

Bakewell Corporation [462]
8820 Ladue Road
St. Louis, Missouri 63124

Cider Bluff Inc. [486]
5445 Holmes
Kansas City, Missouri 64110

Ener-Tech, Inc. [541]
1924 Burlewood Drive
St. Louis, Missouri 63141

Engel Industries, Inc. [110]
8122 Reilly Avenue
St. Louis, Missouri 63111

Gary Glenn, AIA Architects [570]
10 South Euclid
St. Louis, Missouri 63108

Peter H. Green & Associates [579]
Architecture—Planning—Urban Design
25 South Bemiston Avenue
St. Louis, Missouri 63105

Haverstick & Associates [589]
Architects—Planners—Designers
Lake Tekakwitha
Pacific, Missouri 63069

Hunter Hunter Associates [599]
8990 Manchester Road
St. Louis, Missouri 63144

Londe—Parker—Michels, Inc. [649]
7438 Forsyth, Suite 202
St. Louis, Missouri 63105

Missouri Department of Natural Resources
Division of Energy
P.O. Box 176
Jefferson City, Missouri 65102

Monsanto Company [213]
800 North Lindbergh Boulevard
St. Louis, Missouri 63166

New Day Builders [680]
Route 2, Box 140-D
Doniphan, Missouri 63935

New Life Farm, Inc. [932]
Drury, Missouri 65638

Arthur Hall Pedersen [693]
34 North Gore Avenue
Webster Groves, Missouri 63119

PSI Energy Systems, Inc. [244]
A Division of Pipe Systems Inc.
1533 Fenpark Drive
Fenton, Missouri 63026

Ross & Baruzzini, Inc. [713]
7912 Bonhomme
St. Louis, Missouri 63105

Simmons & Sun, Inc. [730]
Solar Earth Consultants, Inc.
P.O. Box 1497
High Ridge, Missouri 63049

Solar-Air by Jira [620]
Jira Heating & Cooling, Inc.
Route 5, Box 400
Columbia, Missouri 65201

Solar Building Corporation [737]
1004 Allen
St. Louis, Missouri 63104

GEOGRAPHICAL INDEX 467

Solar Greenhouse Association [1023]
34 North Gore Avenue
Webster Groves, Missouri 63119

Terra-Dome Corporation [808]
14 Oak Hill Cluster
Independence, Missouri 64057

Tona-Midsouth, Ltd. [405]
P.O. Box 1160
Poplar Bluff, Missouri 63901

MONTANA

Central Montana Human Resources
  Development Council [870]
District VI
604 W. Main
Lewiston, Montana 59457

Independent Power Developers Inc.
  [909]
P.O. Box 1467
Noxon, Montana 59853

Montana Department of Natural
  Resources and Conservation
Energy Division
32 South Ewing
Helena, Montana 59601

Montana State University [923]
Mechanical Engineering Department
220 Roberts Hall
Bozeman, Montana 59717

National Center for Appropriate
  Technology [928]
P.O. Box 3838
Butte, Montana 59701

Ryniker Steel Products Co. [267]
P.O. Box 1932, N.P. Industrial Site
Billings, Montana 59103

NEBRASKA

Allied Industries International [12]
East Highway 275, Industrial Area
Fremont, Nebraska 68025

American Energy Savers, Inc. [17]
912 St. Paul Road, Box 1421
Grand Island, Nebraska 68801

Architectural Concepts [454]
3140 O Street
Lincoln, Nebraska 68510

Hemphill, Vierk & Dawson [592]
908 Terminal Building
Lincoln, Nebraska 68508

Ionic Solar Inc. [612]
8934 J Street
Omaha, Nebraska 68127

LI-COR [194]
4421 Superior Street, Box 4425
Lincoln, Nebraska 68504

Nebraska Solar Office
Nebraska Hall
University of Nebraska
Lincoln, Nebraska 68588

The Schemmer Associates Inc. [717]
Architects, Engineers, Planners
10830 Old Mill Road
Omaha, Nebraska 68154

Lee Schriever—Architect [722]
RFD
Bennet, Nebraska 68317

Solar, Incorporated [314]
008 Sunburst Lane
Mead, Nebraska 68041

NEVADA

Nevada Department of Energy
400 West King Street
Carson City, Nevada 89710

Richdel, Inc. [260]
1851 Oregon Street, P.O. Box A
Carson City, Nevada 89701

Southwest Ener-Tech, Inc. [362]
3020 Valley View Boulevard
Las Vegas, Nevada 89102

S. Wiel & Associates [831]
940 Matley Lane, Suite 8
Reno, Nevada 89502

NEW HAMPSHIRE

Community Builders [493]
Shaker Road
Canterbury, New Hampshire 03224

Contemporary Systems, Inc. [61]
The CSI Solar Center, Route 2
Walpole, New Hampshire 03608

Erie Scientific Company [116]
Division of Sybron Corporation
Portsmouth Industrial Park
Portsmouth, New Hampshire 03801

Hollis Observatory [157]
One Pine Street
Nashua, New Hampshire 03060

Kalwall Corporation [179]
Solar Components Division
P.O. Box 237
Manchester, New Hampshire 03105

Natural Power, Inc. [222]
Francestown Turnpike
New Boston, New Hampshire 03070

New Hampshire Governor's Council
  on Energy
2-½ Beacon Street
Concord, New Hampshire 03301

Solar Age Magazine [1021]
Church Hill
Harrisville, New Hampshire 03450

Solar Sauna [758]
P.O. Box 466
Hollis, New Hampshire 03049

Solar Survival [966]
Cherry Hill Road, Box 275
Harrisville, New Hampshire 03450

Philip S. Tambling, Architect [803]
42 Middle Street
Portsmouth, New Hampshire 03801

Total Environmental Action, Inc.
  [815]
Church Hill
Harrisville, New Hampshire 03450

Urethane Molding [414]
RFD, Box 190
Laconia, New Hampshire 03246

NEW JERSEY

Abacus Controls Inc. [2]
80 Readington Road
Somerville, New Jersey 08876

Berry Solar Products [38]
P.O. Box 327, Woodbridge at Main
Edison, New Jersey 08817

Calmac Manufacturing Corp. [43]
Box 710
Englewood, New Jersey 07631

GEOGRAPHICAL INDEX 469

Canfield Group [44]
1000 Brighton Street
Union, New Jersey 07083

Chemplast Inc. [51]
150 Dey Road
Wayne, New Jersey 07470

Creighton Solar Concepts, Inc. [67]
Kennedy Boulevard
Woodbine Airport, Building No. 4
Woodbine, New Jersey 08270

EASCO Aluminum [89]
P.O. Box 73
North Brunswick, New Jersey 08902

Edmund Scientific Co. [90]
7897 Edscorp Building
Barrington, New Jersey 08007

Edwards Engineering Corp. [92]
101 Alexander Avenue
Pompton Plains, New Jersey 07444

Eisler Engineering Company [93]
750 South 13th Street
Newark, New Jersey 07103

Environmental Consulting & Testing
  Services [843A]
P.O. Box 3521
Cherry Hill, New Jersey 08034

Ergenics [114]
681 Lawlins Road
Wyckoff, New Jersey 07481

Friedrich & Dimmock, Inc. [128]
P.O. Box 230
Millville, New Jersey 08332

H&H Precision Products [153]
Division of Emerson Electric Co.
25 Canfield Road
Cedar Grove, New Jersey 07009

Ista Energy Systems Corp. [171]
29 South Union Avenue
Cranford, New Jersey 07016

Kathabar Systems [180]
Ross Air Systems Division
Midland-Ross Corporation
Box 791
New Brunswick, New Jersey 08903

Kelbaugh & Lee Architects [625]
240 Nassau Street
Princeton, New Jersey 08540

New Jersey Institute of Technology
  [681]
Center for Technology Assessment
323 High Street
Newark, New Jersey 07102

New Jersey State Energy Office
Office of Alternative Technology
101 Commerce Street
Newark, New Jersey 07102

RCA Corporation [948]
RCA Laboratories
Princeton, New Jersey 08540

Sealed Air Corporation [272]
Corporate Office
Park 80 Plaza East
Saddle Brook, New Jersey 07662

Solar Energymaster [302]
P.O. Box 592
Florham Park, New Jersey 07932

Solar Industries, Inc. [316]
Monmouth Airport Industrial Park
Farmingdale, New Jersey 07727

Solar-Trol [343]
Division of Trolex Corporation
1078 Route 46
Clifton, New Jersey 07013

Sunworks Division of Sun Solector
   Corp. [390]
P.O. Box 3900
Somerville, New Jersey 08876

Unisol Energy Corporation [816]
36 N. Day Street
Orange, New Jersey 07050

NEW MEXICO

Adobe Solar Ltd. [434]
930 Matador S.E.
Albuquerque, New Mexico 87123

Anachem Inc. [850]
Princeton Park of Commerce
3300 Princeton N.E.
Albuquerque, New Mexico 87107

Aztech International Ltd. [35]
2417 Aztec Road, N.E.
Albuquerque, New Mexico 87107

Green Horizon [581]
Route 7, Box 124 MS
Santa Fe, New Mexico 87501

Los Alamos Scientific Laboratory [915]
Solar Energy Group, Q-11
P.O. Box 1663, MS 571
Los Alamos, New Mexico 87545

Stephen Merdler Associates [662]
300 Calle Sierpe
Santa Fe, New Mexico 87501

New Mexico Energy and Minerals
   Department
P.O. Box 2770
Santa Fe, New Mexico 87501

New Mexico Solar Energy Association
   [1012]
P.O. Box 2004
Santa Fe, New Mexico 87501

Pittman Earthworks [700]
P.O. Box 5031
Santa Fe, New Mexico 87502

Shelton Energy Research [957]
P.O. Box 5235
Santa Fe, New Mexico 87502

Solar Resources, Inc. [333]
Box 1848
Taos, New Mexico 87571

Solar Sustenance Team [967]
P.O. Box 733
El Rito, New Mexico 87530

Solar Thermal Test Facilities
   Users Association [968]
First National Bank Building East
Suite 1204
Albuquerque, New Mexico 87108

Tetra Corporation [980]
P.O. Box 4369
Albuquerque, New Mexico 87196

NEW YORK

Advance Energy Technologies, Inc.
   [7]
P.O. Box 387
Solartown
Clifton Park, New York 12065

Aldermaston Inc. [11]
P.O. Box 34
Locust Valley, New York 11560

Alternate Energy Industries
   Corporation (AEIC) [443]
420 Lexington Avenue, Suite 1628
New York, New York 10017

The Amcor Group Ltd. [16]
350 Fifth Avenue, Suite 1907
New York, New York 10001

GEOGRAPHICAL INDEX 471

Bio-Energy Systems, Inc. [39]
221 Canal Street
Ellenville, New York 12446

Brookhaven National Laboratory
  [864]
Upton, Long Island, New York 11973

Cary Arboretum of the New York
  Botanical Garden [867]
Box AB
Millbrook, New York 12545

Clairex Electronics [53]
560 South Third Avenue
Mt. Vernon, New York 10550

Conklin & Rossant [495]
251 Park Avenue South
New York, New York 10010

DAS/Solar Systems [502]
188 Flatbush Avenue, Ext.
Brooklyn, New York 11201

Dubin-Bloome Associates, P.C. [516]
42 West 39th Street
New York, New York 10018

Edge Research [522]
RD 1, Box 394 B
Kingston, New York 12401

W. S. Fleming & Associates Inc. [556]
840 James Street
Syracuse, New York 13203

W.S. Fleming & Associates Inc. [557]
3 Computer Drive
Albany, New York 12205

Ford Products Corporation [125]
Ford Products Road
Valley Cottage, New York 10989

Four Seasons Solar Products Corp.
  [126]
910 Route 110
Farmingdale, New York 11735

Grumman Energy Systems Inc. [139]
10 Orville Drive
Bohemia, New York 11716

Hitachi Chemical Co. America Ltd.
  [155]
437 Madison Avenue
New York, New York 10022

Irvine Engineering [614]
P.O. Box 246
Suffern, New York 10901

Jan Johnson [1006]
Solar Greenhouse Consultant
Box 383-B
High Falls, New York 12440

Kingston Industries Corp. [182]
205 Lexington Avenue
New York, New York 10016

Lipe-Rollway Corporation [913]
806 Emerson Avenue
P.O. Box 1397
Syracuse, New York 13201

Long Island Lighting Co. [914]
250 Old Country Road
Mineola, New York 11501

Mechanical Technology Inc. [918]
968 Albany-Shaker Road
Latham, New York 12110

Robert Mitchell [666]
Solar Systems Design Inc.
RD 3, Box 239
Selkirk, New York 12158

472 SOLAR CENSUS

National Semiconductors Ltd. [218]
331 Cornelia Street
Plattsburgh, New York 12901

National Stove Works [221]
Howe Caverns Road
Cobleskill, New York 12043

New York Institute of Technology
 [1013]
Center for Energy Policy and Research
Old Westbury, New York 11568

New York State Energy Office
Solar Program
2 Rockefeller Plaza
Albany, New York 12223

Parkway Windows [941]
127-17 20th Avenue
College Point, New York 11356

Edward Pedersen [694]
Architect/Planner
109 Haffenden Road
Syracuse, New York 13210

Polytechnic Institute of New York
 [944]
Solar Energy Applications Center
333 Jay Street
Brooklyn, New York 11201

Revere Solar and Architectural
 Products, Inc. [258]
P.O. Box 151
Rome, New York 13440

Rigidized Metals Corporation [261]
658 Ohio Street
Buffalo, New York 14203

Rochester Institute of Technology
 [953]
Institute for Applied Energy Studies
1 Lomb Memorial Drive
Rochester, New York 14623

Paul Rosenberg Associates [712]
330 Fifth Avenue
Pelham, New York 10803

Solar Fab, Inc. [308]
151-23 34 Avenue
Flushing, New York 11354

Solar Structures Inc. [760]
Wenning Associates
221 Mill Street
Poughkeepsie, New York 12601

Solar Sunstill [965]
15 Blueberry Ridge Road
Setauket, New York 11733

Solartek [763]
P.O. Box 298
Guilderland, New York 12084

Solation Products Inc. [350]
111 West Road
Cortland, New York 13045

Solex Solar Energy Systems, Inc.
 [355]
444 Bedford Road
Pleasantville, New York 10570

The Stein Partnership [778]
588 Fifth Avenue
New York, New York 10036

Sunmaster Corporation [375]
12 Spruce Street
Corning, New York 14830

Sunpower Research & Development
 Corp. [976]
200 Madison Avenue
New York, New York 10016

Total Energy [406]
55 Knickerbocker Avenue
Bohemia, New York 11716

Vegetable Factory, Inc. [419]
100 Court Street
Copiague, New York 11726

Lawrence Wittman & Co. Inc. [425]
1395 Marconi Boulevard
Copiague, New York 11726

NORTH CAROLINA

Carolina Solar Systems [46]
527 Hillsborough Street
Raleigh, North Carolina 27602

Energy Control Systems [100]
3324 Octavia Street
Raleigh, North Carolina 27606

Integrated Energy Systems [610]
301 North Columbia Street
Chapel Hill, North Carolina 27514

North Carolina Department of
    Commerce
Energy Division
430 North Salisbury Street
Raleigh, North Carolina 27611

Solar Products Information &
    Engineering (Solar P.I.E.) [757]
P.O. Box 506
Columbus, North Carolina 28722

Sunshelter Design [791]
1209 Hillsborough Street
Raleigh, North Carolina 27603

Turner Greenhouses [409]
Highway 117 South
Goldsboro, North Carolina 27530

NORTH DAKOTA

Engineers—Architects P.C. [543]
408 First Avenue Building
Minot, North Dakota 58701

University of North Dakota [934]
Engineering Experiment Station
P.O. Box 8103
University Station
Grand Forks, North Dakota 58202

North Dakota Energy Management
    and Conservation
1533 North 12th Street
Bismarck, North Dakota 58501

R. C Solar Co. Inc. [252]
Box 248
Richardton, North Dakota 58652

Roggensack Insulation & Solar Inc.
    [710]
55 Prairiewood Drive
Fargo, North Dakota 58103

Dan Smith, Architect [735]
164 East Main, Box 536
Valley City, North Dakota 58072

Solar Dakota [742]
Box 1394
Minot, North Dakota 50701

OHIO

Alpha Solarco Inc. [13]
Suite 2230, 1014 Vine Street
Cincinnati, Ohio 45202

Angel, Mull & Associates, Inc. [449]
3049 Sylvania Avenue
Toledo, Ohio 43613

Architectural Bureau [452]
1900 E. Broad Street
Columbus, Ohio 43209

Architectural Collective [453]
Architronics, Inc.
4615 W. Streetsboro Road
Richfield, Ohio 44286

Architekton, Inc. [457]
700 Walnut Street
Cincinnati, Ohio 43209

Battelle Memorial Institute [861]
Columbus Laboratories
505 King Avenue
Columbus, Ohio 43201

Howard Bell Enterprises, Inc. [466]
P.O. Box 413, 5931 East Low Road
Valley City, Ohio 44280

W. J. Black, AIA/Architect [471]
1122 Westwood Avenue
Columbus, Ohio 43212

Robert J. Bregar Associates [475]
22700 Shore Center Drive, Suite 303
Cleveland, Ohio 44123

Environmental Design Alternatives
  [545]
1951 Brookview Drive
Kent, Ohio 44240

John M. Evanoff, Architect [551]
3405 River Road
Toledo, Ohio 43614

Forest City Dillon, Inc. [561]
A Subsidiary of Forest City
  Enterprises, Inc.
10800 Brookpark Road
Cleveland, Ohio 44130

Fusion Incorporated [129]
4658 East 355th Street
Willoughby, Ohio 44094

General Solar Systems Division [134]
General Extrusions Inc.
P.O. Box 2687
Youngstown, Ohio 44507

Glass-Lined Water Heater Company
  [135]
13000 Athens Avenue
Cleveland, Ohio 44107

Harris Architects, Inc. [588]
3821 Wales Road N.W.
Massillon, Ohio 44646

M. R. Immormino [605]
23950 Hazelmere Road
Cleveland, Ohio 44122

J.F.A. Services, Inc. [619]
7340 Kingsgate Way, Suite 230
West Chester, Ohio 45069

Libbey-Owens-Ford Company [193]
Solar Energy Systems
1701 East Broadway Street
Toledo, Ohio 43605

H. D. Luther Manufacturing Co.
  [196]
5297 Southway Avenue S.W.
Canton, Ohio 44706

James A. Martis, Jr., Architects
  [657]
28790 Chagrin Boulevard, No. 250
Cleveland, Ohio 44122

Mitchell & Jensen [667]
4247 Philadelphia Drive
Dayton, Ohio 45405

James A. Monsul & Associates [669]
642 Brooksedge Boulevard
Westerville, Ohio 43081

Mor-Flo Industries, Inc. [214]
18450 South Miles Road
Cleveland, Ohio 44128

Ohio Department of Energy
30 East Broad Street, 34th Floor
Columbus, Ohio 43215

Olmon & Hutchinson Architects [684]
9-½ South Market Street
P.O. Box 339
Troy, Ohio 45373

Olympic Solar Corporation [233]
208 15th Street, S.W.
Canton, Ohio 44707

Owens-Illinois Sunpak Division [234]
P.O. Box 1035
Toledo, Ohio 43666

David Panich, Architect [690]
Solar Designs
Route 5, Box 88C
Athens, Ohio 45701

Rom-Aire Solar Corporation [265]
121 Miller Road
Avon Lake, Ohio 44012

Roth Eckert & Jacobson, Inc. [714]
227 West Crawford Street
Findlay, Ohio 45840

Solar Energy Engineering [751]
1838 Alverne Drive
Poland, Ohio 44514

Solar Home Systems Inc. [753]
8732 Camelot Drive
Chesterland, Ohio 44026

Solartec Inc. [339]
250 Pennsylvania Avenue
Salem, Ohio 44460

Solar Usage Now, Inc. [347]
420 East Tiffin Street, P.O. Box 306
Bascom, Ohio 44809

Steed/Hammond/Paul [776]
10 Court Street
Hamilton, Ohio 45011

R. D. Strayer, Inc. [781]
5490 Ashford Road
Dublin, Ohio 43017

R. S. Subjinske & Associates, Inc. [783]
1826 South Main Street
Akron, Ohio 44301

Sunpower Inc. [975]
6 Byard Street
Athens, Ohio 45701

Teledyne Mono-Thane [393]
1460 Industrial Parkway
Akron, Ohio 44310

Thermal Engineering & Design Co. [398]
44 N. Summit Street
Akron, Ohio 44308

Turbonics Inc. [408]
11200 Madison
Cleveland, Ohio 44102

United States Solar Industries [412]
25 W. Fifth Street
London, Ohio 43140

Zaugg & Zaugg Architects [841]
60 South Main Street
Mansfield, Ohio 44902

OKLAHOMA

Big Five Community Services, Inc. [996]
215 North 16th, P.O. Box 371
Durant, Oklahoma 74701

Coating Laboratories Inc. [55]
3133 East Admiral Place
Tulsa, Oklahoma 74110

General Solar Corp. [568]
P.O. Box 15835
Tulsa, Oklahoma 74112

Joe Hylton & Associates [600]
566 Buchanan Street
Norman, Oklahoma 73069

University of Oklahoma [936]
Division of Economics
Norman, Oklahoma 73019

Oklahoma Department of Energy
4400 North Lincoln Boulevard
Suite 251
Oklahoma City, Oklahoma 73105

Oklahoma State University [937]
Architectural Extension
118 Architecture Building
Stillwater, Oklahoma 74078

Oklahoma State University [938]
School of Mechanical Engineering
Stillwater, Oklahoma 74078

Solartronics Inc. [344]
3101 East Reno
Oklahoma City, Oklahoma 73117

OREGON

Amity Foundation [853]
P.O. Box 7066
Eugene, Oregon 97401

Paul F. Bogen, Architect [474]
2350 Columbia Street
Eugene, Oregon 97403

Miller & Sun Enterprises Inc. [664]
P.O. Box 19151
Portland, Oregon 97219

University of Oregon [939]
Solar Energy Center
Department of Architecture &
  Allied Arts
Eugene, Oregon 97403

Oregon Department of Energy
Labor and Industries Building
Room 111
Salem, Oregon 97310

Oregon Solar Institute [1015]
215 S.E. 9th, Room 21
Portland, Oregon 97214

Rusth Industries [266]
P.O. Box 1519
Beaverton, Oregon 97075

Sundex Solar Equipment Sales [368]
677 Valley View Road
Talent, Oregon 97540

Sun Harvesters of Oregon [786]
P.O. Box 5481
Eugene, Oregon 97405

Sun Life Systems Inc. [374]
6210 N.E. 92nd Drive, Suite 101
Portland, Oregon 97220

Technical Ventures [805]
P.O. Box 06175
Portland, Oregon 97206

Western Solar Utilization Network
  (Western SUN) [1033]
921 S.W. Washington Street, Suite 160
Portland, Oregon 97205

GEOGRAPHICAL INDEX 477

Ray A. Wiley, Architect/Engineer [832]
541 Willamette, Suite 308
Eugene, Oregon 97401

PENNSYLVANIA

Ametek, Power Systems Group [24]
1025 Polinski Road
Ivyland, Pennsylvania 18974

Burt Hill Kosar Rittelmann Associates [478]
400 Morgan Center
Butler, Pennsylvania 16001

Doucette Industries [83]
P.O. Box 1641
York, Pennsylvania 17405

Energy Development Company [894]
179 East Road 2
Hamburg, Pennsylvania 19526

Free Energy Systems Inc. [127]
Holmes Industrial Park
Holmes, Pennsylvania 19043

IBE Energy [601]
656 West Main Street
Palmyra, Pennsylvania 17078

Lancaster County Community Action Program [636]
Solar Project Program
630 Rockland Street
Lancaster, Pennsylvania 17602

Mehrkam Energy Development Co. [204]
R.D. 2
Hamburg, Pennsylvania 19526

Mototron Inc. [925]
292 Montgomery Avenue
Bala Cynwyd, Pennsylvania 19004

NRG Company [228]
P.O. Box 306
Ardmore, Pennsylvania 19003

Pennsylvania Governor's Energy Council
1625 North Front Street
Harrisburg, Pennsylvania 17101

The Pennsylvania State University [943]
Department of Engineering Science and Mechanics
227 Hammond Building
University Park, Pennsylvania 16802

Pittsburgh Corning Corporation [240]
800 Presque Isle Drive
Pittsburgh, Pennsylvania 15239

Rectech Inc. [950]
421 East Beaver Avenue
P.O. Box 177
State College, Pennsylvania 16801

Solar Patents [964]
P.O. Box 21
Turtle Creek, Pennsylvania 15145

Sonnewald Service [360]
Route 1, Box 1508
Spring Grove, Pennsylvania 17362

Sunearth Solar Products Corp. [371]
352 Godshall Drive
Harleyville, Pennsylvania 19438

Sven Tjernagel Solar Systems [404]
477 Woodcrest Drive
Mechanicsburg, Pennsylvania 17055

Viking Energy Corporation [819]
4709 Baum Boulevard
Pittsburgh, Pennsylvania 15213

RHODE ISLAND

Aeolian Kinetics [9]
P.O. Box 100
Providence, Rhode Island 02901

Beckman, Blydenburgh & Associates [465]
P.O. Box 100
Providence, Rhode Island 02901

Brown University [865]
Division of Engineering
Providence, Rhode Island 02912

Crystalite Embedments Inc. [879]
6 Industrial Drive
Smithfield, Rhode Island 02917

Elmwood Sensors, Inc. [98]
1655 Elmwood
Cranston, Rhode Island 02907

Eppley Laboratory, Inc. [113]
12 Sheffield Avenue
Newport, Rhode Island 02840

Independent Energy, Inc. [167]
42 Ladd Street, P.O. Box 860
East Greenwich, Rhode Island 02818

Megatherm [203]
803 Taunton Avenue
East Providence, Rhode Island 02914

Research & Design Institute [951]
P.O. Box 307
Providence, Rhode Island 02901

Rhode Island Governor's Energy Office
80 Dean Street
Providence, Rhode Island 02903

Solar Homes Inc./Suntrol [313]
275 Harborside Boulevard
Providence, Rhode Island 02905

Suntree Solar Company [388]
20 Austin Avenue
Greenville, Rhode Island 02828

SYenergy Methods, Inc. [391]
1367 Elmwood Avenue
Cranston, Rhode Island 02910

Vulcan Solar Industries, Inc. [421]
6 Industrial Drive
Smithfield, Rhode Island 02917

SOUTH CAROLINA

Energy Associates [532]
Box 6602
Greenville, South Carolina 29606

Energy Designs/Architects [535]
201 Woodrow Street
Columbia, South Carolina 29205

Helio Thermics Inc. [150]
1070 Orion Street
Donaldson Center
Greenville, South Carolina 29605

R.P. Woodcrafters [715]
1614 Westminster Drive
Columbia, South Carolina 29204

South Carolina Division of Energy Resources
1122 Lady Street, 11th Floor
Columbia, South Carolina 29201

USDA-SEA-AR [984]
Rural Housing Research Unit
Box 792
Clemson, South Carolina 29631

GEOGRAPHICAL INDEX 479

SOUTH DAKOTA

Dunham Associates [886]
Symcom Inc.
528 Kansas City Street
Rapid City, South Dakota 57701

Mass Energy Systems, Inc. [430C]
P.O. Box 1311
Rapid City, South Dakota 57709

Passive Solar Alternatives [691]
302 Denver, No. 203
Rapid City, South Dakota 57701

South Dakota State Solar Office
Capitol Lake Plaza Building
Suite 102
Pierre, South Dakota 57501

The Spitznagel Partners, Inc. [771]
1112 West Avenue, North
Sioux Falls, South Dakota 57104

Synago Corporation [801]
P.O. Box 444
Rapid City, South Dakota 57709

TENNESSEE

AFG Industries Inc. [10]
P.O. Box 929
Kingsport, Tennessee 37662

Corporate Energy Developments Inc.
[878]
P.O. Box 332
Hermitage, Tennessee 37076

Dixie Royal Homes Inc. [80]
460 East 15th Street
P.O. Box 805
Cookeville, Tennessee 38501

Energy Design Corporation [101]
P.O. Box 34294
Memphis, Tennessee 38134

Ionic Solar [612]
148 8th Avenue North
Nashville, Tennessee 37203

W. L. Jackson Manufacturing Co.
[173]
1205 East 40th Street
P.O. Box 11168
Chattanooga, Tennessee 37401

Monroe Modular Homes, Inc. [212]
Route 5, Industrial Park
Madisonville, Tennessee 37354

Oak Ridge National Laboratory
[935]
P.O. Box X
Oak Ridge, Tennessee 37830

Lee Ragsdale Associates [247]
2325 Sutherland Avenue
P.O. Box 10605
Knoxville, Tennessee 37919

State Industries, Inc. [365]
Ashland, Tennessee 89015

University of Tennessee [979]
Solar Program
Energy, Environment and Resources
  Center
329 South Stadium Hall
Knoxville, Tennessee 37916

Tennessee Energy Authority
Capitol Building, Suite 707
226 Capitol Boulevard
Nashville, Tennessee 37916

Tennessee Valley Authority [807]
Architectural Design Branch
400 Commerce
Knoxville, Tennessee 37902

TEXAS

Alternative Energy Resources Inc. [14]
1155K Larry Mahan Drive
El Paso, Texas 79925

American Solar King Corp. [21]
P.O. Box 7399
Waco, Texas 76710

Ana-Lab Corp. [851]
2600 Dudley Road
Kilgore, Texas 75662

The Center for Maximum Potential
  Building Systems [869]
8604 FM 969
Austin, Texas 78724

Cole Solar Systems, Inc. [56]
440A East St. Elmo Road
Austin, Texas 78745

Croft & Company [499]
Engineering Division
708 W. 10th Street
Austin, Texas 78701

Devices & Services Co. [79]
3501 A Milton Avenue
Dallas, Texas 75205

Dodge Products Inc. [82]
P.O. Box 19781
Houston, Texas 77024

Electrolab Inc. [96]
2103 Mannix
San Antonio, Texas 78217

Environment Associates [549]
2777 Allen Parkway, Suite 207
Houston, Texas 77019

Ronnie Freeman [564]
Solar Energy Consultant
Box 215
Trent, Texas 79561

Holt + Fatter + Scott, Inc. [594]
2525 Wallingwood, No. 501
Austin, Texas 78746

Lennox Industries Inc. [190]
P.O. Box 400450
Dallas, Texas 75240

Modupac Inc. [922]
8309 Monroe
P.O. Box 12897
Houston, Texas 77017

Moreland Associates [673]
904 Boland
Fort Worth, Texas 76107

Navarro College [1010]
P.O. Box 1170
Corsicana, Texas 75110

Northrup Inc. [225]
Subsidiary of Atlantic Richfield
  Company
302 Nichols Drive
Hutchins, Texas 75141

Radiation Research Associates,
  Inc. [946]
3550 Hulen Street
Ft. Worth, Texas 76107

R&D Enterprise [253]
Route 3, Box 258
Carthage, Texas 75633

Research Institute for Advanced
    Technology [952]
U.S. Highway 190 West
Killeen, Texas 76541

Harold A. Smith & Associates [280]
P.O. Box 42431
Houston, Texas 77042

Solar Design Consultants Inc. [744]
2540 Walnut Hill Lane, Suite 166
Dallas, Texas 75229

Solar Kinetics, Inc. [318]
8120 Chancellor Row
Dallas, Texas 75247

Texas A&M University [1027]
Center for Energy and Mineral
    Resources
College Station, Texas 77843

Texas Energy and Natural Resources
    Advisory Council
411 West 13th Street
Room 800
Austin, Texas 78701

The Texas Solar Energy Society, Inc.
    [1028]
1007 South Congress, No. 359
Austin, Texas 78704

Texas State Technical Institute
    (TSTI) [1029]
Solar Energy Mechanic Program
Route 3
Sweetwater, Texas 79556

UTAH

Cascade Solar Technics [868]
216 Paxton Avenue
Salt Lake City, Utah 84101

Cover Pools, Inc. [65]
117 West Fireclay Avenue
Salt Lake City, Utah 84107

Devon D Inc. [991]
P.O. Box 227
Fillmore, Utah 84631

TIOS Corporation [814]
740 South 300 West
Salt Lake City, Utah 84101

University of Utah [985]
Mechanical and Industrial
    Engineering Department
Salt Lake City, Utah 84112

Utah Energy Office
231 East 400 S., Suite 101
Salt Lake City, Utah 84111

VERMONT

Brattleboro Design Group [842]
113 Main Street, P.O. Box 235
Brattleboro, Vermont 05301

Dimetrodon [509]
Route 1, Box 160
Warren, Vermont 05674

Earth Services Inc. [88]
Route 30, Box 99
Pawlet, Vermont 05761

Enertech Corporation [109]
P.O. Box 420
Norwich, Vermont 05055

Green Mountain Homes, Inc. [137]
Royalton, Vermont 05068

North Wind Power Co., Inc. [226]
Box 315
Warren, Vermont 05674

Vermont State Energy Office
4 East State Street
Montpelier, Vermont 05602

VIRGINIA

American Gas Association [995]
1515 Wilson Boulevard
Arlington, Virginia 22209

Helios International Corp. [149]
Sunstar Group
2120 Angus Road
Charlottesville, Virginia 22901

Intertechnology/Solar Corp. [170]
100 Main Street
Warrenton, Virginia 22186

Martin Processing Inc. [200]
P.O. Box 5068
Martinsville, Virginia 24112

North American Solar Development
  Corp. [223]
2800 Juniper Street
Fairfax, Virginia 22030

Solar Energy Design, Inc. [749]
3822 Prince William Drive
Fairfax, Virginia 22031

Solar Spectrum Inc. [336]
Seahorse Plastics Corporation
4680 Shoulders Hill Road
Suffolk, Virginia 23435

Sunshine Development Corporation,
  Inc. [382]
A Subsidiary of RISE Inc.
1641 East Main Street
Salem, Virginia 24153

Survival Consultants [799]
P.O. Box 21
Rapidan, Virginia 22733

Virginia Energy Information and
  Services Center
310 Turner Road
Richmond, Virginia 23225

Virginia Polytechnic Institute
  and State University [986]
Agricultural Engineering Department
311 Seitz Hall
Blacksburg, Virginia 24061

Westinghouse Electric Corporation
  [828]
Solar Heating & Cooling Systems
5205 Leesburg Pike, Suite 201
Falls Church, Virginia 22041

WASHINGTON

Balance Associates [463]
201 Summit Avenue East
Seattle, Washington 98102

Ecotope Group [520]
2332 East Madison
Seattle, Washington 98112

E&K Service Co. [94]
16824 - 74th N.E.
Bothell, Washington 98011

ENERAD Inc. [893]
P.O. Box 3982
Bellevue, Washington 98009

National Institute of Creativity,
  Inc. [929]
P.O. Box 44067
Tacoma, Washington 98409

F. C. Radice [947]
Research & Development
13515 N.E. 70th Street
Redmond, Washington 98052

GEOGRAPHICAL INDEX 483

Space/Time Designs, Inc. [770]
P.O. Box 4229
Bellevue, Washington 98009

University of Washington [987]
Aerospace and Energetics Research
  Program FL-10
Seattle, Washington 98195

Washington State Energy Office
400 East Union Street, 1st Floor
Olympia, Washington 98504

WEST VIRGINIA

West Virginia Fuel and Energy Office
1262-½ Greenbrier Road
Charleston, West Virginia 25311

WISCONSIN

Advance Air Control [435]
1506 Terry Andrae Court
Sheboygan, Wisconsin 53081

Advanced Energy Systems Inc. [5]
Sunstone
P.O. Box 194, 126 Water Street
Baraboo, Wisconsin 53913

Affiliated Engineers Inc. [437]
Fuad & Associates
P.O. Box 5039
625 North Segol Road, Suite C
Madison, Wisconsin 53705

Blake—Huettenrauch—Schuyler
  Associates, Inc. [472]
330 West Silver Spring Drive
Milwaukee, Wisconsin 53217

Community Action Commission for
  the City of Madison and the
  County of Dane, Inc. [492]
1045 East Dayton Street
Madison, Wisconsin 53703

Creative Energy Products [66]
1053 Williamson Street
Madison, Wisconsin 53703

Dale and Associates Inc. [70]
Distributor Division
1401 Cranston Road
Beloit, Wisconsin 53511

Eland Electric Corp. [892]
1841 Morrow Street
Green Bay, Wisconsin 54304

Erie Manufacturing Co. [115]
4000 S. 13th Street
Milwaukee, Wisconsin 53221

The Evjen Associates, Inc. [552]
Box 152
Hudson, Wisconsin 54016

Glenn F. Groth, Architect [583]
2619 Camelot Boulevard
Sheboygan, Wisconsin 53081

Hedland Products [146]
2200 South Street
Racine, Wisconsin 53404

HSR Associates, Inc. [598]
100 Milwaukee Street
La Crosse, Wisconsin 54601

Arlan Kay & Associates [623]
5685 Lincoln Road
Oregon, Wisconsin 53575

LaFont Corporation [185]
1319 Town Street
Prentice, Wisconsin 54556

Microcontrol Systems, Inc. [205]
6579 North Sidney Place
Milwaukee, Wisconsin 53209

484 SOLAR CENSUS

Natural Energy Workshop, Inc. [678]
Route 1
Box 130
North Freedom, Wisconsin 53951

PRADO [702]
Box 1128
Madison, Wisconsin 53701

Research Products Corporation [256]
1015 E. Washington Avenue
P.O. Box 1467
Madison, Wisconsin 53701

Scientific Building and Energy
   Consultants, Inc. [956]
Route 1
Box 314X
Jefferson, Wisconsin 53549

Silicon Sensors, Inc. [277]
Solar Systems, Inc., Division
Highway 18 East
Dodgeville, Wisconsin 53533

Solar Unlimited, Inc. [346]
240 East Main
P.O. Box 63
Sun Prairie, Wisconsin 53590

Sun Stone Company, Inc. [385]
P.O. Box 138
Baraboo, Wisconsin 53913

West CAP (West Central Wisconsin
   Community Action Agency) [827]
525 2nd Street
Glenwood City, Wisconsin 54013

University of Wisconsin—Madison
   [990]
Solar Energy Laboratory
1500 Johnson Drive
Madison, Wisconsin 53706

Wisconsin Division of State Energy
101 South Webster, 8th Floor
Madison, Wisconsin 53702

WYOMING

Community Action of Laramie
   County [999]
Cheyenne Community Solar
   Greenhouse
1603 Central Avenue, No. 400
Cheyenne, Wyoming 82001

Park Energy Company [235]
Star Route, Box 9
Jackson, Wyoming 83001

Wyoming Energy Extension
   Service
P.O. Box 3965
University Station
Laramie, Wyoming 82070